D1810984

Rational pesticide use

Rational pesticide use

Proceedings of the
Ninth Long Ashton Symposium

Edited by

K. J. BRENT

and

R. K. ATKIN

Long Ashton Research Station
University of Bristol

The right of the
University of Cambridge
to print and sell
all manner of books
was granted by
Henry VIII in 1534.
The University has printed
and published continuously
since 1584.

CAMBRIDGE UNIVERSITY PRESS

Cambridge

New York   New Rochelle

Melbourne   Sydney

Published by the Press Syndicate of the University of Cambridge
The Pitt Building, Trumpington Street, Cambridge CB2 1RP
32 East 57th Street, New York, NY 10022, USA
10 Stamford Road, Oakleigh, Melbourne 3166, Australia

© Cambridge University Press 1987

First published 1987

Printed in Great Britain at the University Press, Cambridge

Library of Congress cataloguing in publication data available

British Library cataloguing in publication data
Long Ashton Symposium (9th)
Rational pesticide use: proceedings of the Ninth Long Ashton Symposium
1. Pesticides
I. Title   II. Brent, K. J.   III. Atkin, R. K.
632'.95   SB951

ISBN 0 521 32068 2

CONTENTS

SECTION 4

Forecasting and Pest Management

FOREWORD

In recent years the use of pesticides in agriculture and horticulture has increased greatly throughout the world, both in the frequency of applications and in the total crop areas that are treated. Undoubtedly this application of pesticides has contributed to the growth in crop production that has occurred in most countries, and has decreased substantially the risks of disastrous crop losses both at the regional level and on the individual farm.

At the same time, public concern over the amounts of agrochemicals that are being applied to the land, and their possible side-effects on human and animal health and on the countryside, has risen sharply. Moreover, we have entered an era of agricultural over-production in Europe, the USA and other major agricultural areas; economic pressures are increasing and farmers and advisers are seeking ways of improving their efficiency of production and competitiveness in world markets.

Thus, the topic of the Ninth Long Ashton Symposium - 'Rational Pesticide Use' - is one of increasing urgency. The basic question at issue is to what extent can we make pesticide applications safer, more effective and precise, and more economic than they are today? This in turn raises such further questions as how do we assess the safety of pesticide applications, how can we restrict their use to situations of real necessity, and how can we avoid the upsurge of pesticide-resistant insects, pathogens and weeds? International experts in many aspects of crop protection research or advisory work gathered in Bristol to address this question, and articles based on their contributions are published in this volume in order to reach a wider audience. Many personal views are expressed and, apart from general editing in the interest of clarity and readability, no attempt has been made to establish uniformity of style or approach. Rather, we have deliberately tried to preserve the different 'flavour' of each individual presentation. It is hoped that the whole gives a balanced and interesting picture of the whole topic of the rational use of pesticides as seen in the mid-1980s.

The chapters cover four major issues: environment - how far is current pesticide use affecting public health, wild-life and the landscape, and are changes required in crop protection practice or research?; application - what are the aims and prospects of research done to improve application methods and to develop new ones, in relation to the effectiveness and precision of pesticide treatments?; resistance – how serious are the problems caused by build-up of resistant forms of pests in crops, and can we avoid or delay failures of control by adjusting pesticide use?; and forecasting and pest management – are existing schemes of risk assessment acceptable as useful aids to decision-making, and is current research coupled with computer technology likely to stimulate new developments in rational pest management?

Of course, these topics intertwine and their boundaries are arbitrary. Professor Haskell in his opening Chapter based on the inaugural Douglas Wills Lecture identifies and discusses the 'selection pressures' that have shaped the whole pattern of world-wide crop protection practices and which are continuing to work to produce further evolutionary change. We feel his analysis of the concepts of integrated pest management and his views on overall priorities for research, information transfer and education deserve particular attention.

Because contributors focus on the principles that might guide the development and implementation of more rational control strategies, we believe that this book will be of value to the many scientists, advisers, teachers, students and others who are concerned with the continued and yet ever-changing struggle to protect crops against their natural enemies. We thank all the authors for their contributions and for their forebearance during unforeseen delays in publishing, and are indebted to the staff of Cambridge University Press for their support at all times. It is a pleasure to record our gratitude to our colleagues, especially Drs. E.C. Hislop, D.J. Royle and B.D. Smith, for their editorial help. We thank too Mr. E.G.R. Chenoweth for his assistance with the illustrations, and Mrs. J.F. Mizen, Miss H.L. Clarke, Mrs. J.M. Llewellyn and Mrs. S.E. Child for typing standard camera-ready copy from the submitted typescripts.

K.J. Brent   Long Ashton Research Station,
R.K. Atkin   January, 1987.

INAUGURAL LECTURE (DOUGLAS WILLS LECTURE)

1. NATURAL SELECTION IN CROP PROTECTION

P.T. Haskell
University College of Wales, P.O. Box 78, Cardiff, CF1 1XL,
Wales

INTRODUCTION
Some people will read the title of this lecture with quite
understandable apprehension, expecting to be subjected to a discourse on
Darwinian theory as applied to crop protection. Let me set their minds
at rest; what I know about natural selection can be written on the edge
of a postage stamp and I was never one to demonstrate my ignorance in
public.

However, one element in the theory of natural selection, the survival of
the fittest, is relevant to my purpose, because amongst other things I
want to consider if and how the selection pressures engendered by, and
acting upon, the several groups involved in pesticide use – for example,
agrochemical companies, farmers and growers, and the public at large –
have affected research, development, operations and the present use of
pesticides and have indeed produced the "best" or "fittest" policies and
practice. If anyone doubts the power of such pressures to affect work
on pesticides just consider the fact that virtually no effective control
techniques exist for one of the most damaging group of pests in world
agriculture – birds. Admittedly there are difficult technical problems,
but nevertheless the great strength of the world bird lobby has
effectively nullified progress towards development of efficient and
practical control measures (Wright, 1980).

There is no escaping some conflict between the many selection pressures
that exist; even if you are a dedicated chemical control protagonist,
you may also be a bee-keeper and hence be aware of the problems caused
in that sphere by pesticide use. Over and above that, you are a member
of the public and hence subject to public opinion as moulded and formulated by government, industry, the media and special interest groups.

The theory of natural selection demonstrates that in a complex system
the best answer lies in compromise; if you are a predator, for example,
it is an obvious advantage to be able to move faster than your prey and
your fellow-predators, but this can only be done at the expense of
increased energy requirements and requires greater efficiency in prey-
catching.

Thus, in my opinion the central problem of rational pesticide use – or
indeed any rational policy on anything at all, whether it be the
agricultural policy of the Common Market or the best length of a yo-yo

sting - is to reconcile the differing general and specific interests of
a heterogeneous number of groups in the community.

Before trying to define and discuss some of these interests in relation
to pesticide usage, it is necessary to have some definition of "rational
use". I propose, for the purposes of argument, that a rational policy
is one based on conscious reason rather than unconscious motivation, one
that is sensible and moderate and neither foolish or extreme - which
hence in principle could be agreed to and supported by a majority of the
total community involved.

It is clear that such a policy can never be - or certainly should not be
- immutable, because it is dealing with a biological system of great
complexity, which is forever changing and evolving in all its aspects,
whether these be biological, commercial, social or political.

One advantage of this definition is that it keeps us within the realms
of practical propositions and politics by emphasising the need to avoid
extremes. If you do not think that is important, consider how the scene
has changed during the last three decades. In the 1940s, the
astonishing success of the newly discovered organo-chlorine and organophosphorus compounds in combating arthropod disease vectors and later
agricultural insect pests gave rise to the belief that all such problems
could be solved by the use of such chemicals; hence, there was a rush by
all concerned to climb on the broad-spectrum chemical bandwagon and
virtually all research was directed to this end.

This is illustrated by the fact that during the years 1940 - 1950, the
percentage of papers published in international journals on insecticide
testing rose from about 55% to 80%, while those on general biology fell
from 30% to 15%, and on biological control from 10% to 8%.

Thus, the moderate and balanced view was temporarily lost, with
unfortunate consequences in the shape of the appearance of resistance on
an increasing scale, environmental pollution and loss of control
efficiency with recurrence of secondary pest outbreaks.

The backlash generated by this - the "Silent Spring" syndrome - was
equally extreme, and as a result there is now a body of opinion which
claims that the only rational policy for pesticide use is to stop it
altogether.

Such violent swings of opinion generally only impede development of a
rational policy, and hence my insistence in stressing the need for any
policy to be moderate and sensible.

FACTORS AFFECTING PESTICIDE DEVELOPMENT AND USAGE
If you accept this, let us look at the various selection
pressures generated by, or acting upon, three major groups involved in
the pesticide scene - agrochemical companies, users and the public.

There is of course one factor dominating them all – the fact
that pests, diseases, weeds cause world-wide health problems and losses
in agricultural production of up to 40%. This situation is widely
accepted as untenable and, since continued pesticide use can demon-
strably reduce these losses, this is again accepted, albeit reluctantly,
by a majority. But clearly this generalised picture is subject to great
geographical variation; in the United States and Europe, for example,
there is over-production of food, and insect pest problems in
agriculture are less imporant than weeds and diseases. It is presumably
the latter problems that have led to fungicides and herbicides being the
growth areas in pesticide development and use in Europe, whereas over-
production has assisted the development of anti-pesticide lobbies.

In developing countries, where increased agricultural production is
vital to survival and where insect vectors of diseases such as malaria,
filariasis and trypanosomiasis cause enormous health problems, pesticide
use is accepted as necessary and beneficial but even here circumstances
alter priorities. In Sri Lanka, for example, although there is only one
vector of malaria, such is the toll of that disease that the Government
has enacted legislation reserving the use of several insecticides solely
for malaria control. This has removed some of the best weapons for
agricultural pest control from general use and has put a premium on the
development of biocontrol and integrated pest management (IPM) in that
country; it has also produced a black market selling the banned
pesticides to farmers. Thus, policies of rational pesticide use will
vary world wide and relate not only to the local biological situation
but also to the local socio-economic organisation.

Viewpoint of the Agrochemical Industry
Let us now turn to some of the selection pressures acting on
the agrochemical industry. Braunholtz (1977) listed three – "commercial
opportunity, business constraints and technical capability".

Commercial opportunity means satisfying user requirements and is
perceived through an understanding of farmers and their needs; but
business constraints, which are financial, social and political, reduce
the possible response spectrum because companies have to make a profit.
This means that by and large they can only be concerned with major pest
problems in major crops on a world basis, and this cuts out small
products use and weighs against the development of highly selective
compounds. Other major constraints relate to the public demand, often
enshrined in government legislation, for certain standards of safety in
use and in regard to the environment; meeting these standards now
requires very high expenditure.

Total costs for a new product from discovery to marketing are now of the
order of £10-15 million and cost recovery has to be made over 10-15
years. Such a situation is bound to affect the reaction of the chemical
industry to systems such as IPM which have as one objective, a reduction
in pesticide usage. Industry's response to IPM has been somewhat
cautious, to say the least; but against that they have to take into
account the growing public concern over environmental issues; in other
words, they have to compromise.

Viewpoint of the users

What of the selection pressures operating through and on the users, farmers and growers? They must use pesticides to prevent crop losses, but only within the limits of cost effectiveness. This is only one aspect of the economics of the situation. A big problem for farmers is the annual variation of yield and hence of income; this is largely due to weather, but pests and diseases are another major factor, and in the UK for example, the use of fungicides has tended to stabilise cereal crop yields, with consequent improvement in forward planning and use of resources, storage and marketing (Martin, 1981). A further bonus for any pest control technique is that it can save labour, and avoiding the great costs of hand-weeding has been a major factor in promoting the use of herbicides (Martin, 1981). In addition, the increasing complexity of modern farming has created a demand for pest control systems which are safe, simple and easy to apply (Smith, 1983).

Viewpoint of the public and special interest groups

The primary pressure generated by the general public in relation to pesticide use is the demand for agricultural products, where cost and availability are important. However, this pressure is indirect. It is only rarely equated with pest control, as for example when the media report some disease epidemic which temporarily reduces availability and hence increases price. It would seem that an equally important public selection pressure is the demand for cosmetic perfection in produce, which has now reached a ludicrous level and can lead to larger applications of pesticide to ensure that produce qualifies for the highest grade. As Southwood (1979) has pointed out, "it is sometimes claimed that as pest control is expensive, economic forces regulate the number of applications of pesticide; this argument fails to take account of the large economic penalties associated with grading failure compared with costs of an application of pesticide". Thus, even in 1971, Class I apples sold for £165 per tonne, while Class II fetched only £57. Since the cost of application of pesticides per tonne of apples was only £1.40. The temptation - the need, growers could claim - to use extra applications to secure upgrading is understandable.

Western society is also much concerned with the safety risk and safe use of pesticides, and with all aspects of environmental problems caused by them. Also some developing countries, e.g., Kenya and India, have now established environmental protection policies.

The special-interest groups in our societies generate specific pressures, many of which cast pesticides as the villains of the piece. The 'bird lobby' mentioned previously is very powerful, but there are many other influential groups with interests ranging from nature conservation and environmental concern, through wildlife protection and animal welfare, to more specific subjects such as beekeeping. It is one sign of a healthy society that such groups exist but often their policies are extreme and inoperable; they too need to produce rational policies, in which unconscious motivation, arising from emotion and sentimentality, is replaced by reason.

These are only some of the multiplicity of selection pressures operating in respect of pesticide usage, but even so they emphasise both the complexity of the problem and the need to follow moderate and sensible policies.

THE CHANGING RESPONSE IN CROP PROTECTION PRACTICE
To deal with all these factors rationally requires a basic philosophy of approach which is sufficiently flexible to accommodate the variations in pest problems induced by climate and geography, which can respond to changes in circumstances and pressures, whether biological and socio-economic, and which can produce pest control systems which are effective, efficient, relatively stable, environmentally-acceptable and practical.

It is now widely, though not universally accepted, that the concept of IPM offers such a philosophy; it is my personal opinion that it offers the best framework on which to develop effective pest control systems and, in our present context, a rational background to pesticide use.

I said that it was not acceptable universally as the best approach and I ascribe this in some measure to the overzealousness of its proponent pressure groups in claiming too much for it. Complete multidisciplinary IPM systems covering animal pests, weeds and diseases, are rare because they are time consuming and demand considerable specialised inputs during the development stage. Moreover, costs are not always cheaper than conventional chemical control, as Table 1 shows. In this case, the system developed in The Netherlands is complex but its advantages are that the system is stable, long lasting and environmentally acceptable, all essentially unquantifiable in economic in economic terms (de Wilde and Leemans, 1981).

Table 1. Comparative annual costs in £ (sterling) per hectare of pest control in apples using IPM and conventional chemical control (data from IOBC Special Report on Integrated Pest Management, 1981).

	IPM	Conventional
Germany, 1972	176	214
Switzerland, 1974	318	484
The Netherlands, 1980	100	88

The agrochemical industry is not greatly impressed with IPM. One of its aims, that of reducing pesticide use, is directly contrary to industrial interests and the IPM concept of selective chemicals for limited use is unprofitable due to high development costs. Integrated Pest Management does not offer a new technology to exploit or sufficiently attractive investment opportunities and it demands knowledge of technologies, e.g. the production and use of bio-control agents, which most companies do not possess. Nevertheless, since IPM offers some prospect of delaying the build-up of pesticide resistance and of producing systems which are relatively stable and environmentally acceptable, industry has to go along with it, albeit somewhat half heartedly.

Farmers, at least in temperate countries where pesticide resistance is not such a pressing problem as in the tropics, have also been cautious about IPM, mainly because it involves higher financial risk and larger inputs of time and knowledge than conventional control. Against this, the increasing cost of pesticides and their application, and the safety and environmental lobbies, incline them towards "supervised control", the half way house to IPM. Hence, much farmer interest is focused on forecasting and timing of application, reduced dosage and more efficient application. Improvements in the last factor depend on purely practical developments, such as the technique of driving a sprayer through a crop on "tramlines", in which the overall economies offset the crop losses caused by this method. Attention to practicality is vital; it is essential to produce systems which are convenient and can be used by farmers with a minimum of knowledge, preparation and equipment. Smith (1983) listed 26 conditions under which farmers should not apply pesticides; some of my colleagues thought certain of these were facetious, e.g. "do not apply pesticides before reading instructions on the can". However, all his points were serious. If label instructions are not clear, simple and legible they will not be followed properly and the whole control system may suffer.

MAXIMISING THE USE OF PESTICIDE RELATED DATA

We have a responsibility to develop systems which are practical to the last detail. Here I believe that our scientific training in plant protection at all levels does not place sufficient emphasis. I recently chaired a Civil Service selection board to recruit some plant pathologists for ADAS; my two colleagues were ADAS experts of long experience and we were all appalled not so much by the lack of practical experience, which in new graduates is understandable, but by the lack of appreciation of the importance of practical issues. More-over, plant protection is only one of a number of inputs to overall crop protection, such a choice of cultivar, water and fertiliser regimes and overall management. It should not be regarded as an entity in itself, and yet many of those making specialist inputs on the plant protection side have insufficient knowledge and experience of crop production.

A further problem of the integrated approach, made much of by critics, is that because a programme has to be developed in relation to a

particular crop/pest complex in a particular eco-system, the finished
product will be usable only in a restricted area. This overlooks the
fact that much of the data-base is common to many eco-systems, but it
does emphasise the need for a large and accessible data-base. Because
the agrochemical industry must seek, to recall Braunholtz (1977),
"commercial opportunities based on farmers' needs", it must maintain
such a data-base, to discover and test the trends of the selection
pressures from farmers and growers. This is also done by ADAS, and to
some extent by crop consultant firms and by public-sector research
laboratories. Between them, these organisations must have enormous
stores of information and it seems wasteful, to say the least, that
these are not fully utilised.

Hirst (1979) drew attention to this, remarking "I am left with the
impression that nationally we waste a good deal of information by not
more effectively pooling the chemical, biological and crop data
available from all the expensive field trials that are conducted".

Now 1984 has been designated by the UK Government as "Information Year",
is there not a strong case for setting up a National Crop Protection
Data Base to store centrally, analyse and correlate the vast amount of
information gathered annually about crop protection, to be made
available to all on demand? Agrochemical companies might worry about
confidential results, but clearly they could still, while protecting
essential information, contribute a great deal to such a base. Progress
in forecasting and pest management depends on a large accurate data-base
in as many crops and eco-systems as possible. A number of small but
relevant data-bases already exists in the UK, such as the Biological
Records Centre at Monkswood, the ECDIN (Environmental Chemicals Data
Information Network), and the Data Bank of the Department of Health and
Social Security dealing with residues in food, and the Toxicology Data
Bank.

Combining these with the wealth of information available through
industry, ADAS, AFRC and other similar sources would be relatively
simple and cheap, and the potential benefits in relation to developing
rational policies on pesticide use would be great. Such a national
data-base would eventually help us to answer the question "What are
farmers' needs in crop protection?" more precisely, in terms of current
losses both by crop and geographically. We might also be better able to
address the question "How can the need to mitigate these losses be
translated into action?" in the sense of priorities of investments in
research and development programmes by industry and the public sector.

RESEARCH OPPORTUNITIES
          Considering first the general requirements for the future we
must ask "Who is now looking at the future of agriculture in Britain in
the year 2000 and the demands this will make on scientific research?" I
can give you one view; I was at an Agricultural Show recently and the
local National Farmers' Union Secretary told me "Farmers are facing a
recession; they do not want to be told how to improve on the things they

are already doing, because many of them feel they will not be doing them
in a few years time. They want to be told of new approaches,
alternatives to their present ones and new technologies which can help
them in the future".

There seems to me to be a great opportunity here, but it can only be
seized if support for what is now called somewhat disparagingly
"curiosity based research" can be stepped up. The trend, unfortunately,
is the other way; in our own discipline in the UK the recent decision to
close the Weed Research Organization and the Letcombe Laboratory are
examples of the current drive not only to reduce public spending but to
increase privatisation and consumer-orientated research. Of course,
consumer research has an important role to play, but the tying of
research to a particular crop or commodity does not permit work on
systems of wide application. Universities might fill part of this gap
but their resources are also being cut on an inexorable annual basis.
On the other hand, I suspect there is a lot of repetitive work carried
out by industry and government on trials of varieties and systems which
could be rationalised and the money saved spent on modest but essential
basic research.

For example, I mentioned earlier the impasse reached, almost entirely
due to public pressure, on the development of bird pest control - one of
the biggest pest problems world-wide. Given the strength of public
opinion, the only feasible approach seems to be that of developing
chemical repellents in some form or other. A few exist, but some are
toxic compounds, some produce unacceptable side-effects and all are
highly variable in their performance.

My approach to this would be to institute a programme of research on the
chemical basis of bird feeding behaviour, similar in principle to that
carried out in the last 10 - 15 years with insects, which has produced
such worthwhile results. When we know the chemical and behavioural
reasons why certain bird species attack certain fruits at certain times
we could well be on the way to develop a repellent, either as an
externally-applied chemical or by altering the crop plant chemistry
(e.g. by diminishing an essential phagostimulant or nutrient which is
attracting the bird in the first place).

Following the same line of reasoning, I would like to see more work on
the role of plant biochemistry in the pest/host relationship and in the
action of growth stimulants, growth retardants, germination activators
and so on.

Research into the structure and mode of action of toxins produced by
fungi and bacteria also seems essential. Microbial pesticides have been
with us for some time but the latest developments, such as the new
formulations of Bacillus thuringiensis H14 are extraordinarily effective
against a number of pests, are safe to use and environmentally
acceptable. They offer the enormous possibility of bacterial strain
selection to tailor the product to a specific requirement and they are
relatively cheap. Of course, like all compounds they have disadvantages

- in this case slow action and the need to be ingested by the pest for
activation. To solve these problems we need to know more about the
detailed structure of the toxins and their mode of action. They are one
of the classes of pesticide that are relatively compatible with the
natural enemy complex, and selectivity towards this is a subject badly
needing more research.

Behaviour modifying chemicals, the semio-chemicals, were hailed on their
discovery as offering a cheap and environmentally-acceptable method of
pest control. Once again the development of practical systems has
proved to be the problem and it seems to me it would be sensible to stop
identifying yet more and more of these compounds and put extra effort
into their practical use. It is accepted now that they offer special
advantages in forecasting (Haskell, 1981).

One particular application of semio-chemicals seems to me to have great
potential. I refer to the work of Prokopy and Lewis and their
colleagues (Prokopy et al., 1978; Lewis et al., 1975, 1977) on chemicals
used by parasites and predators to home in on their prey, because this
may offer the chance of improving the performance of a natural enemy
complex in a given eco-system. Further field studies (Lewis et al.,
1979) showed that the use of synthetic kairomones could indeed increase
predation, but there was a limit to this because the whole system was
density dependent. However, there is also the exciting prospect of
using plant-produced chemicals in situ, to attract predators to the crop
area, where they then can detect and follow the prey.

This and other work points the way towards increasing the efficiency of
natural enemy populations, a subject which must be one of the most
important in relation to pesticide use in the future.

Once we have got our chemicals, we need to deliver them into or on to
the target and here, as we all know, there is much to be done to
decrease overall dosage of active ingredient while still depositing a
lethal dose on to the target, to reduce drift and overall losses and to
improve sticking power and penetration. In this connection, there is
much research interest in spray droplet size, controlled droplet
application and the use of electrostatics. However, it has always
surprised me that more research has not been done on systemic
application - one of the most elegant, safe and effective methods of
transferring pesticide to target.

Crisp (1971) described the possible use of precursor compounds which are
modified by the metabolic processes in the plant to produce the active
compound during or after translocation. Wain (1978) described a
possible approach to the development of systemic fungicides by combining
the latter with a sugar molecule, which transported it within the plant
and into the pathogen.

By combining such approaches with that of controlled release, would it
be possible to produce a capsule which, implanted in the soil, releases
at periods related to attack by pests or diseases, compounds trans-

located by the plant which can deter such attacks? Such a system is no
more complex than some of those which have evolved in nature; for
example, when an insect lands on its specific host plant, it receives a
chemical stimulus which causes it to produce an attractant sex pheromone
so that mating and oviposition takes place on the host. Further
possibilities exist by putting into or on to the plant, precursors of
the desired compound, which are activated by the effect of light or
temperature. Work at Rothamsted has shown this is possible with
pheromones and I think it is a technique worth further exploitation.
Harnessing plant physiological processes to time the release of an
active compound in relation to insect attack could help in damping down
losses early on, when they are least detectable.

An obstacle to the wider use of systemic compounds against insect pests
has been that they affect only sucking insects, but our knowledge of
pest/host plant interaction and its chemical basis must be reaching the
point where we can design and develop systems, using the techniques
mentioned above, to produce chemicals on plant surfaces which deter both
insect pests and plant pathogens.

It would seem that approaches of this nature would be supported by
selection pressures from industry, users and the public; they would
overcome costly and environmentally difficult application methods, they
would be simple and cheap to apply, they would be safe and non-
polluting. But to develop these would need a multidisciplinary effort
by entomologists, chemists and plant physiologists which may be beyond
the capacity of any one organisation.

Is not this a case for the development of cost-sharing R and D
programmes between industry and government? Many people have both
advocated and been involved in such collaboration, but it has always
been on a small scale and generally the result of pressures by
individuals. The Government-backed organisation set up to facilitate
such work in the UK, the British Technology Group, has not had much
success in this direction and some new approach is necessary to harness
the experience and expertise which exists in numerous organisations
throughout the UK. The lack of practical joint industry/government
effort seems to me to constitute a severe obstacle to development of a
rational policy of pesticide use.

PESTICIDES AND THE ENVIRONMENT
        I turn finally to the question of pesticides and the
environment. If ever a subject required a rational view, this is it,
and yet unfortunately, it is the one most beset by unconscious
motivation in the form of fear, sentiment, emotion and tradition.
Although objective, overall cost/benefit analyses, such as that of
Pimentel et al. (1980) show a positive return of 4 to 1 favouring
pesticide use, this can seem irrelevant in the face of the core of truth
on which current public attitudes are based.

Pesticides are toxic compounds deliberately introduced into the environment, where they can cause death and disease in man and other non-target organisms and produce biological changes in such effects as pesticide resistance and shifts in wild-life populations.

These effects of pesticides, potential and actual, are emphasised by a number of pressure groups, several of them international, and ranging in influence from political parties with considerable public support to lunatic fringe bodies. Included in their ranks are many organisations with rational objectives cogently argued, but there are others where emotion, exaggeration and exploitation predominate.

It is my personal opinion that this area of concern forms one exception to the rule that extreme views are ultimately self-defeating, because in general and on balance, pressures generated by environmentalists in the last decade have stimulated and exacted responses from the crop protection discipline, in the shape of more concern for, interest in, and research on environmental effects, safety measures and non-chemical control techniques, which have been of overall benefit.

But there seem to me to be two outstanding problems: the first is that in relation to many environmental questions we lack both reliable information and the support to carry out the necessary research to get it, and the second is that we lack both means and mechanism to mount a really objective and informative campaign to put the public in possession of an objective balance sheet on pesticide use. Environmental research needs to be continued over a time span sufficient to detect lasting ecological effects; this would be another field materially assisted by the establishment of a National Crop Protection Data Base.

The lack of a mechanism to mount an effective public relations campaign provides an interesting example of the interaction of selection pressures. Clearly both the agrochemical industry and users are concerned to demonstrate that they have at heart the safety of special sections, like spray operators, as well as that of the public at large, that their products do not represent a public health threat and that their operations overall do not damage the environment to an unacceptable degree.

They are hampered in their defence of current pesticide use because the public image of pesticides has been directed by the anti-chemical lobby precisely towards these problems, which are real, whereas the public can never see direct evidence of the benefit derived from pesticide use. Although this could be remedied, such an effort has been neglected by the major companies, and by trade associations such as GIFAP, in favour of an over-emotive response of their own to the more extreme opinions of the environmentalist lobby, such as for example the total cessation of pesticide use. This has never been proposed by the bulk of the lobby, as has been made clear on several occasions (e.g. Bull, 1982) but a lot of the industry response has dealt with the problems of food supply and public health, which would ensue if pesticides were to be abandoned and even here, as Pimentel et al. (1980) have pointed out, exaggerated claims have been made.

The farming community, while not over-reacting so much, has also shown unwarranted defensiveness on a range of environmental issues. Of course, the development of agriculture world wide has been the cause of huge changes in the environment, some acceptable and some not. Farmers have had traditional and emotional support from the general public, but this has decreased of late because of their increasing awareness of environmental issues and, in Europe, because of the current overproduction of crops.

There are problems in developing a telling public relations campaign, since all the media rate the headline "Man dies of pesticide poisoning" far more newsworthy than "Integrated pest management reduces pesticide use". However, I think if serious thought was given to this by industry, governments and users, an informative campaign could be developed to redress the present balance of selection pressures. The British Crop Protection Council, a unique body representing the agrochemical industry, growers and advisers, government and universities, has a current information and training programme, including such schemes as BASIS, which is good but needs to be much expanded to be really effective. Industry ought to support such an expanded effort, otherwise further and probably unnecessary restrictions and regulations on the registration and use of pesticides, much more draconian than those now about to be introduced in the UK, will be demanded by growing public pressure.

CONCLUSION
Rational use of pesticides is necessary - but rational for who - agrochemical industry or users or the public? There is clearly no simple solution because we see that some of the selection pressures acting on these groups are not conductive to rational pesticide use and some are acceptable only to one or other of those groups. I think that to reconcile these pressures so that a rational policy can be developed requires that more attention be paid to the socio-economic background of pesticide use. This means investigation of how the manifold selection pressures, a few of which I have described, are generated, how they interact and how that interaction affects the practical policies followed by industry and government. Such investigation will not be a 'one-off' study. Attitudes both within and towards agriculture are changing rapidly, especially in Europe. We must ensure that the all crop protectionists identify it, and respond to the new challenges.

REFERENCES
Braunholtz, J.T. (1977). The crop protection industry: products in prospect. Proceedings of the 9th British Insecticide and Fungicide Conference, 3, 659-670. British Crop Protection Council Publications.
Bull, D. (1982). A Growing Problem: Pesticides and the Third World Poor, pp. 192. Oxford: Oxfam.
Crisp, C.E. (1971). The molecular design of systemic insecticides and organic functional groups in translocation. Proceedings of

the 2nd International Congress of Pesticide Chemistry $\underline{1}$, 211-264.

Haskell, P.T., Campion, D.G. and Nesbitt, B.F. (1981). Behaviour modifying chemicals in tropical pest management. Mededelingen van let Faculteit Landbouwwetenschappen Rijksuniversiteit Gent $\underline{46}$/1. 27-38.

Lewis, W.J., Jones, R.L., Nordlund, D.A. and Sparks, A.M. (1975). Kairomones and their use for management of entomophagous insects. I. Evaluation for increasing rates of parasitisation by Trichogramma spp. in the field. Journal of Chemical Ecology $\underline{1}$, 343-347.

Lewis, W.J., Nordlund, D.A., Gross, H.R., Jones, R.L. and Jones, S.L. (1977). Kairomones and their use for management of entomophagous insects. V. Moth scales as a stimulus for predation of Heliothis zea (Boddie) eggs by Chrysopa carnea (Stephens) larvae. Journal of Chemical Ecology $\underline{3}$, 483-487.

Lewis, W.J., Beevers, M., Nordlund, D.A., Gross, H.R. and Hagen, K.S. (1979). Kairomones and their use for management of entomophagous insects. IX. Investigations of various kairomone treatment patterns for Trichogramma spp. Journal of Chemical Ecology $\underline{5}$, 673-680.

Martin, J.S. (1981). The use of chemicals in modern farming - a farmer's view. Proceedings of the 11th British Crop Protection Council Conference - Pests and Diseases $\underline{3}$, 675-683. British Crop Protection Council Publications.

Pimentel, D., Andow, D., Dyson-Hudson, R., Gallahan, D., Jacobson, S., Irish, M., Kroop, S., Moss, A., Schreiner, I., Shepherd, M., Thompson, T. and Vinzant, B. (1980). Environmental and social costs of pesticides; a preliminary assessment. Oikos $\underline{34}$, 127-140.

Prokopy, R.J., Zeigler, J.R. and Wong, T.T.Y. (1978). Deterrence of repeated oviposition by fruit-marking pheromone in Ceratitis capitata. Journal of Chemical Ecology $\underline{4}$, 55-63.

Smith, P.R. (1983). Farming - partly an art and partly a science. Proceedings of the 10th International Congress of Plant Protection $\underline{3}$, 56-60.

Southwood, T.R.E. (1979). Pesticide usage: prodigal or precise. Proceedings of the 10th British Crop Protection Council Conference - Pests and Diseases $\underline{3}$, 603-619. British Crop Protection Council Publications.

Wain, R.L. (1978). Crop protection - some developments and future prospects. Report of Long Ashton Research Station for 1977, 224-235.

de Wilde, J. and Leemans, A. (1981). Cost-benefit aspects of integrated pest management. IOBC Conference on Integrated Pest Management, Bellagio, 1980, 42-49.

Wright, E.S. (1980). Bird Problems in Agriculture, pp. 212. Croydon, England: British Crop Protection Council Publications.

Section 1. ENVIRONMENT

2. EFFECTS OF PESTICIDES ON WILDLIFE AND PRIORITIES IN FUTURE STUDIES

J.P. Dempster
The Institute of Terrestrial Ecology, Monks Wood
Experimental Station, Abbots Ripton, Huntingdon,
Cambridgeshire, England

INTRODUCTION

This paper describes some of the ways in which pesticides have affected wildlife over the past 30 years, and considers the types and the uses of pesticides which are most likely to cause environmental problems. It also attempts to identify those gaps in our knowledge that most urgently require research. This is a huge field of study which contains numerous problems which deserve attention, and so my remarks are based primarily on experience in Britain. Furthermore, the assessment of research priorities which I give is very much a personal one.

Since pesticides are designed to kill living organisms and the great majority of pesticides are non-specific poisons, it is almost inevitable that any application of a pesticide will kill some non-target species. Of course, many individual animals and plants will be killed by a vast range of human activities. However, it is not the fate of individuals which concerns most ecologists when considering the impact of pesticides, but rather it is the fate of populations. High death rates do not necessarily imply harmful effects to populations, since a high mortality may be countered by density-dependent changes in subsequent survival, reproduction or dispersal. In other words, the fact that the population has crashed may lead to increased degrees of survival, reproduction or immigration. The impact of a pesticide will depend also upon its degree of toxicity to other species which affect the population, as competitors, enemies or prey, as well as to the pest species itself (Dempster, 1975). Chemicals vary greatly in their spectra of toxic action on different species, and in their likelihood of coming into contact with each species. Thus, a laboratory LD_{50} for a given chemical and one species tells us little about the likely effects of that chemical in the field (Moriarty, 1983).

Another inevitable, indirect, effect of a pesticide application, is the removal of a large part of the food supply of species feeding in the crop. Thus, an insecticide inevitably reduces the food supply of insectivores, whilst a herbicide will remove food plants and weed seed needed for many herbivores. In this respect, pesticides are no different from any non-chemical means of pest control, they are simply more effective.

It is clear, then, that the effect of any one pesticide on the crop flora and fauna (and on the flora and fauna of surrounding land, if the pesticide is allowed to reach it) is extremely complicated and difficult to predict. Generally, insufficient will be known about the population ecology of non-target species to predict the effects of pesticides on them, and one has little alternative but to do large-scale field trials with the chemical and to monitor the impact. Nevertheless we can still learn from past experiences, and make generalisations about the likelihood that certain types and uses of pesticide will cause wildlife problems.

The effect of a pesticide on non-target species, in the majority of cases, is likely to be local and temporary. Most populations will probably recover, either by reproduction or by recolonisation. A species with a very local distribution will be especially susceptible to damage. Even a widespread species may be harmed if the proportion of its population affected is sufficiently large to prevent normal recovery. This is much more likely to happen if the pesticide is applied over a very large area, or if it is present for a long time due to its persistence or repeated application. A further consideration is that harmful sublethal effects on populations can occur with some, perhaps all, pesticides, particularly if the chemical is very persistent.

ORGANOCHLORINE INSECTICIDES

Public concern over the effects of agricultural pesticides on wildlife originated largely from the use of one group of chemicals, the organochlorine insecticides. Besides being toxic to a wide range of organisms, these chemicals have three main properties which contribute to their impact on wildlife. They are chemically very stable, so that they can persist in the environment, more or less unchanged, for many years, and can be moved considerable distances, to areas remote from agricultural land, in air or water, or in the bodies of migrating animals, They are soluble in fat, so that they tend to accumulate in the fatty tissues of animals and can be transferred along food chains, from prey to predators. They can affect the reproduction of certain species, notably birds, at sublethal concentrations.

The problems caused by the organochlorine insecticides came to light in the UK when large numbers of seed-eating birds were killed in the springs of 1956-60, as a result of feeding on cereal seed which had been treated with aldrin, dieldrin or heptachlor, to control wheat bulbfly. Bird casualties were locally so large that the use of these materials in spring was discontinued by agreement between Agrochemical Industry and Government. During the period when they were still in use, several birds of prey decreased greatly in numbers. The sparrowhawk (Accipiter nisus) and peregrine falcon (Falco peregrinus), both bird-feeding raptors, were particularly affected (Ratcliffe, 1963, 1980; Prestt, 1965), but lesser or more local declines occurred in several other birds of prey. A survey in 1964 showed that the sparrowhawk, formerly one of the commonest and most widely distributed diurnal birds of prey, had been virtually exterminated from much of southern and eastern England. Deaths had been caused by these predators feeding on birds which had been poisoned by eating cereal seed dressed with organochlorine pesticides.

There was some resurgence of populations of both the peregrine and sparrowhawk in the north and west of England, within a few years of the bans on the use of aldrin, dieldrin and heptachlor as seed treatments but recovery in the intensively arable areas further south has been very slow. In fact, the sparrowhawk is still absent as a breeding bird from Lincolnshire, Cambridgeshire, and much of Essex, Norfolk, Suffolk and Kent (Newton and Haas, 1984).

Since the time of the population crash, reproduction of these species has been impaired as a result of their laying a high proportion of thin shells eggs. This has led to egg breakage and a high embryo mortality. The thin eggshells have been caused by a continuing contamination of these birds by another organochlorine, DDE, a persistent metabolite of DDT. Decreased reproduction has probably slowed down the recovery of both the peregrine and sparrowhawk, although it has not prevented it.

Many seed-eating species of bird also declined around 1960, but these shorter-lived species had a sufficiently high reproductive rate to recover more quickly than the raptors.

Since the era of the organochlorines, there have been few reports in the UK of major wildlife incidents involving pesticide poisoning. This is partly the result of the greater vigilance provided by the Pesticides Safety Precautions Scheme and the quick response by the Ministry of Agriculture, Fisheries and Food (MAFF) in recommending the stoppage in use of pesticides which appeared harmful. However, one particular series of incidents is worth a mention as it demonstrates the difficulty in predicting problems.

ORGANOPHOSPHATES

With the phasing-out of the organochlorine insecticides, two organophosphates, carbophenothion and chlorfenvinphos were used as cereal seed treatments to control wheat bulbfly. This use of carbophenothion, led to a series of incidents of poisoning of geese (particularly Anser spp.) feeding on recently-sown fields (Stanley and Bunyan, 1979). The first occurred in November 1971, and involved the death of approximately 500 greylag geese (Bailey et al., 1972). Deaths of both greylag and pinkfooted geese also occurred in the winter of 1974-75 (Hamilton et al., 1976). Game birds, pigeons, corvids, rabbits and rats feeding on the treated grain on the same sites appeared to be unaffected. Laboratory studies on the toxicity of carbophenothion to five species of geese indicated that grey geese (Anser spp.) were very susceptible to poisoning, more so than geese of the genus, Branta (MAFF, 1978). Although the deaths of geese were restricted to a few areas in southern Scotland and Humberside, a very high percentage of the world population of pinkfooted geese overwinters in Britain and could have

been at risk. MAFF therefore advised farmers to use alternative
chemicals in those areas where overwintering geese fed.

Two points are particularly worth noting about these incidents. First,
the method of application of the pesticide, to grain which could be left
on the surface of the soil, brought seed-eating birds directly into
contact with high concentrations of carbophenothion (as had also
happened with the organochlorine seed dressings). Secondly, it is
unlikely that the usual toxicity tests done on laboratory animals would
have identified the exceptional sensitivity of grey geese to
carbophenothion.

OTHER PESTICIDES
Although there have been a number of other incidents of
pesticides killing birds and mammals (Hardy and Stanley, 1984) these
have all been of a very local nature. It is unlikely that any direct
toxic effect causing deaths of vertebrates would have been overlooked,
but far less is known about the impacts of agricultural pesticides on
invertebrates. Birds appear to be particularly sensitive to pesticides,
and public interest is such that large kills of birds receive immediate
publicity. Kills of smaller, less "popular" animals would almost
certainly go unnoticed. One exception is the honey bee, and large kills
of bees have been observed repeatedly over the years (Stevenson et al.,
1978), generally as a result of spraying a crop at the time of
flowering. Honey bees usually return to the hive to die and so high
mortalities are readily noticed. The four-fold increase over the past 5
years (1979-1984) in the UK area of oilseed rape has resulted in deaths
of bees due to the spraying of this crop with triazophos, an
organophosphate insecticide, to control Ceuthorynchus and Dasineura.
The effects of these sprays on other pollinating insects are likely to
be comparable to those on honey bees, but no quantitative data are
available.

FACTORS AFFECTING SEVERITY OF PESTICIDE SIDE-EFFECTS
We have seen that chemical persistence was central to the
problems caused by the organochlorine insecticides. Similar effects
might also be expected if a less persistent pesticide were applied
repeatedly to a perennial crop, or to an annual crop continuously grown
on the same land. The effect on non-target species would be all the
greater if the crop were grown over very large areas, since this would
increase the likelihood of species being eliminated, with little chance
of successful recolonisation by untreated organisms. Little research
has been done to determine the effects of scale of use on pesticide
impact but, in my opinion, change in the scale of use of pesticides is
the most worrying aspect of agriculture in Britain today, particularly
with respect to the control of cereal pests.

In 1983, almost 80% of tilled land in Britain, or about 4 million
hectares was under cereals, mainly wheat and barley. Since 1945 there
has been a revolution in cereal growing with yields per hectare more

than doubling, due to the development of high-yielding and disease-resistant cultivars and an increased use of fertilisers and pesticides. Often break crops are no longer grown; instead, cereals are grown continuously on the same land. Reduced cultivation and the increasing tendency to sow in autumn, rather than spring, have greatly increased the need for pesticides, so that a large number of applications of different chemicals are now applied routinely to cereals each year. Figure 1 shows the number of applications of pesticides applied to cereals in 1977 (Steed et al., 1979), where it can be seen that on average four applications of pesticide were made. Some of these would have been mixtures of more than one chemical, such as an insecticide plus a fungicide, or a fungicide and herbicide mixture. Forty-four per cent (over one million ha) of the area under wheat and barley received five or more applications (Fig. 2.), although the very large numbers of applications made to some crops were mainly spot treatments to control small areas of particular pests, such as weeds. The increase in the use of pesticides on cereals has been rapid. For example, field applications of insecticides to cereals increased from 44,000 ha in 1974 to 534,000 ha in 1977, i.e. an 11-fold increase in three years (Sly, 1977; Steed et al., 1979) and there has been a similar large increase in the use of fungicides more recently. All told, a huge number of toxic materials is being applied to a vast area of land, but we have very little idea of their impact on our flora and fauna.

ASSESSING PESTICIDE EFFECTS: CURRENT STUDIES

It is difficult to assess the effects of a large number of pesticides, repeatedly applied to a piece of land. We can test the toxicity of individual chemicals to a range of organisms in the laboratory, but this tells us little about the likely effect of mixtures of chemicals, applied against a background of other activities associated with crop husbandry, on the total flora and fauna occupying the crop and its immediate surrounds. Added to this, agricultural techniques are changing rapidly; for example, the current swing towards winter cereals carries with it a change in pesticide use which is still evolving. In my opinion, the only way of studying the impacts of pesticides in cereals is to monitor as much as possible of the total cereal ecosystem. Any change that is identified would then require experimental study to establish the causes.

Two current research projects in the UK go some way towards doing this: the work of the Game Conservancy in West Sussex and the MAFF experiment at Boxworth EHF in Cambridgeshire. The Game Conservancy's study has shown large downward trends in the numbers of many plants and insects in cereal fields in the study area since 1970 (Potts, 1984). The decline in the grey partridge (Perdix perdix) has been a direct result of the loss of insect food caused by the widespread use of pesticides in cereals, and especially by herbicides which remove insect food plants (Potts, 1980). The MAFF experiment has begun more recently, and aims to determine the total effects of cereal husbandry on pest incidence, crop yield, and the flora and fauna, both within the crops and in adjacent habitats. Within the experiment, three pesticide strategies are being

Figure 1. Number of different pesticides applied to wheat and barley in 1977 (from Steed _et al._, 1979).

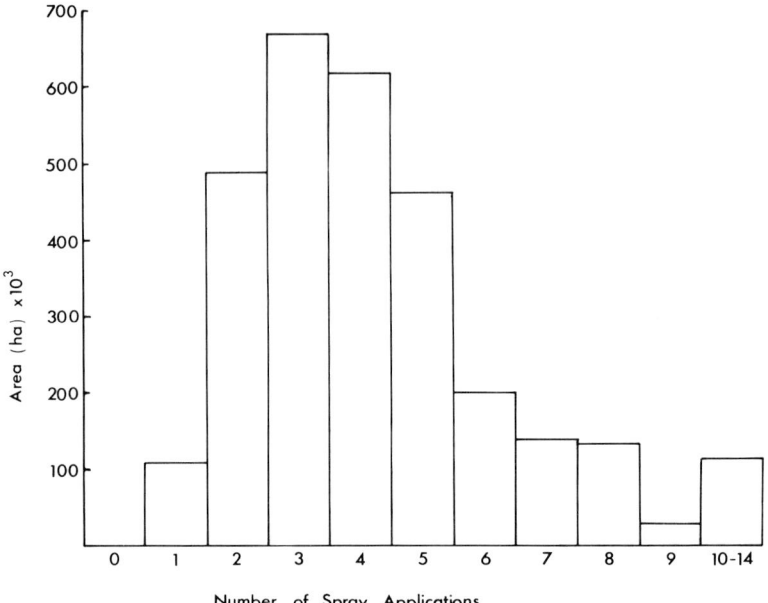

Figure 2. Area of wheat and barley receiving different numbers of pesticide applications in 1977 (from Steed _et al._, 1979).

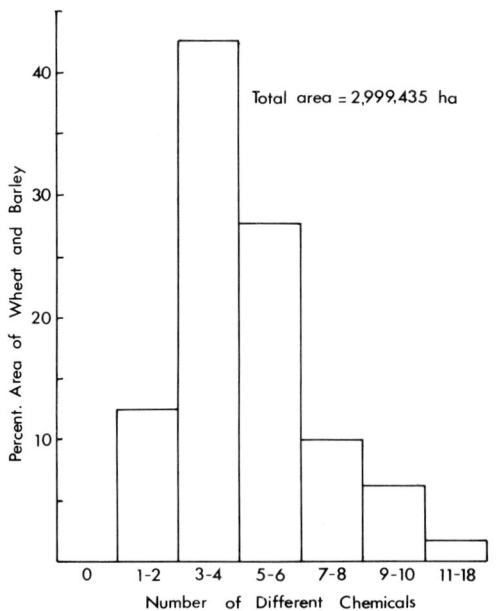

compared, (i) high 'insurance' pesticide input, (ii) a 'reduced' pesticide input, based on pest and disease monitoring, and (iii) an 'integrated' approach in which changes in husbandry will be exploited to minimise the need for pesticides (Stanley and Hardy, 1984).

Many of the insects shown by the Game Conservancy's study to be adversely affected by pesticides used on cereal crops, are beneficial to the farmer, particularly as predators of cereal aphids. This highlights the problems of impacts of pesticide use which are directly harmful to the farmer himself, and opens up the question of the long-term sustainability of crop production using present methods of crop protection and nutrition. This is too big a subject to elaborate upon here, but I would stress one point. The impacts of crop husbandry and pesticide use on soil flora and fauna and very poorly known, largely because we know so little about the precise 'roles' of many soil inhabitants. Although there have been studies of the effects of certain individuals pesticides on major arthropod groups inhabiting the soil (Edwards, 1984), little research has been done on the effects of regular applications of the mixtures of chemicals routinely applied to cereal crops. There is an urgent need for more research on this subject.

Just as one might expect the effects of repeated applications of non-persistent pesticides, on plant and animal populations, to be comparable to the effects of persistent chemicals, it is possible that repeated exposure may lead to sublethal effects which can affect populations. Again we just do not know. A characteristic effect of organophosphorus insecticides on vertebrate animals is to inhibit the activity of a range of esterase enzymes, including acetylcholine esterase, which is important in the transmission of impulses across nerve synapses. Long-term impacts of esterase inhibition on the behaviour of animals in the laboratory and the field requires study.

CONCLUDING COMMENTS

The main points made in this paper can be summarised as follows:

1. Ecological effects are brought about by the impacts of pesticides upon populations, not upon individuals. This makes it necessary to assess effects against a knowledge of the population dynamics of organisms.

2. Species differ from one another in their suceptibility to toxic chemicals, and there are limitations to the value of laboratory tests of toxicity in predicting effects in the field.

3. The major declines in wildlife populations, known to have occurred as a result of the use of pesticides, have involved highly persistent chemicals. The repeated application of less persistent pesticides could theoretically have similar effects to the persistent ones – research to determine whether this is so is urgently needed.

4. The recent changes in the scale of pesticide use on
cereal crops are likely to have an increased impact on our
flora and fauna. Insufficient research is being done to
monitor changes, particularly in soil organisms.

5. Our inability to predict the likely impact of any single
pesticide, when applied in the field, stems largely from our
ignorance about the basic ecology of most species, and how
populations interact one with another. Even in simplified
agricultural ecosystems, much basic ecological research
needs to be done.

6. I have confined my remarks to research required to
understand the environmental impacts of pesticides.
Complementary research to find ways of reducing the usage of
pesticides, by alternative control measures, is also
urgently needed.

7. Although this paper concentrates on the impacts of
pesticides, these are only one of the tools of agriculture.
To understand their impacts upon the inhabitants of crops,
studies must include other agricultural practices, since
these too will be affecting wildlife species.

REFERENCES

Bailey, S., Bunyan, P.J., Hamilton, G.A., Jennings, D.M. and Stanley,
P.I. (1972). Accidental poisoning of wild geese in
Perthshire, November 1971. Wildfowl 23, 88-91.

Dempster, J.P. (1975). Effects of organochlorine insecticides:
persistent organic pollutants. In Organochlorine
Insecticides: Persistent Organic Pollutants, ed. F.
Moriarty, 231-248. London: Academic Press.

Edwards, C.A. (1984). Changes in agricultural practice and their impact
on soil organisms. In Agriculture and the Environment, ed.
D. Jenkins, 56-65. Cambridge: Institute of Terrestrial
Ecology.

Hamilton, G.A., Hunter, K., Ritchie, A.S., Ruthven, A.D., Brown, P.M.
and Stanley, P.I. (1976). Poisoning of wild geese by
carbophenothion treated winter wheat. Pesticide Science 7,
175-183.

Hardy, A.R. and Stanley, P.I. (1984). The impact of the commercial
agricultural use of organophosphorus and carbamate
pesticides on British wildlife. In Agriculture and the
Environment, ed. D. Jenkins, 72-80. Cambridge: Institute of
Terrestrial Ecology.

MAFF (1978). Pest Infestation Control Laboratory Report, 1974-76.
234-239. London: HMSO.

Moriarty, F. (1983). Ecotoxicology, the Study of Pollutants in
Ecosystems. London: Academic Press.

Newton, I. and Haas, M.B. (1984). The return of the sparrowhawk.
British Birds 77, 47-70.

Potts, G.R. (1980).   The effects of modern agriculture, nest predation
        and game management on the population ecology of partridges
        (Perdix perdix and Alectoris rufa).   Advances in Ecological
        Research 11, 2–79.
Potts, G.R. (1984).   Monitoring changes in the cereal ecosystem.   In
        Agricultural and the Environment, ed. D. Jenkins, 128–134.
        Cambridge: Institute of Terrestrial Ecology.
Prestt, I. (1965).   An enquiry into the recent breeding status of some
        of the smaller birds of prey and crows in Britain.   Bird
        Study 12, 196–221.
Ratcliffe, D.A. (1963).   The status of the peregrine in Great Britain.
        Bird Study 10, 56–90.
Ratcliffe, D.A. (1980).   The Peregrine Falcon.   Vermillion, South
        Dakota: Butes Publishers.
Sly, J.M.A. (1977).   Review of usage of pesticides in agriculture and
        horticulture in England and Wales, 1965–1974 (Survey Report
        8).   Pinner: MAFF.
Stanley, P.I. and Bunyan, P.J. (1979).   Hazards to wintering geese and
        other wildlife from the use of dieldrin, chlorfenvinphos and
        carbophenothion.   Proceedings of the Royal Society, London B
        205, 31–45.
Stanley, P.I. and Hardy, A.R. (1984).   The environmental implications of
        current pesticide usage on cereals.   In Agriculture and the
        Environment, ed. D. Jenkins, 66–72.   Cambridge: Institute of
        Terrestrial Ecology.
Steed, J.M., Sly, J.M.A., Tucker, C.G. and Cutler, J.R. (1979).   Arable
        Farm Crops, 1977 (Survey Report 18).   Pinner: MAFF.
Stevenson, J.H., Needham, P.H. and Walker, J. (1978).   Poisoning of
        honeybees by pesticides.   Investigations of the changing
        pattern in Britain over 20 years.   Report of Rothamsted
        Experimental Station for 1977 2, 55–72.

3. INTEGRATING CHEMICAL CONTROL WITH THE ACTIVITY OF
   BENEFICIAL ORGANISMS

S.A. Hassan
Institute for Biological Pest Control, Heinrichstrasse 243,
D-6100 Darmstadt, West Germany

INTRODUCTION

Beneficial arthropods, parasites and predators of agricultural pests reduce the population of their host or prey and help to limit damage caused by the pest. Modern plant protection recommends therefore the reduction of the use of chemical pesticides to a minimum. If however the use of pesticides is indispensable, selective pesticides that would control the pests without destroying their effective natural enemies should be chosen. In recent years there has been considerable interest in the development of progammes that assure a more compatible use of chemical and biological methods of pest control. Among the most useful of these programmes are some in which success depends upon the selection of pesticides that are less toxic to the most important natural enemies of pests in a crop. Unlike the more complicated practices involving special timing or placement of toxicants to avoid pesticide contact with the natural enemies, this type of programme requires less supervision and less knowledge of the differences in pest and natural enemy behaviour.

SEARCH FOR SELECTIVE PESTICIDES

In order to choose selective pesticides that can be used in integrated control programmes, pesticides available in the market should be tested on the most important natural enemies of pests in the different crops. In an attempt to encourage international co-operation in this field of work, the Working Group "Pesticides and Beneficial Organisms" of the International Organisation for Biological Control (IOBC), West Palaearctic Regional Section (WPRS) was formed in 1974. Standard testing methods agreeable to the members in the different countries were developed and joint pesticide testing programmes were organised. Laboratory, semi-field and field test methods were developed and the testing of 60 pesticides was completed in 1984. The natural enemies tested by members of the Working Group at the present time are: Trichogramma cacoeciae (Chalcidoidea, Hym.), by Hassan at Darmstadt; Coccygomimus turionellae (Ichneumonidae, Hym.) by Bogenschütz at Stegen-Wittental; Phygadeuon trichops (Ichneumonidae, Hym.) by Naton at Munich; Chrysopa carnea (Chrysopidae, Neur.) by Bigler at Zurich; Leptomastix dactylopii (Hym.) by Viggiani at Naples; Drino inconspicua (Tachninidae, Dipt.) by Huang at Göttingen; Amblyseius potentillae, Typhlodromus pyri (Phytoseiidae, Acari) by Overmeer and van Zon at Amsterdam; Phytoseiulus persimilis, Typhlodromus finlandicus (Phytoseiidae, Acari) by

Vanwetswinkel at Sint-Truiden; Syrphus vitripennis (Syrphidae, Dipt.) by
Rieckmann at Hannover; Encarsia formosa (Aphelinidae, Hym.) by Oomen at
Wageningen; Cryptolaemus montrouzieri (Coccinellidae, Col.) by Brown at
Kent; Anthocoris nemorum (Anthocoridae, Heteropt.) by Firth also at
Kent; Opius sp. (Braconidae, Hym.) by Ledieu at Littlehampton;
Lepthyphantes tenuis (Linyphiidae, Araneae) by Inglesfield at Kent;
Aleochara bilineata (Staphylinidae, Col.) by Samsøe-Petersen at Lyngby;
Pterostichus cupreus (Carabidae, Col.) by Edwards at Bracknell;
Bembidion lampros (Carbidae, Col.) by Chiverton at Uppsala; Cales noacki
(Hym.) by Vivas at Valencia; and the entomopathogenic fungus
Verticillium lecanii (Moniliaceae, Hyphomycetes) by Tuset also at
Valencia.

Forty pesticides were tested on 13 beneficial organisms by group members
in six countries. The results of these tests as well as information on
the testing methods and the addresses of the testing members of the
group were published by Franz et al. (1980) and Hassan et al. (1983). A
further publication involving another 20 pesticides and 15 beneficials
is in preparation. Table 1 shows a list of 23 pesticides (8
insecticides/acaricides, 11 fungicides and 4 herbicides) as well as 2
plant growth regulators. They were found to be less harmful to most of
the tested beneficial organisms. These preparations can therefore be
recommended for use in integrated control. Although the group members
test the pesticides on as many beneficial organisms as is practical,
there is a general agreement that the beneficials chosen for a routine
testing should be relevant to the crop on which the particular pesticide
is to use used. Table 2 gives some examples of the choice of
beneficials in different crops.

BENEFICIAL ARTHROPODS DEVELOPING RESISTANCE TO PESTICIDES
        Although integrated control programmes that are based on the
use of true selective pesticides are more reliable, in some cases it was
possible to make use of beneficial arthropods that are resistant to
pesticides. Predatory mites seem to develop tolerance or resistance to
pesticides more easily than other beneficial arthropods. Several
predatory mites are known to be key natural enemies of spider mites on
fruit orchards, vineyards and on glasshouse crops. Resistance of
predatory mites to commonly used pesticides was recorded by Hoy and Knop
(1979) with Metaseiulus occidentalis, by Schulten et al. (1976) and
Hassan (1982) with Phytoseiulus persimilis and by Croft (1976) with M.
occidentalis, Amblyseius fallacis and Typhlodromus pyri and by Englert
and Kettner (1983) with T. pyri and successful integrated control
programmes were developed.

CLEVER USE OF NON-SELECTIVE PESTICIDES
        Varying method, time and dose of application of pesticides
can reduce its toxicity to beneficial arthropods. The use of
insecticides in the soil, for example to control the cabbage root fly,
is known to kill predatory Carabidae and Staphylinidae. It was reported
that these losses can be reduced to a minimum by restricting the use of

Table 1. Pesticides that are less harmful (or less persistent) to beneficial arthropods in the order of their increased toxicity.

Insecticides and Acaricides	Fungicides	Herbicides
Dipel (Bacillus thuringiensis)	Nimrod (bupirimate)	Illoxan (diclofop-methyl)
Torque (fenbutatin-oxide)	Bayleton (triadimefon)	Semeron (desmetryne)
AAzomate (benzoximate)	Ronilan (vinclozolin)	Betanal (phenmedipham)
Dimilin (diflubenzuron)	Orthocid 83 (captan)	Kerb 50 W (propyzamid)
Tedion V 18 (tetradifon)	Cercobin-M (thiophanate-methyl)	
Kelthane (dicofol)	Ortho Difolatan (captafol)	
Spruzit-Nova-fl. (pyrethrum + pip.butoxide)	Derosal (carbendazim)	Plant growth regulators
Pirimor (pirimicarb)	Daconil 500 (chlorothalonil)	
	Plondrel (ditalimfos)	Cycocel (chlormequat)
	Pomarsol forte (thiram)	Rhodofix (1-naphthylacetic acid)
	Dithane Ultra (mancozeb)	

Table 2. Choice of beneficials when the side effects of pesticides are to be tested - some examples.

Field crops, e.g. wheat:

1 general predator (e.g. Chrysopidae, Coccinellidae)
1 aphid parasite (e.g. Aphidiinae)
1 soil living predator (e.g. Carabidae, Staphylinidae)

Vegetables, e.g. brassica crops:

1 aphid predator
1 aphid parasite
1 Lepidoptera egg parasite (e.g. Trichogramma)
1 Lepidoptera larval (pupal) parasite (e.g. Tachinidae, Braconidae)
1 soil living predator
1 soil living parasite (i.g. Cynipidae, Ichneumonidae)

Vegetables under glass, e.g. cucumber:

The predatory mite, Phytoseiulus persimilis
The whitefly parasite, Encarsia formosa
1 aphid predator
1 aphid parasite

Fruit orchards, e.g. apple:

1 general predator
1 aphid parasite
1 predatory mite (Typhlodromus), the mite predator Anthocoris
1 Lepidoptera egg parasite
1 Lepidoptera larval (pupal) parasite

Vineyards

The predatory mite, Typhlodromus pyri
1 Lepidoptera egg parasite

Forest

1 Lepidoptera larval (pupal) parasite
1 Lepidoptera egg parasite
1 general predator
1 predatory mite

---

the insecticide to the soil immediately around the plants and applying
it under the surface of the soil (Wright et al. 1960; Hassan, 1969).
Broadcasting of insecticide granules was shown by Wright et al. (1960)
to kill predators and induce more damage by the pest.

Knowledge of the behaviour of pests and natural enemies and the
development of suitable methods to monitor the time of their occurence
in the field will help in establishing the most appropriate time for
pesticide treatment. A peak population of a pest is usually followed by
a peak population of natural enemies. If applied at the appropriate
time, a short-lived non-selective pesticide can reduce the pest
population and allow the natural enemies to develop and control the
remainder of the major pest as well as other minor pests. In an
experiment on corn where chemicals for aphid control were integrated
with the egg parasite, Trichogramma evanescens, to control the European
corn borer, the effectiveness of the parasite was reduced to different
degrees depending on the time of the chemical treatment. The
effectiveness of Trichogramma was reduced by 27.3, 51.8 and 50.7%, when
Pirimor, Metasystox and Nexion-stark respectively, were applied three
days after the emergence of the parasite and by 19.0, 32.9 and 25.2 %
when the same chemicals were applied 10 days after the parasite's
emergence (Hassan et al. 1984).

The integration of biological and non-selective pesticides requires
experience and intensive supervision. In co-operation with specialised
plant protection advisers, farmers may be able to make use of such
programmes.

REFERENCES
Croft, B.A. (1976). Establishing insecticide-resistant phytoseiid mite
            predators in deciduous tree fruit orchards. Entomophaga 21,
            383-399.
Englert, W.D. and Kettner, J. (1983). Nebenwirkungen von Pflanzen-
            behandlungsmitteln aus Spinnmilben und Raubmilben.
            Mitteilugen uas der Deutsche Gesellshaft fur Allgemeine und
            Angewandte Entomololgie. (II. European Entomology Congress
            Kiel) 4, 89-91.
Franz, J.M., Bogenschütz, H., Hassan, S.A., Huang, P., Naton, E., Suter,
            H. and Viggiani, G. (1980). Results of a joint pesticide
            test programme by the Working Group: Pesticides and
            Beneficial Arthropods. Entomophaga 25, 231-236.
Hassan, S.A. (1969). Observations on the effort of insecticides on
            coleopterous predators of Erioischia brassicae (Diptera:
            Anthomyiidae). Entomologia expermentilis et applicata 12,
            157-168. Amsterdam: North-Holland Publishing Co.
Hassan, S.A. (1982). Relative tolerance of three different strains of
            the predatory mite Phytoseiulus persimilis A.-H (Acari,
            Phytoseiidae) to 11 pesticides used on glasshouse crops.
            Zeitschrift für angewandte Entomologie 93, 55-63.
Hassan, S.A., Bigler, F., Bogenschütz, H., Brown, J.U., Firth, S.I.,
            Huang, P., Ledieu, M.S., Naton, E., Oomen, P.A., Overmeer,

            W.P.J., Rieckmann, W., Samsøe-Petersen, L., Viggiani, G. and
            van Zon, A.Q. (1983). Results of the second joint pesticide
            testing programme by the IOBC/WPRS-Working Group "Pesticides
            and Beneficial Arthropods". Zeitschrift für angewandte
            Entomologie 95, 151-158.
Hassan, S.A., Stein, E. and Dannemann, K. (1984). Integration des
            Einsatzes von Trichogramma evanescens zur Bekämpfung des
            Maiszünslers Ostrinia nubilalis mit Insektizidspritzungen
            gegen Blattläuse. Gesunde Pflanzen 36, 268-273.
Hoy, M.A. and Knop, N.F. (1979). Studies on pesticide resistance in the
            Phytoseiid Metaseiulus occidentalis in California. In
            Recent Advances in Acarology, Vol.I, 89-94. New York:
            Academic Press Inc.
Schulten, G.G.M., van de Klashorst, G. and Russell, V.M. (1976).
            Resistance of Phytoseiulus persimilis A.-H. (Acari:
            Phytoseiidae) to some insecticides. Zeitschrift für
            angewandte Entomologie 80, 337-341.
Wright, D.H., Hughes, R.D. and Worrall, J. (1960). The effect of
            certain predators on the numbers of cabbage root fly
            (Erioischia brassicae (Bouché)) and the subsequent damage
            caused by the pest. Annals of Applied Biology 48, 756-763.

4. ENVIRONMENTAL TOXICOLOGY: ITS ROLE IN CROP PROTECTION

C.F. Wilkinson
Institute for Comparative and Environmental Toxicology,
Cornell University, Ithaca, New York, 14853, USA

INTRODUCTION

We live in a world in which we have become heavily dependent on chemicals. Indeed, if we stop to consider how and where we live, the variety of foods available to us, and the range of drugs that cure our ills, we will appreciate the enormous impact that chemicals have on all of us. Chemicals have become an integral part of our lives, and of the technology that surrounds us, and on balance they have improved immeasurably the quality of our lives.

Although a variety of chemicals were employed to control agricultural pests even by the late 1800s, it was not until about 1940, when DDT exploded on to the world scene, that the real pesticidal potential of synthetic organic chemicals became apparent. DDT was immediately hailed as a panacea for insect control; its discoverer, Paul Müller, was awarded the Nobel Prize, and the new "miracle" chemical was quickly picked up and used with enthusiasm. During the next few years, the worldwide use of DDT resulted in enormous economic benefits through the effective control of a wide variety of agricultural pests, and decreased worldwide human suffering dramatically through the control of mosquitos and other insect vectors of human disease. Spurred on by the success of DDT, the chemical industry began an intensive search for other synthetic organic pesticides, and a steady stream of new insecticides, herbicides, fungicides and other pesticidal products began to appear on the market; it has continued to this day, and the US Environmental Protection Agency currently estimates that there are approximately 600 basic pesticide chemicals marketed in some 45,000 to 50,000 formulations.

The use of pesticide chemicals is an essential adjunct to modern agriculture, indeed, many modern farming practices - increased automation, new cultivation techniques, monocultures, the development of new high-yielding crop varieties - are only possible because of the availability and use of pesticides. Pesticides are synonymous with modern agriculture and provide the most effective, and economically efficient means of controlling the many thousands of species of insects, weeds, fungi, and nematodes that compete for our food and fibre. Without chemicals we could not continue to enjoy, at reasonable cost, the variety, quality and quantity of the fruit, vegetables and other agricultural commodities to which we have now become accustomed. The estimated 30% losses from pests that would occur in the absence of pesticides, would spell economic and human disaster for many developing

countries around the world, where even now, millions of people exist on
a suboptimal calorific daily intake. The latest figures for the world
population - 4.7 billion and expected to rise to 10 billion in the next
40 years - emphasise the continuing need to increase agricultural
productivity (and distribution) worldwide.

THE CHANGING IMAGE
During the first 20 years of this pesticide revolution,
emphasis was placed on the positive, beneficial aspects of chemical use.
As with any new technology there was overuse and misuse, and despite the
appearance of pest resistance and recognition of some adverse effects on
fish, birds and other non-target species, little serious thought was
given to the potential long-term consequences of pesticide use in terms
of human health or the environment.

And then, in 1960, came Rachel Carson and "Silent Spring" and almost
overnight the balance shifted. In her book "The Apocalyptics", Edith
Ephron identifies Rachel Carson as the first of a number of individuals
like Paul Ehrlich, Barry Commoner and George Wald, all noted biologists
and ecologists, who vociferously expressed the view that planet earth
was a finite entity, and that Man and the whole global biosphere was
doomed, unless somehow he learned quickly to control his technological
abilities; some indicated that it was already too late. Rachel Carson's
emphasis in "Silent Spring" was, of course, on pesticides, which she
used symbolically to illustrate her view of the dire consequences of
Man's continued efforts to master Nature through technology. Despite
it's numerous scientific inaccuracies and broad unsubstantiated
conclusions, "Silent Spring" had an enormous impact, at least in the
USA, on the way in which pesticides are viewed. For the first time, we
were all made aware of the hidden costs of pesticides, and their
potential for causing adverse effects on human health and the environment. While the benefits were still obvious in terms of crop
protection, we began to realise that we must balance these more carefully against the risks.

Pesticides thus became the initial focus of the US environmental movement of the 1960s and, in retrospect, we benefited from the increased
awareness that resulted. Since that time, it seems we have become
increasingly concerned with the potential risks with which pesticides
are, or might be, associated - indeed many believe that the pendulum has
swung too far. Pesticides continue to receive a measure of "bad press"
that belies their contribution to the overall problem of environmental
pollution, and at the same time typically underestimates their benefits
to society.

For the last decade and a half, there has existed in the USA an
emotional atmosphere, often verging on hysteria, that society is being
not-so-slowly poisoned by pesticides and other products of modern
chemical technology. "We are becoming a nation of healthy
hypochondriacs" says Dr. Lewis Thomas of the Sloan-Kettering Memorial
Cancer Center. Aided and abetted by an eager, sensation-hungry Press,

the general public has become fearful, angry and confused; it is
suspicious of the profit-related motives of chemical industry and
mistrustful of government efforts to protect it. Carried along by
antichemical, antitechnological public opinion, US legislation directed
towards the protection of human health and/or the environment from
chemicals of all kinds, increased greatly during the 1970s. Regulations
continue to become more complex and difficult to interpret. In the case
of pesticides, chemicals that are introduced deliberately into the
environment, the public feels there can be no excuse for not developing
a regulatory policy that will prevent the release of those materials
that might pose a threat to health or the environment.

ENVIRONMENTAL TOXICOLOGY
Up until about the early 1960s, toxicology was considered a
branch of pharmacology and was restricted primarily to the medical
profession; it's main concern was with the diagnosis and treatment of
human poisonings from drugs and natural products.

Because of the claims of the environmental movement that we were
poisoning ourselves and the biosphere with pesticides and other
chemicals, toxicology began to expand in new directions. It acquired a
new identity and became known more generally as environmental
toxicology. Scientists, whose comparative studies were previously
restricted to identifying differences between rats, mice and
occasionally guinea pigs, started to look at the effects of pesticides
on fish, birds, and other wildlife, and to investigate the problems of
biomagnification associated with many of the chlorinated hydrocarbon
insecticides. At the same time, our analytical capabilities increased,
so that we could detect traces of pesticides and their metabolites, and
follow their movement, disposition, and biodegradation in the environment. And for the first time, we sought to devise appropriate animal
tests, to evaluate the potential toxicological effects of pesticides on
humans.

Toxicology/environmental toxicology has advanced during the last two
decades, to emerge today as a bona fide multidisciplinary science, that
attracts the attention of chemists, biochemists, physiologists,
geneticists, pathologists and others. It is a science that studies the
adverse and interactive effects of chemicals on living organisms, and
assesses the probability of their occurrence. Although our understanding of many basic aspects of toxicology - pharmacokinetics,
metabolism, chemical interactions with genetic material and other
targets - has progressed remarkably in recent years, it is disquieting
to realise that, with only a few exceptions, we still know very little
about how chemicals exert their toxic effects, and have little capacity
to predict the adverse effects of a given chemical in an intact animal,
particularly a human animal.

Similarly impressive advances have been made in the subdiscipline of
ecotoxicology that seeks to determine the fate and disposition of
chemicals in the biosphere; but again, we are still unable to predict

with any certainty the long-term impact of pesticides on natural populations, communities or ecosystems.

ENVIRONMENTAL TOXICOLOGY AND CROP PROTECTION

As we have seen, it was the widespread use and misuse of pesticide chemicals that first triggered the environmental movement, that in turn challenged the pesticide industry to change its ways, and demanded the enactment of legislation to better protect human health and the environment. The discipline of environmental toxicology emerged, in part, in response to the many questions raised as a result of this changing philosophy. Now, after almost 20 years, we can begin to evaluate the role of environmental toxicology in crop protection, examine some of its successes and shortcomings, and take a look into the future.

One of the first things that became evident was that the organochlorine insecticides – DDT, and the cyclodienes such as aldrin, dieldrin, and heptachlor – all relatively inexpensive, highly effective, broad-spectrum materials that had been used extensively for many years, were no longer acceptable. The stability and persistence of these materials in the environment, their accumulation in the tissues of living organisms, and their lack of selectivity, were major factors in the development of pest resistance and in their deleterious impact on beneficial species. Their ability to persist and be biomagnified through food chains, and to concentrate at high levels in certain primary species of raptorial birds and fish, clearly placed some of these species in jeopardy. Furthermore, significant organochlorine residues were also found in human adipose tissue, and tests showed evidence of carcinogenicity in some strains of laboratory animals. Clearly, these chemicals had to go. DDT, registered for use on some 334 crops and agricultural commodities in 1961, was banned in the USA in 1972, and after lengthy hearings, the uses of most other chlorinated hydrocarbon insecticides were either banned or severely restricted during the next decade. At about the same time, the 1972 amendments to the Federal Insecticide, Fungicide and Rodenticide Act (FIFRA), the major US statute under which pesticides are regulated, mandated more rigorous testing of new pesticides, and a careful review of those already registered.

Initially, these actions were greeted with concerned indignation and even ridicule by many in the pesticide industry. For most, they were a signal of hard times ahead, and some, unable to bear the cost didn't make it. However, many in the pesticide industry responded positively, and began to reorganise their resources to meet the challenge of finding replacements, for what had been considered indispensable pesticides. No longer would it be sufficient to find a chemical that was an effective pesticide per se; in addition to efficacy, the chemicals of the future would have to be environmentally acceptable and pose an acceptably low level of risk to humans. New terms like biodegradation began to appear, and older ones like selective toxicity, took on a new significance. An industry lulled into quiescence by its early successes took on a new life; and it was focused heavily on environmental toxicology.

PESTICIDE DISCOVERY

One of the major areas that changed was the methodology used to discover new pesticides. While the empirical screening of chemicals for pesticidal activity still remains an important means of identifying new leads, the search for new pesticides has become increasingly more sophisticated. Long-term basic research programmes have been established, to study the structure and function of new potential pesticide targets, many of which are unique to various groups of organisms, in the hope that a better understanding of such targets might pave the way for the design of new highly selective biocides. We are still a long way from tailoring pesticides to fit specific target sites, but our better understanding of the comparative aspects of toxicology and molecular biology, has resulted in the more rational, directed synthesis of many new groups of bioactive compounds. Examples of such compounds are dimilin and its derivatives, that interfere with the synthesis of chitin, a key constituent of insect cuticle, the nitrogen heterocycles (imidazoles, triazoles, etc.) that inhibit ergosterol biosynthesis in fungi, and the sulfonylurea herbicides that block the synthesis of essential amino acids in plants. Our increased knowledge of insect endocrinology, and our recognition of the critical role of hormones in controlling insect growth and development, have led directly to the synthesis and sale of the juvenile hormone mimics. New screens have been developed to look for new types of biological activity. We have discovered our ability to interfere with insect feeding and mating behaviour through use of antifeedants and pheromones, and have come to realise that, in some cases, a disorientated, disorganised insect is as good as a dead insect, as far as pest control is concerned. Biologists, biochemists, toxicologists, and synthetic organic chemists now work together to identify biologically-active natural products and to synthesise even more active analogues. For example, synthetic pyrethroids bear little or no resemblance to the natural pyrethrins from which they were derived, and are far more potent as insecticides.

The majority of these new pesticides have been designed and developed as a direct result of our increased knowledge of the interactions of chemicals with living organisms - environmental toxicology. In general, they are much more effective than the earlier pesticides, so that now we can think in terms of application rates of grams of active ingredient/ha instead of kilograms/ha. As a result, although pesticide usage continues to rise, the total amount of chemicals applied for crop protection has actually started to decline during the last few years. Furthermore, most are readily biodegradable, so that undesirable environmental effects are markedly reduced.

Many of the newer pesticides are also far more selective than the older broad-spectrum materials. This has been achieved, in part, through the emphasis on new specific targets, and in part by taking advantage of other species differences that determine the bioavailability of the pesticide at the target. Recent advances in our understanding of the key processes of pesticide absorption, distribution and metabolism that often limit toxicity, provide additional opportunities for improving selectivity.

While we are still dependent to a large extent on many of the older chemicals, our pest control capabilities have been strengthened immeasurably by the availability of new chemicals that have resulted from the more sophisticated, scientific approach to pesticide discovery. Environmental toxicology has played an important role in the evolution of this approach.

ENVIRONMENTAL AND HUMAN HEALTH EFFECTS
Prior to the FIFRA amendment of 1972, the registration of a new pesticide in the USA depended mainly on proof of efficacy, and a demonstration of an acceptable acute toxicity to animals. Since that time, registration requirements have increased enormously and now include additional requirements for data on product chemistry, environmental fate, impacts on wildlife, and a complex series of animal toxicity tests designed to detect potential chronic health effects on humans.

All pesticides registered since 1972 have had to meet all of these latter requirements, and compounds registered prior to 1972 have to satisfy the same criteria prior to re-registration. By the end of 1984, approximately 90 of the 600 pesticides in use (representing about 50% of the pesticide usage) will have been re-registered, and the plans call for review of about 25 chemicals per year. As might be imagined, this has created a considerable task for both industry and government, and there has been an increasing demand for environmental toxicologists to design and conduct the necessary tests, and evaluate the results.

Despite serious attempts by industry to incorporate more selectivity into pesticides, it is necessary to bear in mind that all are biocides, and that at a subcellular level most organisms are dependent on similar biochemical processes. Consequently, it is inevitable that most pesticides will continue to have the potential to cause adverse effects in foe and friend alike. Even those based on natural products or those directed at specific "non-human" targets, cannot be assumed to be free from risk, since they are still biologically reactive and may interact with totally different targets in mammals.

Environmental toxicology has helped in many ways in the preparation of the substantial environmental and health-related data base currently required for pesticide registration. However, since present concerns, at least in the USA, are focused heavily towards assessing the acute and chronic effects of pesticides on human health, a few comments in this area are appropriate.

ACUTE TOXICITY TESTING
A few years ago, toxicology testing almost exclusively emphasised the importance of acute toxicity resulting from single dose exposures. Acute toxicity remains a serious problem for pesticide formulators and applicators, who along with the victims of accident, misuse and suicide, account for most of the estimated 20,000

pesticide-related deaths that occur annually worldwide. There are many who feel that this is a needless toll, that could readily be obviated by replacing a few of the major culprits, e.g. parathion, with less toxic substitutes, and by denying the registration of new compounds with acute LD_{50}s of less than about 50 mg kg^{-1}. However, although testing for acute toxicity is still required, it is seldom of great concern to the general public, who seem able to accept the possibility of a quick, relatively "clean" demise. Perhaps the most important reason behind this general lack of concern for acute toxicity is the fact that according to the fundamental tenet of toxicology, the severity of the response is directly related to the dose. Thus, for acute toxic effects, it is possible to measure a threshold dose below which no adverse effects are likely to occur. The so-called "no observable effect level" (NOEL) has become a useful benchmark for assessment of acute effects and the application of various "safety" factors to this has become an acceptable regulatory procedure. Moreover, the determination of acute effects such as oral and dermal toxicity, and eye and skin irritation effect, do not present serious practical problems; it is relatively easy to count dead noses!

What the public cannot accept, however, is the possibility, no matter how remote, that chronic, low-level exposure to pesticides might ultimately lead to cancer and other effects such as mutagenicity or birth defects, that are considered the ultimate insults to human health. As a result, the assessment of carcinogenic risk has emerged as the very hub of. modern toxicology.

CHRONIC TOXICITY TESTING

The theoretical and practical problems inherent in carcinogenic risk assessment are formidable, particularly when we try to quantify the risk potential in humans. It is fraught by innumerable problems, and scientific uncertainty and controversy exist at every step along the way.

In contrast to the assessment of acute toxicity, where we measure the severity of specific observable adverse effects occurring in individual animals, at high doses, over relatively short periods of time, cancer risk assessment seeks to measure increases in the frequency of occurrence of a low probability event in a population, at low doses, over long periods of time. Since probability statistics indicate that, short of using many thousands of test animals, it is simply not feasible to detect effects occurring at doses realistic of human exposures, it is necessary to compromise. Thus, in practice, most oncogenicity studies employ relatively small groups of test animals and expose them to high doses often at, or approaching the "maximum tolerable level". It is assumed that, although not measurable directly, the effects at low doses can be estimated by extrapolation from those observed at high doses. Furthermore, since it is currently assumed by regulatory agencies that there is no threshold for carcinogenic effects, the dose extrapolation must pass through zero. From this point on, the risk assessment process becomes a statistical problem. It seems that we are becoming increasingly obsessed with developing numerical risk estimates, derived from highly theoretical models based on controversial, unsubstantiated assumptions.

We are attempting to quantify and predict responses that are beyond the realm of biological and scientific certainty, and the toxicologist is being pushed further and further away from the final and most important stages of the risk assessment process. A qualitative estimation of cancer risk is a scientific process; it becomes a political process when it is described in numerical terms. Unfortunately, in the USA we seem to have become truly caught up in the numbers game, and unfortunately the numbers are perceived to be precise and accurate values by a fearful and concerned public. Anyone believing this should carefully consider the fact that the numbers generated during the recent EDB controversy were derived from a single toxicological data point; they bear little or no relationship to the real world, and do not give due consideration to the substantial toxicological data base that exists for this material.

Toxicologists are continually placed in the uncomfortable position of answering questions in the uncertain framework that the current state-of-the-art of cancer risk assessment provides. There are no simple "yes" or "no" answers to questions like "does it?" or "doesn't it?", "is it?" or "isn't it?", and lawyers and the general public often cannot understand why the scientists hesitate and disagree, and why they cannot provide assurance of absolute safety. It is impossible to prove a negative; honest toxicologists can never, ever, provide an assurance of safety − only some low level of risk for even the most innocuous chemical.

It is clear that in the area of assessing the potential long-term chronic health effects of pesticides, environmental toxicology has still a long way to go. Furthermore, toxicologists have allowed the statisticians and politicians to overinterpret and misinterpret toxicological data beyond the point of scientific objectivity. The time has come when toxicologists must insist on a more rational, balanced approach to chronic risk assessment, and must carefully explain the facts, especially the uncertainties, associated with these types of studies.

To those deeply concerned over the potential for pesticides to cause cancer in humans, it should be pointed out that, contrary to popular belief, most types of cancer have not increased during the last 50 years; in fact, with the exception of lung cancer, due mainly to smoking cigarettes, most have either stayed constant or have declined. We are not in the midst of a cancer epidemic, and if we continue to be obsessed with assessing cancer risk, we run a very real risk of squandering our limited testing resources and failing to give proper consideration to other more prevalent and potentially more serious adverse effects.

THE FUTURE

Environmental toxicology will undoubtedly continue to play an increasingly important role in the discovery and development of the pesticide chemicals of the future.

As our knowledge and understanding of the interactions of chemicals with living organisms become more complete, many new opportunities will be provided for developing even more sophisticated materials for use in crop protection. They will be more effective, more selective chemicals that will continue to meet the more stringent requirements associated with our changing perceptions of environmental preservation and the protection of human health. Of necessity, they will be more expensive materials as development costs continue to escalate; but we will continue to accept these costs for what they are - the price of our survival.

5.   PESTICIDE REGISTRATION: BENEFIT OR BUREAUCRACY?

B. Thomas
FBC Limited, Chesterford Park Research Station, Saffron
Walden, Essex, CM10 1XL, England

INTRODUCTION
Imagine if you will the development of a new pesticide in an
ideal world. Such a pesticide would be of insignificant mammalian
toxicity, would have no effect on wildlife species or the environment,
would control the target pest and then disappear quickly leaving no
residues in crops, soil or water; it would also be highly efficacious
and inexpensive. Unfortunately, we live in the real world where such
ideals are seldom, if ever, realised and where pesticides have, in
common with all chemicals, a potential to cause adverse effects. The
purpose of pesticide registration is to assess these potential effects
at a pre-marketing stage, so as to prevent the use of 'unsafe'
pesticides and, by the imposition of appropriate precautions during and
after use, to ensure that any adverse effects from registered pesticides
are minimised.

Over recent years we have seen the establishment of an increasing number
of National Registration Schemes and concomitantly, demands for increasing amounts of data in support of requests for pesticide registration.
The data can conveniently be grouped into four main areas - chemistry,
residues, ecotoxicology and mammalian toxicology - which are aimed at
assessing the potential risks to operators, consumers of treated food-
stuffs and the environment (see Thomas, 1983). The assessment of these
potential risks cannot be considered in isolation but must be balanced
by an appraisal of benefit, and to this end most Registration
Authorities require some evidence of biological efficacy. Others, such
as the United Kingdom Pesticide Safety Precautions Scheme, have felt
that poor efficacy would not be tolerated by normal market forces and
thus do not require such data for registration. However, new pesticide
legislation in the UK now requires the submission of efficacy data.

Both industry and Government undertake risk-benefit analysis, and
pesticide registration and its requirements can be considered in the
light of the different approaches adopted by these two parties.

THE INDUSTRY VIEW
Costs and time
The prime objective of industry is to provide a profitable
enterprise for its shareholders. The road to profit, however, is a
bumpy one and has many risks along the way. Development of a new
pesticide is very expensive and various estimates are quoted depending

on the pesticide being developed and the geographical location.
Complete development, including manufacturing plant, may be well in
excess of £15 million of which a significant proportion, about £4-5
million, is spent on the acquisition of the data needed to achieve
registration. Another facet of capital investment is the time element.
It takes a minimum of 8-10 years, and often even longer after discovery
before a product reaches the market; the realisation of significant
profits is likely to take longer still. During this period a capital
investment of many millions of pounds is involved and, at the same time,
money must be invested in the discovery and development of new compounds
or the extension of use of existing ones.

Thus, there is a major risk in investing capital and time to obtain data
which subsequently show that the product is, from a toxicological or
environmental point of view, unacceptable either to the company itself,
or to a registration authority. it is important to recognise at this
point that many compounds are "killed off" by companies at an early
stage or even at a relatively late stage in their development because
the company itself decides that the compound's properties pose an
unacceptable degree of risk in the proposed use. In this sense, the
company is therefore continuously undertaking its own 'regulatory'
assessment well before any submission to a Registration Authority. This
responsible attitude is often overlooked by those who criticise agrochemical manufacturers, as is the fact that the Pesticide Industry
consists of average people who are also consumers and gardeners and
parents, who wish to live in and see their children brought up in a
healthy and pleasant environment; like the Merchant of Venice they also
bleed when cut!

Variations in data requirements
Because of the high capital costs involved, a world-wide
market is essential for a satisfactory return on investment and
therefore the requirements of Registration Authorities in different
countries must be satisfied. This in itself adds new problems. Thus,
although many "common" data are required by the various National
Registration Authorities, often there are subtle or even major
differences in these requirements or in the ways in which specific data
must be obtained. A typical example is the mammalian long-term feeding
study in rodents which is designed to identify chronic toxicity and
carcinogenic potential which are key items in the toxicological data
package. The cost of a single 2-year study is about £300,000 and
therefore it is generally impracticable to perform more than one study
per species. The one definitive study must thus be designed to suit all
the major Registration Authorities, and a considerable time is spent in
planning a protocol for such a study that is widely acceptable.

Another major problem arises when certain Registration Authorities
require data which reflect national rather than international needs. In
certain cases these somewhat parochial requirements can be justified in
terms of particular environmental conditions; for example in The
Netherlands, information is required on the behaviour of the pesticide
in surface water. Demands for more esoteric data by other Authorities

are, however, less easily justified. In some countries, for example, we have recently seen what can only be described as a 'quantum' jump in environmental data requirements. For instance, results of cellulose degradation and soil enzyme studies are increasingly requested but for no clear purpose, especially in the absence of any agreed protocols or any real ability to interpret the results of such studies. Similarly, it is difficult to accept the need for extensive studies on a wide range of aquatic organisms, such as the effects on Daphnia reproduction, for pesticides which would not intentionally be added to water and which are unlikely to find their way into water.

Some Authorities ask for tests merely because the test systems are available or 'in vogue', whilst at the same time ignoring the problems of interpretation. In such cases, industry might well feel that its limited resources are being used for the purposes of academic research – a role which is clearly more suitable to the public sector.

Delays in registration
Another element of risk facing the Pesticide Industry, and which is highly specific to this particular industry, relates to the seasonal nature of the business. The timing of marketing is critical and the delay of a registration, if only for a month or two, can result in the loss of a whole season's sales and even allow a competitor to obtain a commercial advantage in a particularly competitive market. Here again, Registration Authorities differ quite considerably in their approach. In some countries such as the United Kingdom and Denmark, provided the data package is acceptable, registration takes a matter of 6 to 8 months whereas in other countries such as Italy, Austria and Japan, the process may take several years. Such long delays often appear to result from excessive bureaucracy rather than from any genuine problems that arise during the assessment process.

Amounts of data
Another important difference which exists between Registration Authorities is the amount of data needed before marketing is allowed. In the United Kingdom, France and Spain a staged system of clearance operates and maximum permitted sales are linked to the amounts and quality of the data, and to the nature and intended use of the pesticide. Thus when appropriate, limited sales can be achieved on the basis of residue data, acute toxicity and sub-chronic toxicity studies, interim chronic studies, teratology and mutagenicity studies and evidence that the use of the compound is unlikely to pose any risk to wildlife and the environment. In other countries, e.g. Denmark, Japan and Germany, sales of any kind can only be achieved after submission of the full chronic (2-year) studies which will only be available some 3 years after the completion of the sub-chronic studies. For certain pesticides this demand seems unnecessarily severe because experience has shown that the information available at the 'pre-chronic' stage is generally sufficient to assess the acceptability of the proposed use of the compound and that the subsequent results of the chronic toxicity studies tend to confirm the initial assessment.

Residue requirements also vary significantly between countries, and an additional complication is the practise of many countries, e.g. most European countries, USA, Canada and Australia, to set statutory tolerance levels for residues in food-stuffs. In many countries, demands for residue data are clearly excessive. The scientific literature contains vast amounts of 'negative' residue data and it is tempting to correlate this increase with the introduction and development of the gas chromatograph. Many amounts of data also remain unpublished because they merely indicate the absence of any significant residues. The end result of residue analysis is regarded by many as a 'definite figure' which is thus administratively and legislatively attractive. Residue analysis therefore results in a considerable investment by industry, by National Authorities and by International Organisations such as the Codex Alimentarius, FAO and WHO. Barnes (1973) commented that there was 'not one iota of evidence that any environmental contamination that has been identified as having been due to the use of pesticides in air, food or water has been followed by any detectable change in the incidence of human disease or disability'. One might therefore argue that residue analysis attracts a disproportionate amount of attention and funds, and that some of this effort and money might be better utilised in other areas of investigation.

These and many other factors, in particular the need for local efficacy data, must be taken into account by industry in developing an overall strategic plan, hopefully leading to the successful marketing of a particular compound.

Post-market activities
The registration of a product and its subsequent marketing is not of course the end of the story as industry must continue to keep the field performance of the pesticide under close surveillance. Apart from monitoring its biological performance and in particular any signs of the onset of resistance, industry must also react quickly and efficiently to any claims of adverse effects. Thus, many companies maintain an 'Adverse Effects Register', often internationally based, in which the investigation of both real and imaginary effects, are recorded. This enables any additional measures or amendments to existing precautions to be introduced quickly. The operation of such 'product stewardship' systems is again one which is invariably overlooked by the critics of the pesticide industry.

THE GOVERNMENT VIEW
The essential activity of Registration Authorities, is to consider on behalf of Government the benefits to the farmer and to the consumer and to weigh these against the risks to the operator, the consumer and the environment.

Political influences
In today's society, Government's task of making a purely objective risk assessment is very difficult if not impossible, because of the influence of various pressure groups. Yet it must be recognised

that Government has a democratic duty to take the opinions expressed by
such groups into its considerations. Activities by organisations such
as "Green Peace" and "Friends of the Earth" and the increasingly
influential role played by the "Green Party" in West Germany and more
recently in European politics, indicate the international nature of
these groups and their concern not only with pesticides but with nuclear
power, noise and chemical pollution in general. Additional developments, particularly in the UK, are the political pressures imposed by
the Trades Unions and their greater role in risk assessment arising as a
direct consequence of the Health and Safety at Work Act. This increase
in public awareness of risk must have an effect on the pesticide risk
assessment process. In considering public and to some extent
'organisational' pressure, it is important to distinguish between actual
risk and what might be considered as "perceived risk".

The 'quantitative' approach. The risk of dying from cancer attracts
much public attention today and clearly the general public views cancer
quite differently from other common causes of death such as heart
disease. Cancer is regarded in a 'sinister' context, and any suggestion
of an association between the use of pesticides and an alleged carcinogenic response in humans becomes a very emotive subject. Many
Registration Authorities have attempted to quantify the risk of cancer
by applying various statistical models to the data derived from animal
experiments. In some cases this has led to a pre-occupation with
mathematical estimates of cancer risk at the expense of a more objective
view of the whole data package, and of the validity of the extrapolation
to Man of certain biological effects seen in experimental animals. This
mathematical approach is most commonly employed in the USA where a
cancer risk of "one in a million" has, in recent years, moved from being
acceptable as a very broad rule of thumb towards a very definite 'cut-
off' point below which pesticide registration may be refused or severely
restricted. I cannot enter here into the arguments as to why this
concept of a mathematical 'cut-off', which is administratively
attractive, is scientifically unacceptable. However, is interesting to
consider how this figure of "one in a million" appears to the public.
Pressure groups will probably argue that even a risk of one in a million
is too great, but to a large extent this arises from a conceptual
problem in understanding the size of such a risk. Examples of a "one-in-
a-million" risk of death in some common activities are as follows:

> Travelling 400 miles by air.
> Travelling 60 miles by car.
> Smoking $\frac{3}{4}$ of a cigarette.
> 1 minutes of rock climbing.
> 20 minutes of living for a man over 60 years of age.

Such risks are taken and accepted by many or all of us from time to
time, and yet the same degree of risk when applied to the use of
pesticides is often criticised and by some deemed to be unacceptable.

The 'qualitative' approach. The public perception of risk is also
well demonstrated by a survey conducted among American college students

who were asked to rank the major sources of danger in the American
environment as they 'perceived' them. The result of this survey and the
actual order was as follows:

Survey Ranking	Actual Ranking
Nuclear power	Smoking
Hand guns	Alcohol
Smoking	Motor vehicles
Pesticides	Hand guns

In terms of causes of actual deaths per 100,000 population, nuclear
power ranks 20th and pesticides rank 28th.

Environmental aspects. The concept of 'perceived risk' extends to the
environmental side, where the public's opinion of pesticides in general
is greatly influenced by the extensive publicity given to a specific but
small number of older pesticides such as dieldrin and DDT. Thus, it is
a quite commonly held but mistaken belief that the use of pesticides in
general causes very serious environmental problems, whereas relatively
little account is taken by the public of the effects caused by
de-afforestation in Third World Countries, or on a more parochial level,
the removal of hedgerows in the UK.

Data requirements

Against this background of an increasing, though perhaps
misinformed public awareness of risk and of the consequential
'political' pressure, it is perhaps not surpising that Registration
Authorities have become more demanding in their requirements for
pre-market information on pesticides. Other factors are also at work,
such as the availability of predictive tests that were not available
some years ago, e.g. combined mutagenicity and carcinogenicity screening
tests, more sophisticated toxicological assessments, the availability of
ever increasing data bases, and the more meaningful interpretation of
data which has become available with the long-term and wide-scale use of
some pesticides, e.g. the organochlorine and organophosphorus
insecticides, the phenoxyacetic and triazine herbicides.

There is, however, a very real danger that data requirements may exceed
those necessary to assess risk reliably and extend to levels which
reflect a "check-list" mentality - in other words, the request for some
data merely becomes a bureaucratic exercise. This concern is not merely
a subjective feeling; on numerous occasions companies have submitted
extensive data on some aspect of pesticide behaviour which more than
adequately define the properties of the pesticide, only to be informed
that these data were not sufficient because of a failure to meet exactly
the arbitary requirements of rigid protocols. This attitude is
particularly prevalent with regard to the environmental studies required
by some European Registration Authorities, and reflects the inability or
unwillingness of such Registration Authorities to assess a pesticide on
its own merits. There are clearly occasions where the physical and
chemical properties of a pesticide coupled with relevant environmental

studies will provide sufficient data to enable a risk assessment to be
made and to obviate the need for further studies on the "check-list".
An example based on personal experience relates to a pesticide having an
extremely low aqueous solubility - so low in fact that it was impossible
to introduce sufficient chemical into the test system to have any effect
on the fish and yet one Registration Authority initially insisted on the
need for an LC_{50} 'figure', and considerable effort was required to
persuade them otherwise.

Environmental data required by many European Registration Authorities
have increased over recent years. It thus comes as somewhat of a
surprise to find that the US Environmental Protection Agency, which has
traditionally been regarded as one of the world's most demanding
Registration Authorities has, in its recent guidelines (EPA, 1982),
considerably relaxed its requirements. In particular, the EPA requirements for data on soil metabolism, microbial effects and soil leaching
are now much more pragmatic and recognise the difficulties in the
interpretation of some of these studies.

The new EPA attitude would, however, seem to be the exception rather
than the rule, and whereas toxicological data requirements have
seemingly reached a fair degree of international harmonisation, environmental data requirements are still some distance away from any such
general agreement.

CONCLUSIONS
Some Registration Authorities and the majority of 'pressure
groups' appear to have fallen into the fallacious trap of equating more
data with greater safety. Ultimately there is of course no guarantee of
absolute safety. The vast majority of data used for risk assessment are
based on laboratory animal or in vitro experiments. There must therefore be a finite, albeit very small, area of doubt in predicting hazards
on the basis of these animal data. In past experience however, these
assessments have proved remarkably good. In the United Kingdom, the
number of accidental poisonings due to pesticides is remarkably low,
especially when compared with other common accidents on the farm such as
tractor mishaps, falls and animal attacks. In less developed countries,
the record is not so good but this is normally reflects the misuse of
the pesticide and not its hazard assessment. The solution to this
problem is the education of users, not more data. The position in the
United Kingdom from an environmental point of view is equally reassuring
although there is, as the Royal Commission on Environmental Pollution
(1979) recognised, a need for continued vigilance and monitoring of the
situation. A case can be made that data requirements to assess risks to
man should now take a different direction and that we should be looking
more towards the acquisition of data on the actual exposure of operators
to pesticides. The study conducted in the United Kingdom by the British
Agrochemicals Association is a significant contribution in this
direction (1984). Such data acquisition, however, poses ethical,
scientific and economic problems - problems which will need the
co-operation of Governments, Industry and the Trades Unions if they are

to be solved successfully. Similar scientific and economic restrictions are involved in monitoring the environmental effects of pesticides. I would suggest that we have reached an optimum level of data requirement based on animal and laboratory experiments and that if such requirements continue to increase it will undoubtedly slow down the development of new compounds with all that this implies for the agricultural industry. The preoccupation of many Registration Authorities with rigid and extensive data requirements might well result not in an increase in safety but merely a better world for rats to live in!

REFERENCES

Barnes, J.M. (1973). Toxicology of agricultural chemicals. Outlook on Agriculture 7, 97-101.

British Agrochemicals Association (1984). Spray Operator Safety Study. London: British Agrochemicals Association.

Environmental Protection Agency (1982). Pesticide Assessment Guidelines, Subdivision N - Chemistry: Environmental Fate. US Department of Commerce, National Technical Information Service.

Royal Commission on Environmental Pollution (1979). Seventh Report: Agriculture and Pollution pp. 280. London: HMSO.

Thomas, B. (1983). Pesticide registration in the United Kingdom and Europe. In Pesticide Residues, MAFF Reference Book 347, 211-218. London: HMSO.

Section 2. APPLICATION

6. REQUIREMENTS FOR EFFECTIVE AND EFFICIENT PESTICIDE
   APPLICATION

E.C. Hislop
Long Ashton Research Station, University of Bristol,
Long Ashton, Bristol BS18 9AF, England

INTRODUCTION
      Pesticides have increased dramatically in their potency over
the last 20 years. Although they are usually used to good biological
effect, few would claim that they are used with high efficiency. Some of
the blame for this undoubtedly can be placed on inefficient application
methods.

Requirements for efficient as well as effective application of
pesticides can be defined basically as follows: placement specifically
on the target of a dose just sufficient to produce the desired biological result. Conventional wisdom suggests that definition of a
problem is half-way to solving it. Unfortunately, this particular
definition is exceedingly simplistic and begs many questions. For
example, can we identify the precise target? The ultimate target for a
pesticide is of course the site of its biochemical action – often a
vulnerable enzyme system within the pest organism. The initital target
for spray droplets may be the surface of a weed or a crop plant that is
attacked by pests or pathogens, but deposition on particular parts of
the plant surface, for example stems, leaf axils or leaf under-sides,
may give the best results for particular pest/chemical combinations.
Can we in fact deposit a well controlled dose? How does the dose
requirement alter with time and with different patterns of deposition,
and how do the latter correlate with different modes of action? Because
of all these complexities, and others, spraying – the most common method
of delivery – is inherently inefficient. In most practical situations,
we can never hope to meet fully the requirements for effective and
efficient application as defined above. But we should be able to
improve on the present position.

An obvious approach to more efficient application would be to use
alternative delivery systems. The 'weed-wiping' technique of applying
herbicides to certain target weeds which protrude above the crop is an
example of a potentially effective and efficient alternative delivery
system: it utilises a minimal quantity of chemical with minimal risk to
the environment. Unfortunately, targets are rarely so accessible. It
is possible to conceive direct delivery systems, such as hand-painting
weeds with herbicide, which might have more general application. But
these would be very labour-intensive and inappropriate in developed
agriculture. Thus, the simple definition of requirements which ignores
inputs other than pesticides is deficient. A delivery system has to be

practical for the situation in which it is to be used.   In third-world
agriculture, where labour is cheaper and capital for investment in
machinery is scarce, logistics often favour fairly simple, if tedious,
delivery systems (Matthews, 1983; Durand et al., 1984).   The opposite is
true for developed agricultural situations where pesticide applications
often have to be made as quickly as possible within certain "weather
windows", and where labour and fuel inputs are important considerations;
here, timeliness of application is often the prime consideration.

This chapter examines some of the requirements and constraints for the
effective and efficient application of pesticides, with an emphasis on
broader aspects of crop management and environmental input.   Justifi-
cation for this approach was amply provided by the recommendations for
minimising risks of environmental contamination given in the Seventh
Report of the UK Royal Commission on Environmental Pollution (1979).
Reference will be made mainly to spraying in developed agriculture.
Table 1 illustrates the scale of pesticide usage in UK agriculture and
horticulture, and shows the predominant importance of herbicides (in
cost terms) and the increasing sales of fungicides.   Pesticide usage in
developing agriculture is generally smaller per unit area with a greater
emphasis on the use of insecticides, but the use of all classes of
pesticides is increasing in most countries, as is concern for their
safety and efficiency.   Not surprisingly, these countries want to avoid
making mistakes that have been made in developed countries.   They are
generally placing a greater emphasis on a managed approach to pest and
disease control (Turner, 1983).   But some still are using potentially
poisonous materials in a manner which puts users at a considerable risk.
While operator safety is not considered in detail in the present
discussion, it is of course an important component of overall
efficiency.

Table 1.   Value (£M) of pesticides sold in the United
Kingdom for use in agricultural and horticulture.   (Source:
British Agrochemicals Association Annual Report and
Handbook, 1984-1985.)

	1980	1984	Increase (%)
Herbicides	117.8	167.9	42.5
Insecticides	16.9	31.0	83.4
Fungicides	37.3	90.3	142.1
Growth regulators	3.4	6.7	97.1

MACHINERY PREPARATION AND CALIBRATION

No application method, whether ground or aerial spraying or
seed or granule application, can be fully effective and efficient if the
dose delivered is inaccurate or uneven. Under-dosing can reduce effectiveness, whilst over-dosing is wasteful, environmentally undesirable,
and may be damaging to the crop.

The traditional hydraulic sprayer is essentially a very simple device,
and its performance can be optimised to considerable advantage (Anon.,
1976). Many arable situations are best sprayed with wide angle (e.g.
110°) flat-fan nozzles (Hislop, 1984). These are usually spaced 50 cm
apart along the boom so that a double overlap pattern is achieved with
the nozzles about 40 cm above the crop. This set-up provides the best
opportunity for evenness of application, and the production of more
driftable droplets by 110° than by 80° nozzles is compensated for by the
lower spray height. An 80° nozzle operating at 60 cm above the crop
does not produce a full double overlap pattern and distribution is
therefore less even than that obtained from wider angled nozzles.

Output from hydraulic nozzles should be checked both visually for spray
uniformity and mechanically for throughput. Although sprayer calib-
ration is now more a question of education than of research, it is
relevant to make a few observations here. The sprayer pressure gauge
should not be accepted as accurate. Application volume rates should be
calculated on the basis of measured speed, and the pressure release
values used to make adjustments.

No spray delivery system will perform well on unstable boom systems
(Nation, 1982). Hislop (1984) has discussed this subject for hydraulic
nozzles compared to controlled droplet application (CDA) techniques.
The same paper also discusses the rationale behind the development of
CDA devices. A detailed review of all ground application techniques has
been presented by Combellack (1984); although this deals specifically
with herbicides, many of the comments are also relevant to other
pesticides.

EFFECTIVE PESTICIDE DELIVERY

Most efficacy testing by pesticide manufacturers relates to
conventional methods of application, i.e. the use of fairly large
volumes of spray atomised through hydraulic nozzle systems. The trials
are usually ad hoc and provide little information about how and why
products succeed or fail to produce the desired result (Hislop, 1983).
Some chemical manufacturers have studied alternative spraying
techniques, the most notable of which is CDA (e.g. Bals, 1969, 1975,
1978; Frost, 1978; Norman, 1978; Rutherford, 1978). However, for a
variety of reasons there are at present relatively few recommendations
by chemical suppliers for the use of products by this technique, which
involves much smaller than usual volumes of liquid.

The last few years have seen the exploitation in spray delivery of the
principle of electrostatic charging (Bowen et al., 1964; Law and Bowen

1966; Law, see Chapter 8). Thus, we now have electrodynamic systems (Coffee, 1979; Durand et al., 1984), charged rotary atomisers (Arnold and Pye, 1980) and charged hydraulic nozzle systems (Marchant and Green, 1982; and Pay, 1984a,b). Aspects of the theory and practice of the use of these systems have been discussed previously (Hislop, 1983b, 1984; Hislop et al., 1983), and the problem to be faced now is to try impartially to consider their merits and limitations with particular reference to the efficacy of their performance. Comparative efficacy of usage is considered later.

Table 2 illustrates some data obtained at Long Ashton over several years of spraying cereals with fungicides. The results suggest that several different application systems can produce very acceptable yield benefits which more than outweigh the cost of the treatments. Much of this work was done using manufacturers' recommended doses irrespective of the volumes applied. However, on occasions (see Hislop et al., 1983) reduced doses have also been effective. Essentially similar results were provided by Evans et al. (1984) in farm situations when hydraulic nozzle application systems were compared with CDA systems for fungicide spraying. Similarly, Robinson and Garnett (1984) and Griffiths et al. (1984) have produced encouraging data on the effectiveness of electrostatic spraying systems. Unfortunately, relatively few reports compare all the alternative spray delivery systems so comprehensive judgement of their merits is difficult.

Table 2. Average yields of winter cereals (barley and wheat combined) sprayed with recommended doses of fungicides at c. GS 30 and c. GS 39-40 at Long Ashton from 1982 to 1984.

Method of spraying	Volume $(1 \, ha^{-1})$	Mean yield $(t \, ha^{-1})$	No. of field experiments
Hydraulic nozzles	200	7.55	5
Hydraulic nozzles	100	7.21	5
Electrodynamic	1 or 2	7.21	4
Unsprayed	-	6.22	5

Data on the relative effectiveness of conventional and CDA systems for herbicide applications are more abundant. Nation (1982) reviewed the available data and concluded that CDA sprays often were similar in efficacy to standard methods of application, particularly when systemic materials were used or when applications were made before emergence. These conclusions have been confirmed in both glasshouse and field experiments, where four soil-applied herbicides were tested with a CDA device spraying 30 l ha^{-1} and with hydraulic nozzles using 250 and 400 l ha^{-1} (Addala et al., 1984a,b). However, there are other data indicating that some herbicides, in particular post-emergence contact materials, do not perform so well when applied in smaller volumes of liquid than those usually recommended. Our own data (Table 3) illustrate this point well for applications made in late spring, when the cereal crop was relatively well developed. Occasionally, however, reduced volume CDA applications can be more effective than the standard method, glyphosate treatments being the outstanding example. There are relatively few data on the efficacy of electrostatic applications of herbicides but those that exist (e.g. Parham, 1982; Griffiths et al., 1984) indicate that these too can be as effective as the standard method when weeds are well exposed, but tend to be less effective in situations where the crop competes with weeds as the target.

Table 3. Effect of one application at c. GS 30, of contact herbicides on the control of broad-leaved weeds in winter cereals at Long Ashton.

Method of spraying	Volume (1 ha^{-1})	Weed control (%)		
		1983	1984	
		Winter barley (1)	Winter (2)	wheat (3)
Hydraulic nozzles	200	92^{a}	83^{a}	70^{a}
Hydraulic nozzles	100	87^{a}	56^{b}	75^{a}
Hydraulic electrostatic	200	–	–	70^{a}
Hydraulic electrostatic	70	–	–	45^{b}
Rotary cages	90	61^{b}	69^{ab}	–
Rotary cages	40	57^{b}	36^{c}	–
CDA	40	81^{a}	39^{c}	–
CDA (Half-dose)	40	64^{b}	–	–

(1) Bifenox/mecoprop.
(2) Ioxynil/bromoxynil/mecoprop.
(3) Experimental herbicide.

Figures in the same column with the same superscript do not differ significantly at 95% confidence limits.

The debate on the efficacy of small volume (e.g. 10 to 40 1 ha^{-1}) spraying systems (particularly CDA) has been protracted and often acrimonious. Frequent claims that the CDA technique allows dose reductions of herbicides have not been substantiated. This does not mean that CDA systems do not have merit, but that users have to guard against over optimistic advertising.

Of course, not all conventional herbicide applications made at about 200 1 ha^{-1} are equally efficient. Hollies (1982) provides data showing that better than 95% control of Avena spp. was achieved in more than 50% of treatments, while 40% gave the same design of control of Alopecurus myosuroides (black-grass). However, he points out that nearly 20% of treatments against these weeds produced only medium or poor control, and that these corresponded to a direct cost penalty of £67 ha^{-1} for Avena spp. and £140 ha^{-1} for A. myosuroides. Poor weed control resulting from any method of spray application exacerbates problems in subsequent years. Thus, the slightly inferior control generally given by CDA techniques compared to conventional treatments, although unlikely to be economically important in the current spray year (Bailey et al., 1982), is likely to increase future problems. Further, Bailey's observations on very poor black-grass control with half dose isoproturon applied at 20 1 ha^{-1} at various drop sizes by CDA are a salutary warning.

Orchard spraying is another area where the merits of competing delivery systems are disputed. Comparisons of standard application methods and reduced volume and sometimes reduced dose spraying, are again sparse. Our data (see Hislop et al., 1983, 1984) indicate that insect pest control is often very satisfactory with small volumes and with reduced doses. However, we have consistently failed to control damaging epidemics of applepowdery mildew when we used small spray volumes (e.g. 50 compared with 500 1 ha^{-1}). Confirmation of this is provided by the data of Umpelby (1984) in Table 4 and our own results in Table 5. On the other hand, other results indicate that small volume spraying is a reasonably cost-effective and a more convenient alternative to normal spraying when the incidence of apple mildew is small (Hislop 1983b).

EFFICIENT PESTICIDE DELIVERY
Much of the spray research at Long Ashton for the last few years has been devoted to measurements of where sprays go and correlations of these measurements with biological results (Hislop et al., 1983). There are three reasons behind this approach:

1. If a particular delivery system puts more of the spray on the target plants, there is less waste and less likelihood of undesirable effects.

2. Any system that puts a greater proportion of the spray on target areas or organisms within a crop (where these can be identified) is likely to be both more efficient and effective.

Table 4. Effects of spray volume and fungicide dose on the control apple powdery mildew at Luddington Experimental Horticulture Station (source: Umpelby, 1984.)

Spray volume (1 ha^{-1})	Fungicide dose per ha	Secondary mildew (%) 1982	Secondary mildew (%) 1983	Deposition on foliage (ng cm^{-2})	Coefficient of variation[+] (%)
550	Full	2	1	110	38
550	25%	8	5	–	
55	Full	8	7	220	89
55/110 [++]	25%/50% [++]	22	20	–	

[+]
[++] Analyses done at Long Ashton.
Volume and dose were increased mid-season.

Table 5. Fungicide deposition on apple foliage, and control of apple powdery mildew at Long Ashton in 1984.

Atomiser	Volume (1 ha^{-1})	Mean deposit (ng cm^{-2})	Mean cover by dry deposits (%)	Secondary mildew (%)[+]
Micronair AU7000	500	101(40)[++]	10.4(59)[++]	29[a]
Micronair AU7000	200	63(48)	5.7(71)	37[ab]
'Electrodyn'	1	132(59)	0.3(81)	42[b]
Unsprayed	–	–	–	56[c]

[+]   Treatments with the same superscript do not differ significantly at 95% confidence limits.
[++] Coefficients of variation (%) are given in parentheses.

3. Such measurements might facilitate a better understanding of the deposition required for more efficient use
of chemicals.

Data in Table 6 summarise some measurements of spray deposition in
cereals while other information in Tables 4 and 5 relates to top fruit.
Data in Table 7 refer to the deposition of sprays on weeds in cereals
sprayed in the spring. Some delivery systems clearly deposited more
material in total on the target than others. Criteria described earlier
suggest that these particular systems were more efficient, yet they
rarely if ever produced a better biological effect. The larger mean
deposits produced by spraying reduced volumes of liquid tended to have
relatively large coefficients of variation (Tables 6 and 7). This means
that some sites were underdosed. On the other hand, the larger volume
sprays sometimes deposited less efficiently overall but with less
variation about the mean. In other words, the traditional spray system
tends to deposit sprays moderately well throughout the crop, and this is
probably the reason for its lasting success and reliability in biological performance. A similar conclusion has been reported by Robinson
and Garnett (1984). Griffiths et al. (1984) and Hislop et al. (1983)
working with electrostatic spraying systems, have also demonstrated
improved overall capture of spray by crops. There were indications that
the distribution of deposits within the crop was sub-optimal for maximum
efficacy. However, the significance of the distribution patterns of
fungicides within crops has to be judged with care because large foliar
deposits have given unexpectedly good control of a cereal stem-base
disease (Janicke and Grossmann, 1984; Robinson and Garnett, 1984).

The phenomenon of pesticide redistribution over the surface of plants
after deposition, either by water (Hislop, 1966, 1970) or as a vapour
(e.g. Bent, 1967; Hislop, 1967) cannot be ignored. Similarly, systemic
pesticides may act at or near the site of deposition by protective or
curative action, but often produce these effects at more distant sites
after translocation in the plant. Most systemic pesticides move within
plants by apoplastic transport, i.e. in the direction of water movement,
although some can move also in the opposite direction. For these
reasons, measurements of deposition on particular leaves or on whole
small weeds do not necessarily indicate their likely biological zones of
action. Identification of precise sites of action is often exceedingly
difficult. The general premise that to be effective many pesticides
have to be intercepted by plants (excluding soil-acting herbicides) is a
reasonable working hypothesis when attempting to analyse efficiency of
pesticide delivery. The fraction reaching the soil is generally considered as waste. However, results of Rawlinson et al. (1982) showing
that the foliar fungicide, triadimefon, can persist in the soil for
considerable periods and protect the next generation of plants from
disease, is a warning against oversimple interpretations of cause and
effects. Action of 'foliar sprays' via fall-out on the soil followed by
root uptake, may be more common than we think, and may contribute to the
unexplained 'tonic effects' of certain fungicides.

Table 6. Spray retention by winter barley at GS 31 at Long
Ashton in 1984.

Method of spraying	Volume (1 ha^{-1})	Mean deposit[+] μg/plant	CV[++]	Capture by plants (%)
Hydraulic nozzles	200	6.2	31	50
Hydraulic nozzles	100	6.6	36	53
Hydraulic electrostatic	180	5.3	39	42
Hydraulic electrostatic	70	6.4	46	51
Electrodynamic	2	8.1	91	65

[+] For application at 100 g ha^{-1}.
[++] Coefficient of variation (%).

Table 7. Deposition of herbicides on broad-leaved weeds in
winter cereals sprayed at c. GS 30, at a site in Oxfordshire

Method of spraying	Volume (1 ha^{-1})	Mean deposit[+] (ng cm^{-2}) 1983[1]	1983[2]	1984[3]
Hydraulic nozzles	200	180(27)[++]	136(42)[++]	216(27)[++]
Hydraulic nozzles	100	69(45)	103(39)	239(37)
Rotary cages	90	65(110)	115(68)	93(46)
Rotary cages	40	165(54)	93(50)	62(86)
CDA (200 μm diam. droplets)	40	108(97)	116(61)	147(85)

[+] Adjusted to notional dose of 100 g ha^{-1}.
[++] Coefficient of variation (%) are given in parentheses.
1. Bifenox/mecoprop in winter barley – deposits on field
    pansy.
2. Ioxynil/bromoxynil/mecoprop in winter wheat – deposits
    on field pansy.
3. Ioxynil/bromoxynil/mecoprop in winter wheat – deposits
    on parsley piert.

Alternative delivery systems using a common pesticide dose in different volumes produce deposits which differ qualitatively as well as quantitatively. For example, a dose of 100 g of active ingredient sprayed electrodynamically in 1 litre as 90 μm drops produces a deposit per drop of about 38 ng. The same dose sprayed by a rotary atomiser in 40 litres as 180 μm drops produces a deposit per drop of about 8 ng, one fifth of the former, and this will cover about four times the surface area. The consequence of these effects are largely unknown, although some attempts are being made to understand them in biological terms (e.g. Scopes, 1981; Herrington and Baines, 1983) and in terms of chemical uptake (Baker and Hunt, 1985). These problems are compounded because not only is the dose of active ingredient per drop site different, but so too are the amounts of all of the formulating agents in the spray. Despite these complications, it is a priority that more fundamental data on the interacting effects of drop size, frequency and content are produced to facilitate better use of existing spray systems and the development of alternatives.

Spraying relatively large volumes of liquid is time consuming, and the passage of heavy equipment over soil is often damaging. Reducing spray volume speeds the operation and facilitates optimal timing. No one disputes the importance of spray timing or the advantage of being able to complete a job within a "weather window" (see Smith (1983), for a farmer's view of the problems). A number of manufacturers of pesticides for use in cereals have recognised the problem, and now make recommend-ations for smaller volume sprays using standard equipment. Others have adopted the CDA concept and have made recommendations accordingly. The use of a variety of spray delivery systems on low ground pressure vehicles is another development which may permit more flexibility in spray timing. Ayres and Cussans (1980) examined the influences of volume rate, hydraulic nozzle size and forward speed of a lightweight low ground pressure vehicle on the activity of three herbicides for the control of weeds in winter cereals. They concluded that spraying speeds up to 19 kph and volume rates down to 60 l ha^{-1} were almost as effective as standard applications and that these techniques were of considerable practical promise. The 'little and often' approach to herbicide spray delivery (Madge, 1982; May, 1982; Ayres and May, 1983) is also effective and practical. This technique owes much of its success to hitting weeds when they are most susceptible, while the decreased herbicide doses are better tolerated by the crop plants.

Obviously any consideration of the overall efficiency of competing spray systems cannot ignore operator and environmental safety. Totally closed spraying systems such as the 'Electrodyn' bozzle used in developing countries (Durand et al., 1984) are an obvious advance in safety, and a similar tractor-mounted system would be advantageous for developed agriculture. Safe, much reduced volume spraying has great appeal in developing countries where water is often scarce, terrain difficult and equipment must be moved or carried manually. Matthews (1983) argues convincingly for the CDA-type applications in developing agriculture where the choice might be between their adoption or doing nothing at

all. The concept of direct injection of concentrated pesticides into
otherwise standard sprayers (Schmidt, 1983) is another development which
will probably be used more widely in the future.

Spray drift is a safety and environmental hazard which is usually
associated with the use of small droplets (Elliott and Wilson, 1983).
Hydraulic pressure nozzles are notorious for producing very large
numbers of small driftable droplets and small numbers of large drops
containing considerable volumes of liquid which are poorly retained on
targets. Table 8 shows how nozzle selection and pressure can influence
the volume of spray produced in small and large drop doses as measured
in-flight with a Malvern laser particle analyser. The relative
accuracies of different methods of measuring hydraulic nozzle
atomisation are at present a matter of dispute and alternative sets of
data showing different absolute magnitudes of droplet size but similar
general patterns have been presented by Sheldon (1984). Frost and Lake
(1981) have discussed some reasons for the differing results obtained
with different equipment but as yet the true position remains unclear.

Table 8. Effect of throughput on droplet spectra of four
110° flat-fan hydraulic nozzles (Lurmark 'Kematal' type) as
measured with a Malvern 2200 laser particle analyser.

Nozzle size[+]	Pressure (bar)	Spray volume (1 ha^{-1})	VMD[++] (µm)	Spray volume in classes (%)		
				50 µm	100 µm	350 µm
F 110/01	4	69	98	20	51	0.1
F 110/015	2	73	179	8	22	10
F 110/03	4	208	164	11	28	9
F 110/04	2	196	245	3	11	20

[+] Nozzles spaced at 50 cm on boom; forward speed 8 kph.
[++] Volume median diameter.

Rotary atomisation as used in many CDA systems can largely eliminate the problem of small driftable drops and this is particularly beneficial for the application of herbicides (Elliott and Wilson, 1983) using drop sizes of 150 to 250 μm. Similarly, CDA fungicide sprays containing drop sizes of c. 100 μm should be less drift-prone than some hydraulic nozzle systems. But it has to be remembered that drops issuing from the periphery of most spinning disc systems have no vertical velocity, unlike drops produced from hydraulic nozzles (see Hislop, 1984), so that crop canopy penetration is restricted. Electrostatic atomiser systems such as the 'Electrodyn' and the APE 80 types might also reduce drift. The extent of this effect is largely unquantified, although encouraging results have been presented for electrodynamic spraying (Hill et al., 1983). Similarly, hydraulic-electrostatic nozzles can also reduce the volume of liquid in small spray drop classes (Table 9) (Pay, 1984). Unfortunately, both these sets of data indicate that the biggest effect is seen with unusually low-throughput nozzles, and the effect is almost lost with nozzles that apply a standard application volume (c. 200 l ha^{-1}). Thus, the true practical meaning of the drop spectra data and the actual drift reduction measured by Sharp (1984) is questionable. But whatever the present value of these observations, they indicate a step in the right direction. Endacott (1983) has suggested that the greater attractiveness of charged sprays for a crop and therefore reduced contamination of lower plant parts and the soil should help preserve beneficial predator insects, so reducing an adverse side effect of traditional pesticide applications.

Table 9. Droplet spectra from three 80° flat-fan hydraulic nozzles operated at 3 bar pressure in charged and uncharged modes, using the Spraycare ES electrostatic charging system. Spectra were determined with a Malvern 2200 laser particle analyser.

Nozzle size[†]	Uncharged (−) or charged (+)	Flow rate (l min^{-1})	Volume (l ha^{-1})[†]	VMD[††] (μm)	NMD[††] (μm)	Ratio VMD: NMD	Spray droplets 50 μm diam. (% by volume)
800067	−	0.26	40	134	26	5.1	8.9
				*	*	*	*
800067	+	0.26	40	139	34	4.1	7.0
8001	−	0.39	59	158	31	5.1	6.4
				*	*	*	*
8001	+	0.39	59	159	38	4.1	5.4
8003	−	1.18	178	205	31	6.7	5.0
				ns	ns	ns	ns
8003	+	1.18	178	208	31	6.7	4.9

ns Adjacent values not significantly different (P = 0.05).
*   Adjacent values significantly different (P = 0.05).
†   Nozzles spaced at 50 cm on the boom; forward speed 8 kph.
†† Volume median diameter and numerical median diameter.

Before leaving the topic of efficient pesticide delivery, it is necessary to comment on alternatives to spraying. Granular applications (Jepson, 1976), weed wiping (Norton, 1982; Lutman et al., 1982) and tree injection (Clifford et al., 1984) are examples of potentially very efficient delivery systems. Using these methods, pesticides are placed more or less precisely where they are required, with relatively little chance of dispersion elsewhere. Unfortunately, each has its own problems and they are relatively little used. For example, there are relatively few granular formulations and the existing application equipment can be very difficult to calibrate. Weed wipers of various kinds are available but the transfer of an optimal dose of herbicide is rarely achieved with most (Lutman et al., 1982; Davison and Derrick, 1983).

RATIONAL PESTICIDE USE AND DELIVERY SYSTEMS
Rational pesticide use implies the input of reason or logic in deciding how to tackle a problem. Some of the factors to be considered when deciding how to apply pesticides are discussed above. However, before attempting to answer the often-asked question of "How should I spray?", first we should consider very briefly the question "Should I spray at all?" Farmers often are risk-averse and tend to favour routine prophylactic spraying, despite the cost. On the other hand, many are now realising that if pesticides are used sparingly, they and the environment will be better off and they will not be exacerbating the problems of pesticide resistance (Martin, 1981). Decisions on whether to spray, and how to spray should be based on individual circumstances (Hislop, 1983a). This implies conscious decision-making based where posible on pest and disease forecasts (Fry and Fohner, 1983; Norton and Way, 1983;) and the wider acceptance of integrated management programmes (Geissbuhler, 1981), associated with estimates of likely profit and loss in their broadest sense (Southwood, 1981). In Europe, with over-production in agriculture and increasing environmental pressures, there is an increasing emphasis on efficiency of pesticide inputs rather than on biological effectiveness alone. Thus, spinning-disc electrostatic spraying or electrodynamic spraying would be favoured if they became commercially available, while in the meantime greater use should be made of existing CDA equipment. Large-scale arable farming could rationally justify two spraying machines, one CDA and one conventional. The more precise and convenient CDA delivery system would probably be used most of the time, while the other would be used for higher volume application of contact herbicides and for dealing with more severe problems. These suggestions might not be practical on small farms where only one sprayer is justified and where the cost of changing could be prohibitive. An easier alternative for those who like the simplicity and efficacy of traditional sprayers is to change to lower-throughput nozzle systems, perhaps using those designed to work at low spray pressures. Such an option will reduce drift, increase work rate and improve timeliness. Top fruit growers are probably best advised to continue using existing mist-blowers but to supervise biologically their fungicide programmes (Butt et al., 1983) and to adjust spray frequency, volumes and doses accordingly.

Data on biological efficacy of unconventional pesticide delivery systems are sparse, but encouraging in terms of overall efficiency. The corner-stone of the present discussion is the need to improve timeliness of applications in the hope that this can be translated in practice into reduced pesticide inputs. Unfortunately, there are at present few pesticide label recommendations which accommodate such thinking, and this could seriously hinder a rapid change to more efficient delivery systems. Cynics suggest that this is because it is not in the best financial interests of pesticide manufacturers to encourage the use of fewer applications or lower doses. However, manufacturers do have to be certain that their recommendations will stand the test of time. In effect, their labels are a warranty that their products will generally perform effectively if applied in a certain manner. They are often unsure that they can back this warranty in the case of unconventional systems. Also, the question of the financial return from the sale of a given quantity of pesticide is almost irrelevant because the price will be adjusted to take account of market forces; what the farmer buys is not quantity but an effect on his season's yield. This is perfectly proper because pesticide manufacturers have to be guaranteed a reasonable return from their massive investment (c. £15 million per product) if they are to research and develop new products needed by consumers. The main problem is that while testing for efficacy is relatively cheap, toxicological, environmental and residue work is costly. Chemical manufacturers are generally unwilling to repeat these tests to meet the requirements of new delivery systems which might be superseded and for products which have a limited life under patent. The same sort of arguments apply to products for use on minor crops where the considerable cost of clearance would not be covered except by substantial sales. These are difficult problems which are not easily solved. They might require greater input from the public research sector.

An encouraging feature of recent application research has been the involvement of pesticide manufacturers in research on novel devices. For example, the Crop Protection Division of ICI invested much time and money examining various electrodynamic devices, and preparing specific pesticide formulations for use with them. Likewise a number of major pesticide manufacturers have tested hydraulic electrostatic spraying systems. Unfortunately, for the manufacturers of CDA machines this kind of collaboration during early stages of development did not occur to any great extent, and rapport between them and the chemical companies was hindered by extravagant claims that the use of this type of delivery system would result in large dose reductions. Co-operation between chemical companies and those determined to improve the efficiency of delivery systems is essential, because without it, suitable label recommendations will not emerge. Pesticide formulations should be optimised to match delivery systems. Rarely has this been done where recommendations exist for the use of products with CDA sprayers. Available products are not optimised to work at relatively large doses per drop; they have been tested almost exclusively in traditional spraying systems using larger spray volumes.

Clearly, improvements in the efficiency of pesticide delivery systems require a commitment and enthusiasm for collaboration between research and development workers in the public and private sectors. It also requires a somewhat open-minded approach from those responsible for formulating regulations governing the use of pesticides. For example, progress could be hindered if regulatory authorities spprecise dosages to be used in practice. Fortunately, concern expressed in the UK farming press on this topic seems unfounded since in the UK only the maximum permissible dose rate will be specified by new legislation, and reduced dose applications will not be illegal. However, those who might choose to use smaller or normal doses of pesticides in unconventional delivery systems rightly will need to obtain clearence. It is generally recognised that pesticide application must remain an integral feature of agricultural and public health for the forseeable future. The many farmers who, through careful application and good timing reduce pesticide inputs are not only maximising profitability and the reliability of crop yields but minimising the risk of adverse side effects. Researchers on application methods are striving towards the same objectives, and marked progress may be expected over the next decade.

REFERENCES

Addala, M.S.A., Hance, R.J., and Drennan, D.S.H. (1984a). Effect of application method on the performance of some soil-applied herbicides. I. Glasshouse experiments. Weed Research 24, 99-104.

Addala, M.S.A., Hance, R.J., and Drennan, D.S.H. (1984b). Effects of application method on the performance of some soil-applied herbicides. II. Field studies. Weed Research 24, 105-113.

Anon. (1976). The utilisation and performance of field crop sprayers. Farm Mechanisation Studies No. 29, pp. 32. Pinner, Middlesex: Ministry of Agriculture, Fisheries and Food.

Anon. (1979). Royal Commission on Environmental Pollution. Seventh Report: Agriculture and Pollution (Command. 7644), pp. 280. London: HMSO.

Arnold, A.J. and Pye, B.J. (1980). Spray application with charged rotary atomisers. In Spraying Systems for the 1980s, Monograph No. 24, 109-112. British Crop Protection Council Publications.

Ayres, P. and Cussans, G.W. (1980). The influence of volume rate, nozzle size and forward speed on the activity of three herbicides for the control of weeds in winter cereals. In Spraying Systems for the 1980s, Monograph No. 24, 57-64. British Crop Protection Council Publications.

Ayres, P. and May, M.J. (1983). The potential for repeated low doses of herbicides. Proceedings of the 10th International Congress of Plant Protection 3, 519.

Bailey, R.J., Phillips, M., Harris, P. and Bradford, A. (1982). The results of an investigation to determine the optimum drop size and volume of application for weed control with spinning disc applications. Proceedings of the British Crop

Protection Conference – Weeds 3, 995-1000. British Crop
Protection Council Publications.

Baker, E.A. and Hunt, G.M. (1985). Factors affecting the uptake of
chlormequat into cereal leaves. Annals of Applied Biology,
106, 579-590.

Bals, E.J. (1969). The principles of and new developments in ultra-low
volume spraying. Proceedings of the 5th British Insecticide
and Fungicide Conference, 189-193.

Bals, E.J. (1975). The importance of controlled droplet application
(CDA) in pesticide application. Proceedings of the 8th
British Insecticide and Fungicide Conference, 153-160.

Bals, E.J. (1978). Reduction of active ingredient dosage by selecting
appropriate droplet size for the target. In Controlled
Droplet Application, Monograph No. 22, 101-106. British
Crop Protection Council Publications.

Bent, K.J. (1967). Vapour action of fungicides against powdery mildews.
Annals of Applied Biology 60, 251-263.

Bowden, H.D., Splinter, W.E. and Carlton, W.M. (1964). Theoretical
implications of electric fields on deposition of charged
particles. Transactions of the American Sociey of
Agricultural Engineers 7, 75-82.

Butt, D.J., Jeger, M.J. and Swait, A.A.J. (1983). Supervised control of
apple orchard diseases in England. Proceedings of the 10th
International Congress of Plant Protection 3, 1005.

Clifford, D.R., Gay, C.N., Gendle, P. and Cooke, L.R. (1984). Injection
for the control of tree diseases. Proceedings of the
British Crop Protection Conference – Pests and Diseases 3,
1067-1074. British Crop Protection Council Publications.

Coffee, R.A. (1979). Electrodynamic energy – A new approach to
pesticide application. Proceedings of the British Crop
Protection Conference – Pests and Diseases 3, 777-789.
British Crop Protection Council Publications.

Combellack, J.H. (1984). Herbicide application: A review of ground
application techniques. Crop Protection 3, 9-34.

Davison, J.G. and Derrick, P.M. (1983). Weed response to spray and
'wiper' application of translocated herbicides. Proceedings
of the 10th International Congress of Plant Protection 2,
520.

Durand, R.N., Pascoe, R., Bingham, W. (1984). The hand-held 'Electrodyn'
sprayer: An operational tool for better crop management in
developing countries. Proceedings of the British Crop
Protection Conference – Pests and Diseases 3, 1083-1090.
British Crop Protection Council Publications.

Elliott, J.G. and Wilson, B.J., eds (1983). The Influence of Weather on
the Efficiency and Safety of Pesticide Application: The
Drift of Herbicides, Occasional Publication No. 3, pp. 135.
British Crop Protection Council Publications.

Endacott, C.J. (1983). Non-target organism mortality – a comparison of
spraying techniques. Proceedings of the 10th International
Congress of Plant Protection 2, 502.

Evans, E.J., Slawson, D.D., Harris, P.B., Harrington, T., Lockley, K.D.,
and Phillips, M.C. (1984). Recent results of national

fungicide application trials in cereal crops. Proceedings
of the British Crop Protection Conference - Pests and
Diseases 3, 1049-1056. British Crop Protection Council
Publications.
Frost, A.R. (1978). Rotary atomisation. In Controlled Droplet
Application, Monograph No. 22, 7-21. British Crop
Protection Council Publications.
Frost, A.R. and Lake, J.R. (1981). The significance of drop velocity to
the determination of drop size distributions of agricultural
sprays. Journal of Agricultural Engineering Research 26,
367-370.
Fry, W.E. and Fohner, G.R. (1983). The role of forecasting in plant
disease suppression. Proceedings of the 10th International
Congress of Plant Protection 1, 139-145.
Gardner, A.J. (1984). A review of terminology to describe the equipment
used for spraying and choosing nozzles to achieve the
desired biological result. Proceedings of a Conference on
Crop Protection in Northern Britain, 259-263.
Geissbuhler, H. (1981). The agrochemical industry's approach to
integrated pest control. Philosophical Transactions of the
Royal Society, London B295, 111-123.
Griffiths, D.C., Caley, G.R., Ethridge, P., Goodchild, R., Hulme, P.J.,
Lewthwaite, R.J., Pye, B.J., Scott, G.C., and Stevenson,
J.A. (1984). Application of pesticides to cereals with
charged rotary atomisers. Proceedings of the British Crop
Protection Conference - Pests and Diseases 3, 1021-1026.
British Crop Protection Council Publications.
Herrington, P.J. and Baines, C.R. (1983). Effects of spray factors on
the control of powdery mildew by bupirimate. Proceedings of
the 10th International Congress on Plant Protection 2, 513.
Hill, H., Hawtree, J.N., Chester, G. and Swaine, H. (1983). Prototype
testing for safety in use of vehicles mounted 'Electrodyn'
sprayers. Proceedings of the 10th International Congress of
Plant Protection 2, 673.
Hislop, E.C. (1966). The redistribution of fungicides on plants. II.
Solution of copper fungicides. Annals of Applied Biology
57, 475-489.
Hislop, E.C. (1967). Observations on the vapour phase activity of some
foliar fungicides. Annals of Applied Biology 60, 265-279.
Hislop, E.C. (1970). Local redistribution of fungicide on leaves by
water. Annals of Applied Biology 66, 89-101.
Hislop, E.C. (1983a). Crop spraying: the need for scientific data. Span
26, 53-55.
Hislop, E.C. (1983b). Methods of droplet production in relation to
pesticide deposition and biological efficacy in cereals and
tree crops. Proceedings of the 10th International Congress
of Plant Protection 2, 469-477.
Hislop, E.C. (1984). Crop spraying: theory and practice. Proceedings
of a Conference on Crop Protection in Northern Britain,
240-251.
Hislop, E.C., Cooke, B.K. and Harman, J.M.P. (1983). Deposition and
biological efficacy of fungicides applied in charged and

uncharged sprays in cereal crops. Crop Protection $\underline{2}$, 305–316.

Hollies, J.D. (1982). A survey of commercially grown high yielding wheat and barley crops from 1977 and 1981. Proceedings of the British Crop Protection Conference – Weeds $\underline{2}$, 609–618. British Crop Protection Council Publications.

Janicke, R. and Grossman, F. (1984). Untersuchungen zu applikation-stechnischen Fragen bei der Bekampflung der Halmbruch-krankheit in Winterweizen. Zeitschrift für Pflanzen-krakheiten und Pflanzenshutz $\underline{91}$, 146–158.

Jepson, W.F. (1976). Review of granular pesticides and their use. In Granular Pesticides, Monograph No. 16, 1–9. British Crop Protection Council Publications.

Law, S.E. and Bowen, H.D. (1966). Charging liquid spray by electrostatic induction. Transactions of the American Society of Agricultural Engineers $\underline{9}$, 501–506.

Lutman, P.J.W., Oswald, A.K. and Byast, T.H. (1982). The chemical estimation and biological activity of glyphosate deposited on four plant species by one rope-wick and two roller applicators. Proceedings of the British Crop Protection Conference – Weeds $\underline{3}$, 1001–1008. British Crop Protection Council Publications.

Madge, W.R. (1982). The "little and often" approach for weed control in sugar beet. Proceedings of the British Crop Protection Conference – Weeds $\underline{1}$, 73–78. British Crop Protection Council Publications.

Marchant, J.R. and Green, R. (1982). An electrostatic charging system for hydraulic spray nozzles. Journal of Agricultural Engineering Research $\underline{27}$, 309–319.

Matthews, G.A. (1981). Improved systems for pesticide application. Philosophical Transactions of the Royal Society, London $\underline{B259}$, 163–173.

Matthews, G.A. (1983). Suitability of pesticide application equipment take over for small-scale farmers in tropical countries. Proceedings of the 10th International Congress of Plant Protection $\underline{2}$, 517.

May, M.J. (1982). Repeat low dose herbicide treatment for weed control in sugar beet. Proceedings of the British Crop Protection Conference $\underline{1}$, 79–84. British Crop Protection Council Publications.

Nation, B.J. (1982). Application technology: review and prospects. Proceedings of the British Crop Protection Conference – Weeds $\underline{3}$, 983–994. British Crop Protection Council Publications.

Norman, R.F. (1978). A view of the impact of CDA on the agrochemical industry: Agrochemicals as good as their application. In Controlled Droplet Application, Monograph No. 22, 265–269. British Crop Protection Council Publications.

Norton, A.J. (1982). The control of sugar beet bolters and weed beet by the height selective application of the isopropylamine salt of glyphosate. Proceedings of the British Crop Protection Conference – Weeds $\underline{1}$, 67–72. British Crop Protection Council Publications.

Norton, G.A. and Way, M.J. (1983). Forecasting and crop protection
        decision making – realities and future needs. Proceedings
        of the 10th International Congress of Plant Protection 1,
        131–138.
Parham, M.R. (1982). Weed control in arable crops with the 'Electrodyn'
        sprayer. Proceedings of the British Crop Protection
        conference – Weeds 3, 1017–1023. British Crop Protection
        Council Publications.
Pay, C.C. (1984). Testing the performance of a new electrostatic
        spraying system. Proceedings of the British Crop Protection
        Conference – Pests and Diseases 3, 1013–1019. British Crop
        Protection Council Publications.
Rawlinson, C.J., Muthyalu, G. and Caley, G.R. (1982). Residual effects
        of triademefon in soil on powdery mildew and yield of spring
        barley. Plant Pathology 31, 143–156.
Robinson, T.H. and Garnett, R.P. (1984). The influence of electrostatic
        charging, drop size and volume of application on the
        deposition of propiconazole and its resultant control of
        cereal diseases. Proceedings of the British Crop Protection
        Conference – Pests and Diseases 3, 1059–1065. British Crop
        Protection Council Publications.
Schmidt, M. (1983). The direct injection technique for preparing the
        spray mix – A method of reducing safety and hygiene problems
        in plant protection? EPPO Bulletin 13, 513–520.
Scopes, N.E.A. (1981). Some factors affecting the efficiency of small
        pesticide droplets. Proceedings of the British Crop
        Protection Conference – Pests and Diseases 3, 875–882.
        British Crop Protection Council Publications.
Sharp, R.B. (1984). Comparison of drift from charged and uncharged
        hydraulic nozzles. Proceedings of the British Crop
        Protection Conference – Pests and Diseases 3, 1027–1031.
        British Crop Protection Council Publications.
Sheldon, W. (1984). Physical data on drop spectra from moulded fan
        spray nozzles. Proceedings of a Conference on Crop
        Protection in Northern Britain, 252–258.
Smith, P.J. (1983). Farming – partly an art, partly a science.
        Proceedings of the 10th International Congress of Plant
        Protection 1, 56–60.
Southwood, T.R.E. (1981). Environmental considerations. Philosophical
        Transactions of the Royal Society, London B295 17–18.
Turner, P.D. (1983). Approaches to integrated pest control in some
        tropical tree crops. Proceedings of the 10th International
        Congress of Plant Protection 3, 991–998.
Umpleby, R.A. (1984). Pest and disease control with medium and reduced
        volume spraying of apples, Luddington EHS, 1981–1984.
        Proceedings of the British Crop Protection Conference-Pests
        and Diseases, 3, 1075–1082. British Crop Protection Council
        Publications.

7. THE NEEDS OF APPLICATION RESEARCH IN DIFFERENT COUNTRIES

E. Kersting
Bayer AG, Pflanzenschutzzentrum Monheim,
5090 Leverkusen-Bayerwerk, West Germany

INTRODUCTION

Pesticide application technology embraces a very wide range of methods. The simplest one is the use of a watering can for sprinkling paths with herbicide to keep them free from unwanted growth. At the other end of the scale is the highly sophisticated aerial application technique employed for treating vast areas of forest land, as in Canada. In both cases, the technique is well suited to the particular purpose. We might ask however whether the application method used in a particular situation is the only one possible, whether it can be improved or perhaps even completely re-designed. Satisfying though it may be to the designer to indulge in the conception and development of new approaches to pesticide application, there is usually also the need to meet an economic objective. Furthermore, technical innovation or improvement may be demanded by particular circumstances. For example, the need for a new application technique may result from the development of a new pesticide with a novel mode of action, or from the increased economic importance of a pest or disease pathogen that previously was unknown or only of minor significance. Let me give an example to illustrate this. If coffee rust (caused by Hemileia vastatrix) were to break out on an epidemic scale in Colombia, growers would be confronted with a new application problem. Fungicides effective against coffee rust are available. But how they should be applied under the given topographical and meteorological conditions has yet to be resolved.

On the other hand, techniques that have already proved themselves in other branches of industry and are new to crop protection, may be adopted for pesticide application. A good example of this is provided by the electrostatic charging technique, which has long been employed to coat metal surfaces with paint, e.g. cars. Studies are currently in progress to establish whether, and in what crops, the use of electrostatic charging in its various technical forms can improve pesticide application. The interest that is now being shown in electrostatic charging of sprays as a technique for improving coverage and dynamic retention on targets is not new, but advances in techniques in industrial spraying, combined with economic, environmental and other pressures have prompted researchers in pesticide application technology to re-examine the potentialities of electrostatic charging.

In the field of crop protection, solutions to application problems must not only be technically sound from the viewpoint of the engineer or physicist. They must produce biologically-effective results, and be economical and convenient to use. It may be technically attractive and feasible to deposit, say, one droplet according to predetermined co-ordinates on one square centimetre. If, however, other parameters are necessary to achieve the desired biological effect, then different technical changes must be made. Thus, if the requirement for economic control of a target organism calls for the deposition of one droplet on two square metres, there will then be no point in devising technical procedures to achieve an impaction of one droplet on one square centimetre.

ECONOMIC, BIOLOGICAL AND TECHNICAL ASPECTS OF PESTICIDE APPLICATION RESEARCH IN DEVELOPING COUNTRIES

In West and Central Africa, about 95% of the area of cotton is treated by ultra-low volume spraying with portable, hand-operated, spinning-disc machines. The widths of the cotton rows are similar in the different countries of these regions. The average swath width per pass is six rows, equivalent to about 4.80 metres. The sprays are applied at a volume of about $3 \ 1 \ ha^{-1}$. Initially, the biological results obtained with this swath width and volume were satisfactory. But it has now emerged that both swath width and volume must be adapted to the pest complex. For the control of "obstinate" pests like spider mites and white flies and other sucking species, $3 \ 1 \ ha^{-1}$ may be quite adequate provided the swath width is reduced to three rows, and the delivery rate per unit area is adjusted accordingly. A more recent trend is to spray at only $1 \ 1 \ ha^{-1}$ instead of $3 \ 1 \ ha^{-1}$ but without reducing the swath width. It is conceivable that $1 \ 1 \ ha^{-1}$ will give satisfactory control of some pest species especially in low population density situations. Such use of a very small volume may be economically justified, but there are doubts about its efficacy and about its toxicological safety. It would therefore seem expedient to have a machine designed for spraying volumes of both 1 and $3 \ 1 \ ha^{-1}$. Technically, this is relatively easy to achieve, by simply changing the nozzles. However, for reasons of operator safety, more suitable solutions must be sought and then certain educational problems are bound to arise. It should also be mentioned that the constancy of the power supply for these machines needs to be improved considerably.

Demands are also placed on pesticide formulation technology. Needless to say, the flow rate is determined by the viscosity of a formulation. But can formulation be developed that has the same viscosity at both 20 and 35°C? Depending upon the type of solvent used, the properties of the active ingredient and the dose required, this may be feasible in exceptional cases, but never as a general rule. In other areas of fluid technology, e.g. with lubricating oils, constant viscosity is achieved by the use of certain additives but whether their admixture in crop protection products will be tolerated by plants and at the same time be economical is very questionable. Further, how can a user know what volume is actually applied per unit area in conditions of changing viscosity caused by temperature variations? More research is needed on this aspect of pesticide formulation and spray delivery.

An interesting recent development in the treatment of cotton, and also cowpea, against insect pests is the hand-held 'Electrodyn' sprayer introduced by ICI. Oil-based insecticide formulations, provided in special sealed containers which include the nozzle, are exposed to a high-voltage potential generated from torch batteries. The charged spray droplets give an effective deposit at volumes as low as 0.5 l ha^{-1}.

Cocoa crops can be treated relatively quickly and economically with insecticides at volumes of 3 to 4 l ha^{-1} by thermal fogging. For the application of products formulated for this technique, oil is used as the carrier. It is known that spraying will give better biological results but for the treatment of cocoa, thermal fogging is more rapid and it provides satisfactory levels of control.

The water volumes that are used per unit area for the treatment of rice crops in the different countries of Asia, vary enormously, although the growing conditions are very similar. The volumes range from 800 to 1,500 l ha^{-1} in some countries as against 250 to 600 l ha^{-1} in others. Both approaches have their advocates and strong opinions favouring one system or the other are held. Perhaps more application research could help to settle this controversy. As elsewhere in the world, there is a need to decrease the spray volumes applied in Asia in order to increase the convenience and speed of spraying, and hence to permit more timely application, to lighten the load under difficult soil conditions, and in situations of shortage to save water.

Outbreaks of famine are recurrent in some developing countries of Africa and Asia. During the last few decades, certain countries have succeeded in conquering this scourge by adopting measures such as the use of fertilisers and crop protection chemicals. Today, crops in those countries are treated with the latest generation of agricultural chemicals but whether the products fulfil their potential under these technical circumstances is sometimes questionable. The problem does not so much concern the need to acquire new research findings but rather the correct application of knowledge that is already available. Here, just as in many other cases, it is education and communication rather than research that is required.

In Central and South America, there is also no lack of available technological know-how on the use of crop protection products. The application techniques are comparatively well-known, albeit through imports of machines. Local manufacture of pesticide application equipment generally is not a profitable enterprise, the poor market

making production costs too high. There are a few exceptions, for
example, the development of a rotary atomiser in Argentina, which
discharges the droplets in both a horizontal or a vertical plane. As
elsewhere, the quest here is also for a decrease in volume rates.

PROBLEMS OF APPLICATION TECHNOLOGY IN EUROPE
          A striking feature of application technology in Europe,
particularly in Western Europe, is the enormous variety of machine
designs, and the heavy emphasis on multi-nozzle tractor drawn equipment.

Crop protection chemicals are applied mainly in liquid form in the
different crops. The volume of water as carrier may amount to as much
as 4,000 l ha^{-1}, for example in hops. In an extreme case, this is
equivalent to the application of 120 t of water ha^{-1}. Projected to the
German hop-growing industry alone, this is equivalent to the use of more
than a million tonnes of water per year. There are strong arguments for
reducing the use of such intolerably large volumes of carrier.

The chief reason for using so much carrier relates to the need – common
to nearly all crops – to penetrate a canopy in order to effect the
necessary coverage. Various approaches are being tried by machine
designers in an effort to solve this problem for hops, one being to
raise the height of the fan on mist-blowers. Measurement of the target
surface areas during the different stages of crop growth also provides
an indication of how the water volume can be adjusted more efficiently.
Much more research work certainly needs to be done in this area.
Considerable progress has already been made in this direction in
grape-vine growing. The surface areas of the plant parts that require
treatment at each application are known exactly for some grape
varieties, e.g. 6 ha of plant surface area may have to be treated on
1 l ha of vineyard at one particular growth stage. The availability of
more precise data helps to adapt an application technique to a given
situation. Advances in equipment design will also ensure greater
accuracy in the deposition of crop protection products on targets.

Crops like top fruit, hops and grape vines certainly make special
demands on application technology. However, in field crops like
cereals, the problem of canopy penetration also has become one of
foremost importance in the last few years. How can such a densely-
growing cereal crop be penetrated with chemical in order to obtain
effective control of pathogens that may infect all plant parts, i.e.
stem, leaf and ear.

In the highly intensive cereal-growing areas of Western Europe, the
farmers inspect their cereal fields at regular intervals. If any
symptoms of disease, say mildew (caused by Erysiphe graminis), are
detected an appropriate fungicide will be applied without any delay. If
this treatment turns out to be unsuccessful, the applied product is
usually blamed and its manufacturers asked to look into the complaint.
The idea that failure of the treatment might have been due to an
incorrect application technique is seldom considered.

Let us imagine the placement of 600 vertical rods in an area of one square metre; the rods are each one metre high and fitted with horizontal plates spaced at intervals down their length. We will attempt to wet this theoretical target area from top to bottom with 30 ml of liquid in half a second. The kinetics of the droplets will certainly allow good coverage of the horizontal surfaces to be obtained. But what about the coverage of the vertical surfaces? A pathogen such as E. graminis, infects the stem of the cereal plant as well as leaves. Assuming the parameters to be those mentioned above, the amount of spray deposited on the lower third section of the 'stem' of each 'plant' will be inadequate. This applies especially to the side of the stem facing away from the direction in which the spraying machine is travelling. There are several theories of how to surmount this problem, but a practicable technical solution has yet to be found. Special applicators have been built but their performance has not yet been sufficiently proven for them to be used commercially. The most expedient solution would appear at present to be the use of a machine utilising an air stream as the carrier. In this context, the question also arises whether the use of electrostatic charging without air assistance is at all purposeful, since penetration into the canopy tends to be even less.

ALTERNATIVE APPROACHES

Direct injection

With direct injection techniques, the spray tank serves only as a water container and the crop protection chemical is injected into the spraying system directly at the nozzle. Prototype injection systems have already been developed, for example in Britain and Germany. Whilst the technique is still in its infancy, it will no doubt progress further because the dose rate can be adjusted according to tractor speed and because exposure of the operator to chemicals during mixing is avoided. From ecological viewpoints, there is a pressing need for more research in this area.

Use of granules

This is undoubtedly an elegant method of applying crop protection products. Herbicide granules are usually applied overall at rates of 10 to 20 kg ha^{-1}, Modern applicators can treat 60 ha with a single hopper load, which makes for substantial economies in costs and energy. Quantities of 10 to 15 kg ha^{-1} are required for band applications of granules, for example as with soil insecticides in sugar beet crops, and there are also easy to handle. However, the use of crop protection chemicals formulated as granules raises certain questions. For instance, given a row spacing of 50 cm, is there really any need for granules to be banded over a length equivalent to 20 km ha^{-1}? Would not a spot application by treating individual seeds with sufficient granules for their protection be sufficient? Spot application could be biologically effective especially if used on drilled crops like brassicas and maize. There is still much scope for research on granule application from viewpoints of both biological efficiency and engineering.

Seed treatments
Seed treatment is another attractive means of applying crop protection chemicals. The seed is treated, before it is sown, to ensure unhindered germination and even emergence. At least that is the theory. The increasing withdrawal of mercurial seed treatments from agriculture has focussed attention on the need to improve the effectiveness of alternative seed treatments. Is it not time to re-consider the technical aspects of seed treatment? The use of polymer film coatings is one very interesting approach which offers prospects of greater uniformity of seed treatment, better adhesion of chemicals, and the facilitated application of several types of treatment, e.g. insecticide, fungicide and nutrient, to each seed. However, sophisticated equipment is at present needed for polymer coating, and it is unlikely to be adopted at the farm level.

Output monitoring
Too little attention has been given to the control of output from sprayers equipped with rotary atomisers. Without any data to aid him, the spray operator is placed in a situation similar to that of a pilot flying blind without instruments.

In contrast, the range of instruments available for monitoring speed, flow rate and other parameters on conventional spraying machines is enormous. Without question, such monitoring instruments are useful to the spray operator but is he not overwhelmed by them? Where is the computer-controlled system or control processor that will shut down a spraying machine immediately or not even allow it to begin spraying if the data fed into the system are wrong or exceed the correct values?

OPERATOR SAFETY
For obvious reasons, operator safety is highly important. Chemical and physical properties do not permit every compound to be formulated as liquids, which are easier to measure and to handle than powders. Crop protection chemicals formulated as powders are therefore now being packed on an increasing scale in special forms, for example in water-soluble film, to avoid operator contamination. A still more elegant approach is formulation as water-dispersible granules which are non-dusty and are readily pourable for weighing and loading. The risk of an operator becoming contaminated with these formulations is minimal but if it does happen it usually will be while he is mixing the spray. During this operation, any risk of contamination will be due pre-dominantly to dermal exposure of hands and arms. Some large spraying machines are fitted with accessory devices such as small clean water tanks and suction probes which enable sprays to be mixed with minimal hazard. But why are only large machines equipped with such devices? Why is it not mandatory for many other machines? No matter how much care is taken, accidents are bound to occur occasionally during mixing and application. It would be desirable if many more applicators used in agriculture were equipped with a small water container so that operators can wash their hands if necessary. In the future, we may well see sealed, direct-loading packs of pesticide concentrates being used to

deliver chemicals without handling, straight into specially designed or modified spray equipment.

CO-OPERATION
Novel types of applicators are continually being developed, marketed and used around the world. It is not uncommon for them to break down during the first operation. However, the manufacturers only guarantee that they have delivered the machine in a satisfactory working condition. The manufacturers of a crop protection chemical point out that their product has received official approval only for certain application methods. So it is the customer who loses out because he alone bears the risk. This is a most unsatisfactory situation. The policy in future technological research and development surely must be for public-sector institutions, and manufacturers of machinery and crop protection products, to co-operate closely and trustingly. Through such co-operation, progress in spray application can be achieved more rapidly and effectively.

8. BASIC PHENOMENA ACTIVE IN ELECTROSTATIC PESTICIDE SPRAYING

S.E. Law
University of Georgia, College of Agriculture,
Department of Agricultural Engineering,
Athens, Georgia, 30602, USA

INTRODUCTION
For over 40 years, electric forces of attraction have been successfully utilised to apply both solid and liquid surface coatings in industry (Miller, 1973). Early attempts were also made to exploit this improved particulate-application technology for use with agricultural pesticides (Bowen et al., 1952). Lack of long-term success in most of these pioneering efforts can be attributed to the shortcomings in directly transferring industrial processes to the considerably different operational requirements of agriculture, and to insufficient understanding of the several basic charge-transfers occurring. Research at North Carolina State University in the 1960s, significantly increased the required understanding of basic phenomena active during charged particle applications on to living-plant targets (Webb and Bowen, 1970; Law and Bowen, 1975).

Within the past decade, economic and environmental incentives have prompted renewed efforts in engineering research and development aimed toward perfecting reliable electrostatic processes and prototype machines suited specifically for the rigorous design demands of agricultural crop spraying. Notably included are the studies in the UK by Marchant and Green (1982), Arnold and Pye (1981), and Coffee (1981), in Canada by Inculet et al. (1981), and in the USA by Law (1983). Extensive laboratory and field documentation of the operational characteristics, and the biological efficacy, of these various electrostatic crop-spraying approaches has been published (Law and Mills, 1980, Law and Lane, 1981; Griffiths et al., 1981; Law, 1982; Giles and Law, 1983; Herzog et al., 1983; Hislop et al., 1983; and Sherman and Sullivan, 1983). The objective of this paper is to provide a comprehensive review of the general electrostatics concepts underlying these diverse approaches, to outline basic charged-droplet phenomena which they all encounter, and to indicate the implications which these phenomena have upon the reliability of electrostatic pesticide spraying on to agricultural crops.

FUNDAMENTALS
It has long been recognised that electrostatic forces are uniquely suited for augmenting gravitational and inertial forces in the control of finely divided matter. Felici (1965) pointed out the scientific basis for this experimentally evident behaviour, by

establishing mathematically that, in a given electric field, the ratio
of charge imparted to a particle divided by the particle's mass varies
inversely with particle diameter. In like fashion varies the ratio of
the electrostatic to the gravitational forces acting upon the particle
and determining its motion. It is thus in the usually troublesome
droplet-size range under 100 µm, that electrostatics strongly attain
dominance over gravitational motion for effecting improved spray
delivery and deposition.

Necessary conditions

In order to incorporate electric forces into pesticide spray
application, two physical conditions must be met: (1) each spray droplet
must be given a significant net electrical charge q_p (e.g. $0.5 - 1.5$ x
10^6 electronic charges for a 50 µm droplet); and (2) the charged droplet
must be acted upon by an electric field (E), which may be self-generated
or may be imposed by other nearby charge assemblies including charged
bodies (e.g. metallic electrodes and other charged droplets). Figure 1
illustrates the various electric forces (F), which can act indepen-
dently, or are superimposed to enhance the motion of the negatively-
charged droplet shown.

Figure 1. Electric-force components exploitable for charged-droplet
crop spraying: (a) inverse-square particle-to-particle force; (b)
induced image-charge force; (c) externally applied electric-force; and
(d) spray-cloud space-charge force.

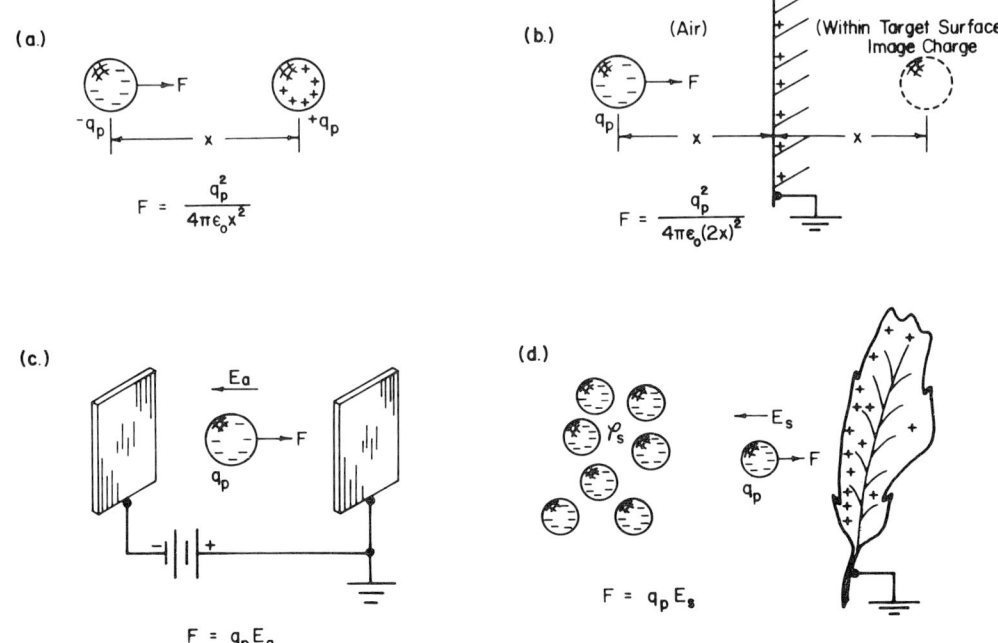

(a.)

$$F = \frac{q_p^2}{4\pi\varepsilon_o x^2}$$

(b.)   (Air)   (Within Target Surface)
Image Charge

$$F = \frac{q_p^2}{4\pi\varepsilon_o (2x)^2}$$

(c.) E_a

$$F = q_p E_a$$

(d.)

ρ_s E_s

$$F = q_p E_s$$

Methods for spray charging

A major portion of the electrostatic crop spraying research mentioned above has been in the development of reliable means for droplet charging. Field-proven methods for imparting the necessary charge (q_p) to pesticide spray particles today are: (1) ionised-field droplet charging of both conductive and non-conductive liquids; (2) electrostatic-induction droplet charging of conductive liquids; and (3) direct electrostatic atomisation and charging of non-conductive liquids. When considered in relation to technical requirements of the vastly differing pesticide-application situations encountered in agriculture, each droplet-charging method possesses its unique advantages as well as disadvantages; they are complementary approaches. Law (1984) presents a detailed consideration of these spray-charging techniques, emphasising their applicability as dictated by the physical properties of the pesticide-liquids to be electrified.

The electrostatic-induction and, especially, the ionised-field charging methods are the ones most widely used throughout many industrial processes. Agricultural charging-nozzle development has also relied primarily upon these two methods in their various embodiments. For ionised-field charging, a D.C. voltage is applied with magnitude sufficient to cause dielectric breakdown of the air surrounding a sharply curved active electrode in the nozzle, such as a metal needle or wire. With a self-sustaining corona discharge current flowing from the point to a surrounding smooth passive electrode, the intermediate gap becomes highly populated with unipolar air ions migrating towards the outer non-ionising nozzle electrode. Either solid or liquid particles, of diameters larger than approximately 0.5 μm, travelling through this ionised-field region can acquire by ion attachment a saturation charge dependent upon the particle's dielectric constant (K), its surface area, and the electrical characteristics of the corona discharge. The fraction (f) of the saturation charge actually attained by the particle depends upon the residence time, and the concentration and mobility of the ions in the charging field. Ionised-field charging theory has been well developed mathematically (White, 1963) and may be used to calculate the net charge imparted to an air-borne particle as:

$$q_p = f \left[1 + 2 \, \frac{K-1}{K+2} \right] 4\pi\varepsilon_o E_o r_p^2 \tag{1}$$

For aqueous-based sprays (K ≃ 80) charged to half-saturation (i.e. f = 0.5) in a typical corona-discharge nozzle, droplet charge in coulombs typically attains a value of:

$$q_p \simeq 6\pi\varepsilon_o E_o r_p^2 \tag{2}$$

with an associated charge-to-mass ratio in coulombs kg^{-1} of:

$$\frac{q_p}{m_p} \simeq \left[\frac{9\varepsilon_o E_o}{2\delta} \right] \frac{1}{r_p} \tag{3}$$

While the ionised-field charging method is routinely used in a variety of commercial and industrial processes ranging from xerography to electrostatic precipitation, great care must be exercised in properly designing the process into agricultural spray-charging devices, in order to maintain long-term charging reliability. Difficulties relate to the fragile nature of the exposed corona electrode, to the elevated ionising voltages required (typically > 15 kV), and to the onset of reverse ionisation from the passive electrode when inadvertently wetted or coated with resistive particles.

Electrostatic induction has proved to be a very satisfactory alternative to the ionised-field method of charging spray droplets for agriculture. In this method, direct charge-transfer to the droplet-formation zone of a liquid jet results from electrostatic induction of electrons on to the continuous jet and in order to maintain it at ground potential the presence of a closely positioned induction electrode of positive polarity is required. Surface densities of free electrons on the order of 10^8 mm^{-2} are typical. Droplets, formed from the surface of this negatively-charged jet, will each depart with net negative charge provided the droplet-formation zone remains subject to the inducing electric field acting between the non-ionising electrode and the jet. Gauss' law indicates that maximum droplet charging should occur for the droplet-production zone located at the region which provides maximum field strength at the terminal surface of the jet. With regard to polarity, this process is completely reversible.

The level of droplet charge imparted by electrostatic induction, depends upon the relative time rate of charge transfer to the droplet-formation zone, as compared with the time required for droplet formation. The charge-transfer capability by conduction from a grounded metal nozzle through the issuing liquid jet, depends upon the electrical properties of the liquid forming the continuous jet. For pesticides, this spray-liquid characteristic may be specified by the charge-transfer time constant (τ), which is a function of the electrical resistivity (ρ), and the dielectric constant (K) of the liquid as:

$$\tau = K \rho \varepsilon_o \qquad (4)$$

If a duration of time, t_f, characterises formation of discrete droplets from the continuous jet, then spray liquids must satisfy the condition $\tau < t_f$, in order to be satisfactorily charged by this induction process. For water-based sprays, Figure 2 indicates that charge-transfer limitation (as defined by $t_f < 5\tau$) would be encountered for resistivities of $\rho > 4.5 \times 10^5$ ohm m. The corresponding value for oil-based sprays (e.g. K = 5) would be $\rho > 7.2 \times 10^6$ ohm m. Spray liquids less resistive than these values should charge satisfactorily by the electrostatic-induction method. Most aqueous tank mixes of pesticides fall within the very favorable $1 - 10^3$ ohm m resistivity range. On-going research in the author's laboratory has verified that with a small percentage of conductivity-enhancing additive, even vegetable-oil carriers of pesticides can be inductively charged.

Figure 2. Charge-transfer time constants
characterising spray liquids as functions of
liquid electrical resistivity and dielectric
constant. (Reproduced with permission from Law
(1978). Copyright, American Society of
Agricultural Engineers.)

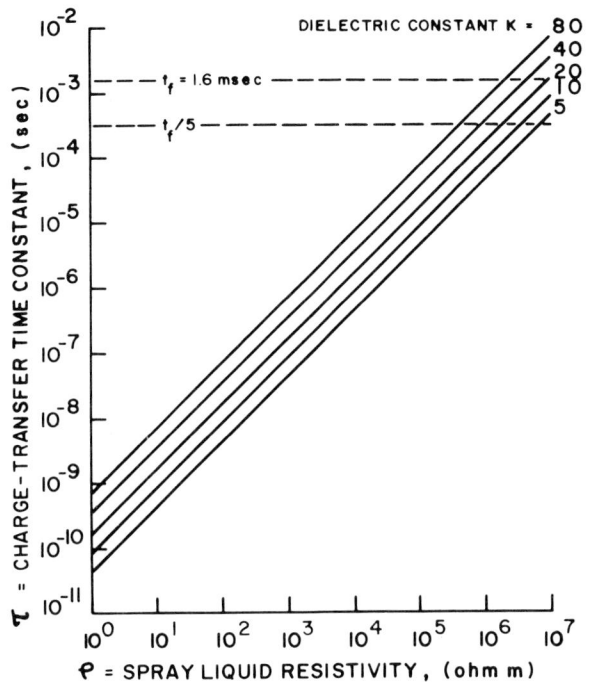

Law (1978) presents details of the engineering design and operational
characteristics of a miniaturised embedded-electrode spray nozzle
developed specifically for induction charging of conductive pesticide
liquids within the $10^{-1} < \rho < 10^{4}$ ohm m range. Typically, sprays of
30–50 μm volume median diameter are electrified to a 10 mC kg^{-1} charge-
to-mass level by charging voltages less than 1 kV and electronic power
consumption less than 25–50 mW. Present efforts are significantly
reducing the 300–400 W power per nozzle required to provide the
pressurised air (e.g. 100–200 kPa) for pneumatic atomisation of the
liquid, and for the turbulent air-entrainment of the charged droplets
necessary to effect their penetration into dense plant canopies.

Comparative analysis of droplet forces

Once charge q_p has been imparted by any of the described
methods to a spray droplet, the force components illustrated in Figure 1
may be brought into action to varying degrees depending upon droplet

size, location relative to the target surface, and other factors.
Instead of directly considering each force's magnitude, their
consequential contributions to droplet terminal velocity (v_t) offers a
truer index for comparing relative effectiveness of the various forces
acting on the air-borne particle in its transit to plant deposition
surfaces. Terminal velocity values for droplets throughout the size
range of interest in pesticide application, can be estimated with
sufficient accuracy by the simple uncorrected form of Stokes' law, as:

$$v_t = \frac{F}{6\pi\eta r_p} \tag{5}$$

It is instructive to determine theoretically how the various terminal-
velocity components depend upon both droplet size and liquid application
rate, and especially to calculate the size range in which electric
forces are predicted to overcome gravity for control of droplet motion.
For example, for water-based sprays corona-charged to the commonly
attained level of half-saturation, Equation (2) and Figure 1 combine to
furnish, along with Equation (5), the following functional relations for
the terminal-velocity components of charged droplets subject to the
stated conditions.

1. Gravitational force. The terminal velocity attained by a droplet
accelerated by gravity is given as:

$$F = \frac{4}{3} \pi r_p^3 \delta g \tag{6}$$

$$v_t = \frac{2\delta g}{9\eta} r_p^2 \tag{7}$$

2. Applied electric-field force. An externally applied driving-field
force can fairly easily be established by maintaining a high D.C.
voltage on a large metal plate positioned above the tops of plants.
Charged droplets released within this applied field (E_a), are driven
toward the top leaves of the grounded plants at a velocity given as:

$$F = 6\pi\varepsilon_o E_o r_p^2 E_a \tag{8}$$

$$v_t = \frac{\varepsilon_o E_o E_a}{\eta} r_p \tag{9}$$

It is practical to maintain a 10 kV m^{-1} driving field by applying 5 kV
to a plate positioned approximately 0.5 m above plant tops, and this
quotient for E_a will be assumed for calculation purposes here.

3. Image-charge force. A charged droplet located near a grounded plant
surface is attracted by an induced image charge of opposite sign
appearing within the leaf. The force of attraction for a droplet-to-
leaf separation (x) is given as:

$$F = \frac{9\pi\varepsilon_o E_o^2}{4x^2} r_p^4 \qquad (10)$$

$$v_t = \frac{3\varepsilon_o E_o^2}{8\eta x^2} r_p^3 \qquad (11)$$

Values calculated from Equation (11), for a charged droplet located at either x = 1 cm and 1 mm spacings from a leaf surface, offer useful insight into the inverse-square behaviour involved, the latter spacing primarily having relevance to the action of this image force for attraction and deposition across target boundary layers.

4. <u>Space-charge force</u>. The movement of a charged droplet within a plant canopy, can be strongly affected by the electric field produced by the space charge associated with all the other neighbouring charged droplets dispersed within the canopy. An approximate mathematical model is provided by considering a charged droplet immersed within a space charge of density ρ_s, uniformly sandwiched between two infinite parallel conducting plates at ground potential. Let the parallel plates, which simulate plant leaves, be located in the X-Y plane at Z = 0 and at Z = d, while the droplet considered is located between the plates at a distance Z. This space-charge field configuration E_s, developed by Bowen et <u>al</u>. (1964) as Equation (12), predicts a charged-droplet force as follows:

$$E_s = \frac{\rho_s Z}{\varepsilon_o} - \frac{\rho_s d}{2\varepsilon_o} \qquad (12)$$

$$F = 6\pi\varepsilon_o E_o r_p^2 \left[\frac{\rho_s Z}{\varepsilon_o} - \frac{\rho_s d}{2\varepsilon_o} \right] \qquad (13)$$

Now an approximation of the average space-charge density (ρ_s) within and surrounding a crop canopy can be calculated as a function of the volume (V_L) of spray liquid dispensed into a target application zone of volume (V_a), and of droplet radius (r_p) as:

$$\rho_s = n_p q_p \qquad (14)$$

$$\rho_s = \frac{3 V_L q_p}{4\pi V_a r_p^3} \qquad (15)$$

For the assumed charging conditions this simplifies to:

$$\rho_s = \left[\frac{9 \varepsilon_o E_o}{2 V_a} \right] \frac{V_L}{r_p} \qquad (16)$$

and the droplet terminal-velocity component, resulting from the interspersed space-charge field within the target confines, is estimated as:

$$v_t = \left[\frac{9 \, \varepsilon_o \, E_o^2}{2 \, V_a \eta} \right] \left[Z - \frac{d}{2} \right] V_L \qquad (17)$$

For illustration, velocity values are calculated from Equation (17) for a charged droplet situated at $Z = 1$ cm from the surface of leaves which are spaced at $d = 10$ cm apart, and for a spray application rate of 9.4 l ha^{-1} (i.e. 1 US gal. ac^{-1}) into a row-crop of 1 m height.

5. Graphical summary. The preceding theoretical analysis provides estimates of the various velocity contributions for droplets charged by the mathematically well defined ionised-field method. While lacking in an explicit functional relation analogous to Equation (1), the electrostatic-induction method and device described earlier, permit droplets to be charged to levels approximately six-fold greater than the half-saturation corona charge. Assuming $q_p \propto r_p^2$ to be valid for both charging methods, then for induction charging the Equations (9), (11), (16) and (17) become increased, respectively, by factors of 6-, 36-, 6- and 36-fold. Figure 3 summarises graphically the values of the terminal-velocity components predicted for inductively charged water-based pesticide sprays, as a function of droplet diameter. For the operational conditions assumed, it is seen that applied-field effects and space-charge field effects at the lowest application rate shown, exceed gravitational effects for droplets smaller than approximately 250 µm and 150 µm, respectively. The importance of image-charge effects is obvious at the very close 1 mm droplet spacing off the target surface, as compared with the negligible contribution at the 1 cm spacing.

Space-charge field effects for droplets smaller than 100 µm are predicted to be dominant within the plant canopy for application rates as low as 9.4 l ha^{-1} (1 US gal. ac^{-1}). As seen from Equation (17) and emphasised in Figure 3, this space-charge contribution is the only case for which the terminal velocity is dependent upon liquid application rate V_L and independent of droplet size. Results of additional theoretical and experimental analyses are presented by Anantheswaran and Law (1981), and these establish the prime importance of space-charge forces for depositing charged droplets within electrostatically-shielded regions such as the interior of plant canopies.

The resultant particle velocity attained in actual practice will, of course, be the vector sum of all combined effects, including turbulent air transport. The preceding calculations and graph are intended to provide theoretical insight into the various particle-size domains in which the different droplet forces predominate, in order to incorporate more rationally the most effective forces into the electrostatic deposition of pesticide sprays.

Figure 3. Terminal-velocity components achieved
by the various force fields acting on an
air-borne charged spray droplet.

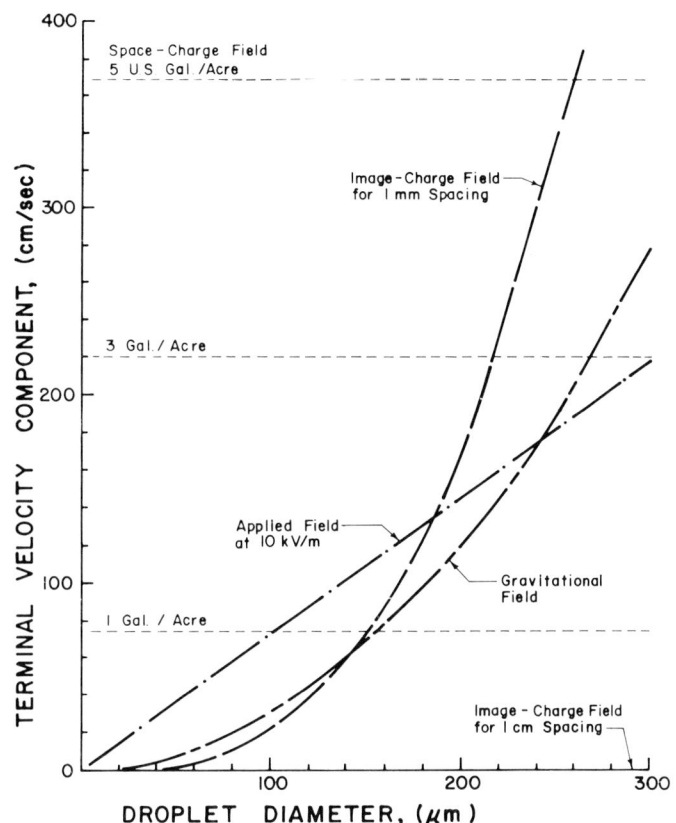

In contrast to the above-mentioned droplet-size independence
which characterises the effect of space-charge upon motion, a detailed
theoretical analysis presented elsewhere (Law and Bowen, 1984) develops
a novel dual particle-specie concept, by which, the space-charge
velocity component of active-specie charged droplets can be greatly
enhanced to achieve their preferential deposition when they are
dispersed with an accompanying inert specie of charged droplets of
judicious size relation. The active-specie velocity amplification
factor which may be achieved is plotted in Figure 4 as a function of the
quantity and the radii ratio of the accompanying inert spray.

AIR-BORNE CHARGE INTERACTIONS
     During flight of charged spray droplets from the dispensing
nozzle to the target object, several basic charge interactions may occur
to affect both the charge retention and the structural integrity of the
droplets. The following sections consider these phenomena and establish
their relative importance with regard to successful practice of the
electrostatic pesticide-spraying methodology.

Figure 4.   Degree of enhancement of
active-specie droplet velocity theoretically achievable by the dual particle
specie concept. (Reproduced with
permission from Law and Bowen (1984).
Copyright, Institute of Electrical and
Electronics Engineers.)

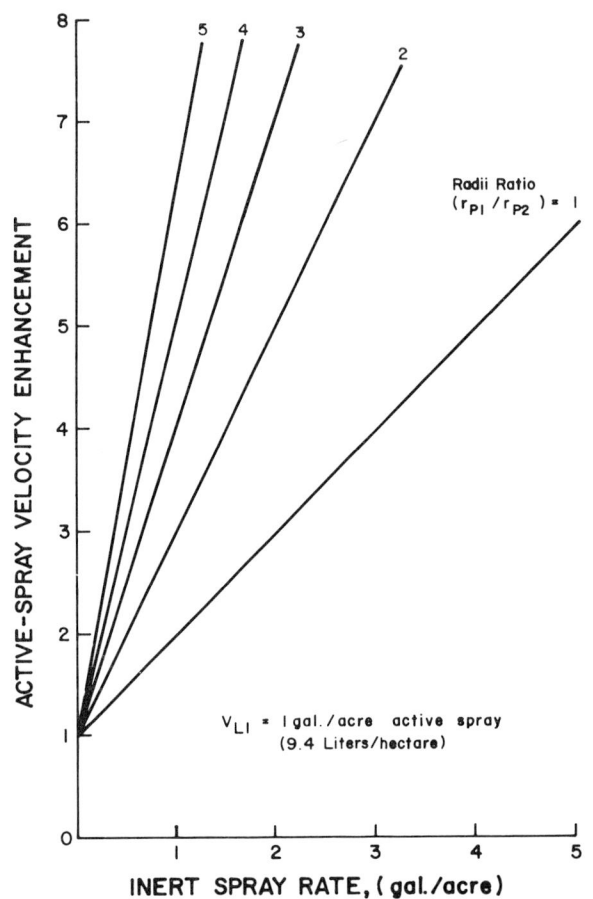

Ambient effects
There exist naturally-occurring free charge and electric
fields in the earth's atmosphere. The following simple analyses
indicate to what degree these ambient conditions may be expected to
interact with air-borne charged spray droplets.

Ambient air ions. In normal air near the earth's surface, ionisation by
cosmic radiation and background radioactivity typically provide a charge
creation rate c_i = 1-7 ion pairs per cubic centimetre per second. The
recombination coefficient α_i for these positive and negative air ions

is approximately 1.6×10^{-6} cm^3 sec^{-1} (Cobine, 1958). Since the rate of ionic recombination depends upon this coefficient as well as upon the concentration (n_i) of both positive and negative species, the air's ion-pair concentration is governed by the relation:

$$\frac{dn_i}{dt} = c_i - \alpha_i n_i^2 \qquad (18)$$

and typically attains an equilibrium value of approximately $1-2 \times 10^3$ ion pairs per cubic centimetre.

These positive and negative air ions of single-electron charge (q_i), can move in response to an imposed electric field with mobilities (μ_i) of 1.4 and 2.1 cm sec^{-1} per V cm^{-1} of field, respectively. Thus, a charged pesticide-spray cloud will encounter some degree of neutralisation by two actions: (1) traversing a region of ionised air; and (2) causing migration of oppositely charged air ions into the region of charged spray. It should be noted, that the lesser degree of droplet-charge neutralisation expected to be inflicted by the $\frac{1}{3}$-slower mobility positive air ions, provides a theoretical basis for the selection of negatively-charged sprays in electrostatic pesticide applications. This is somewhat offset by a slightly greater numerical density of positive air ions near the earth's surface.

In Figure 5, an isolated segment of a moderately charged spray cloud (e.g. $\rho_s = -5$ μC m^{-3}), has been simplified to a spherical charged droplet assembly of 1 m diameter travelling leftward at velocity v_c. As a fraction of the total free charge within the volume occupied by the spray cloud, the positive air ions account for only approximatey 60 ppm. Thus, in order to neutralise the entire cloud, it would have to sweep through a travel length (L) of atmospheric ions estimated by:

$$L = \frac{4 \, r_c \rho_s}{3 \, n_i q_i} \qquad (19)$$

The 10 km value calculated from Equation (19) for the idealised conditions of Figure 5, greatly exceeds the several metres or less of droplet trajection characterising agricultural spraying. Thus, this mode of spray neutralisation can be considered negligible.

If in the second mode of neutralisation, positive ions are considered to be attracted from all directions radially toward the negative cloud, then the following relation indicates the radius (r_i) of the concentric zone of air ions required for complete discharge of the spray to be 12.5 m.

$$r_i = \left[\frac{\rho_s}{n_i q_i} \right]^{1/3} r_c \qquad (20)$$

Of course, the neutralisation process is a dynamic, time-varying one, which initiates at a high rate and lessens exponentially as the

self-generated electric field (E_c) of the charged cloud falls in value
due to the neutralisation already achieved. The transient nature, and
the time required for this neutralisation, depend upon the speed of the
electrical migration of the ions into the charged-spray zone. This
speed at any location r, can be quantified in terms of ionic mobility
(μ_i) as:

$$v_i = \mu_i E_c \tag{21}$$

where, for the zone ($r > r_c$) outside the spray cloud, this becomes:

$$v_i = \left[\frac{\mu_i r_c^3}{3 \, \varepsilon_o} \right] \frac{\rho_s}{r^2} \tag{22}$$

For the conditions depicted in Figure 5, the initial velocity values
describing this attraction of positive air ions are plotted in Figure 6
as a function of distance (r) from the cloud's centre. It should be
noted that these ion-velocity values depend linearly upon the
instantaneous magnitude of spray-cloud charge density, and inversely
upon the square of the distance away from the cloud.

Figure 5. Ambient air-ion neutralisation of the droplet charge carried
on a spherical air-borne spray cloud.

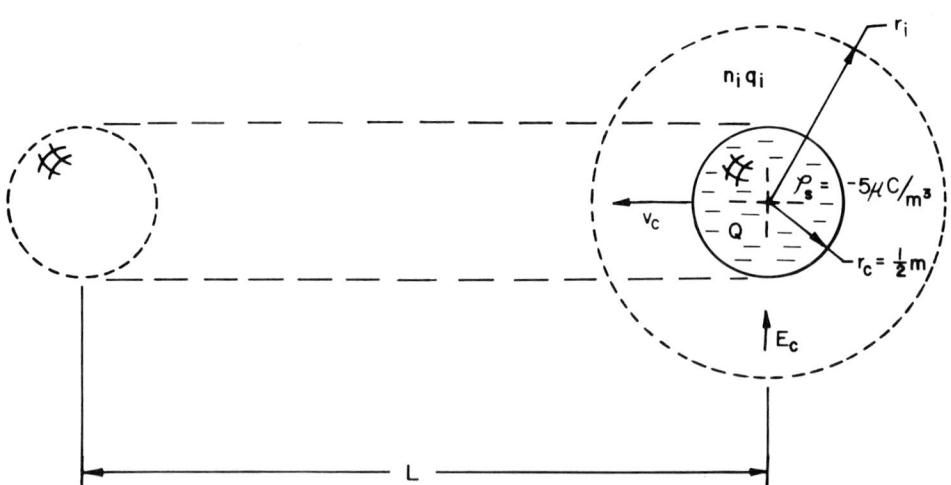

Figure 6.   Initial values for the approach
velocity of positive air ions and for the space-
charge electric field associated with a spherical
charged-spray cloud.

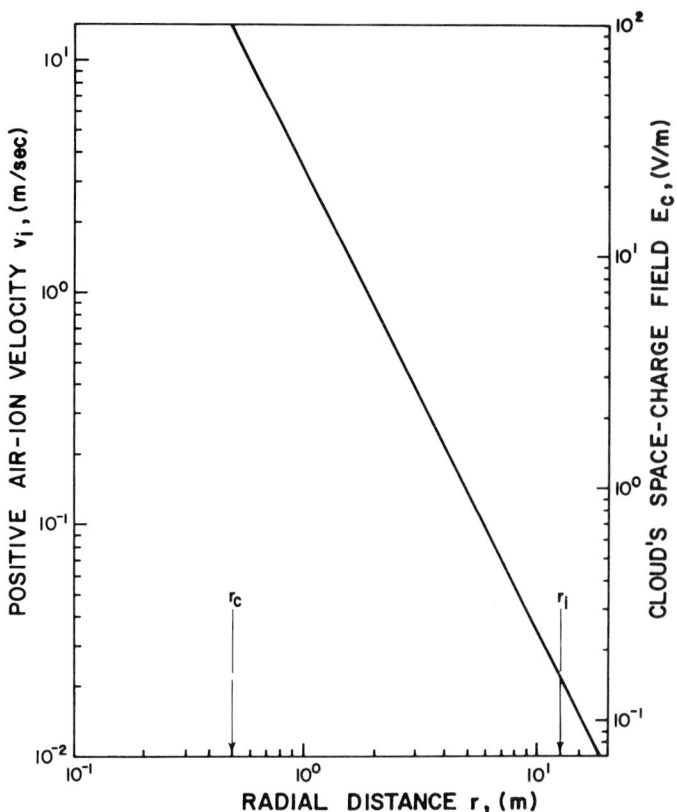

If we disregard ion diffusion in comparison with the
electrical migration, then elementary considerations verify that the
ion-charge flux (i_i) through any spherical concentric shell surrounding
the droplet cloud is independent of radial location.   It has a magnitude
of:

$$i_i = 4\pi n_i q_i \mu_i E_c r^2 \tag{23}$$

and in terms of total cloud charge $Q = (4\pi r_c^3 \rho_s /3)$, it can be equated
at the cloud's boundary r_c to the time rate of neutralisation as:

$$\frac{dQ}{dt} = - \left[\frac{n_i q_i \mu_i}{\varepsilon_o}\right] Q \tag{24}$$

Integration by separation of variables and application of the initial charge condition (<u>viz.</u>, $Q = Q_o$ at $t = 0$), yield the transient equation for spray-cloud neutralisation by ion migration as:

$$Q = Q_o e^{-(n_i q_i \mu_i / \varepsilon_o)t} \qquad (25)$$

This decay process is characterised by the time constant:

$$\tau = \frac{\varepsilon_o}{n_i q_i \mu_i} \qquad (26)$$

which for the conditions assumed in Figure 5 has a value of $\tau = 198$ sec. Approximately 63% of the initial cloud charge would, thus, be neutralised in a period of one time constant or slightly over 3 minutes. As further examples, 21 sec would be required for 10% charge loss, and during the 3 sec period characterising row-crop spraying, a neutralisation of only 1.5% is predicted. Thus, in a practical sense, this mode of spray-charge neutralisation may also be tolerated with little concern (except perhaps in the cases of drift-spraying and orchard airblast spraying with their longer air-borne phases).

The results calculated from the idealised Equation (26) provide a very strenuous, "worst-case" prediction of the amount of time required for droplet neutralisation, which under actual field operation is likely to take considerably longer. A major difference in "real-life" would result from the electrostatic shielding of the charged spray cloud once it was propelled into earthed plant canopies. The action of the field (E_c) would be greatly suppressed, and would extend throughout a zone of ion-attraction much smaller than the radius, (r_i), required for complete cloud neutralisation.

Fair-weather electric field effects. Due to the vertical partitioning of the air-ion constituents of the atmosphere, a naturally occurring electric-potential gradient (E_e) exists near the earth's surface. While its value varies somewhat with time of day, season and location, it typically averages 130 V m^{-1} (Johnson, 1954). With the upper atmosphere of positive polarity and an induced negative charge on the earth's surface, the fair-weather field vector (E_e) points downward towards the earth. The following calculations establish to what degree this ambient field interacts with air-borne charged pesticide droplets in determining their motion.

The -5 μC m^{-3} space-charge density of the spray shown in Figure 5, corresponds to a readily achieved charge-to-mass ratio of approximately -4 mC kg^{-1}. Consider a single isolated droplet of 40 m diameter possessing this charge level. The corresponding electrical-to-gravitational force ratio (F_e/F_g) acting upon the droplet can be estimated by multiplying by the respective field gradients as:

$$\frac{F_e}{F_g} = \frac{q_p E_e}{m_p g} \qquad (27)$$

For this example, the earth's electric field is found to act upon the charged 40 μm droplet with a force only 5% as great as gravitational. While all such droplets would descend to earth, negatively-charged ones would be slowed slightly by E_e, while positively-charged ones would be slightly accelerated. Thus, it is likely that the fair-weather field has only a minor effect upon motion for the typical operational conditions assumed here.

Under certain conditions, however, there may exist a theoretical basis to select, in contrast with the earlier ion-mobility consideration, positive polarity for charged agricultural sprays. If spray is atomised more finely than the 40 μm size specified, then the charge-to-mass achievable should theoretically increase with the reciprocal of diameter, and lift by the earth's field could become appreciable for negative droplets. The lift force on a 4 μm droplet would counteract half its weight (while a 0.5 μm droplet would be levitated), and this could have adverse implications regarding the unwanted air-borne drift of toxic spray from a treated crop area. Conversely, positively charged pesticide droplets would be enhanced in their movement towards earth, with a proportional lessening in drift hazard.

Evaporation of air-borne charged droplets also increases their charge-to-mass value beyond that originally imparted (Law and Bowen, 1975) and, consequently, it would increase their interaction with the fair-weather electric field. For liquids of surface tension Γ, a critical upper limit on droplet charge-to-mass ratio exists (as explained in the following section) such that:

$$\left[\frac{F_e}{F_g}\right]_{max.} = \frac{6 \sqrt{\epsilon_o \Gamma} \; E_e}{\delta \; g \; r_p^{3/2}} \tag{28}$$

This expression predicts it to be impossible to levitate negatively-charged water droplets larger than 32 μm in the earth's field; for surfactant-laden liquid (e.g. $\Gamma \simeq 20$ mN m^{-1}), the limiting diameter would drop to approximately 10 μm.

Several final comments regarding charged-droplet motion resulting from the earth's ambient electric field are given in order to emphasise its primary relevance. During electrostatic application of sprays to crops, the self-generated space-charge electric field of the charged-droplet cloud greatly exceeds the ambient field; it is calculated (Fig. 6) to be over 700-fold greater at the boundary (r_c) of the moderately charged cloud of Figure 5. Also, once the charged spray has penetrated to within the plant canopy, it becomes effectively shielded from external electric fields. Thus, it is the charged droplets which inadvertently stray into air-borne drift above the crop that are likely to become subject to the action of the fair-weather field. As seen, this inter-action can be made, to some degree, either beneficial or detrimental in effecting downward motion for droplet collection depending upon the use

of positively- or negatively-charged spray, respectively. In any
specific engineering design of electrostatic crop-spraying, the
selection of spray-charging polarity should weight the benefits of
positive spray (discussed above) against the reduced ion-neutralisation
benefits likely to be associated with negative spray.

Charged droplet evaporation

The presence of unbound surface charge on an air-borne spray
droplet can drastically alter its surface energy condition. The
theoretically-predicted interactions of this surface charge with droplet
evaporation, and their implications to electrostatic pesticide spraying
have been presented by Law and Bowen (1975). It can be shown
theoretically that charge introduces surface forces, which (1) can
disrupt the structural integrity of the droplet, and (2) tend to reduce
its rate of evaporation. The magnitude and relative importance of these
droplet-charge effects are considered below.

Droplet instability. It has been determined that during evaporation,
vapour leaving a charged droplet carries away no electrical charge (Law,
1968). Thus, as an isolated droplet having a given initial charge
evaporates, the surface charge density must increase. This charge per
unit area cannot increase indefinitely; it approaches a critical level
at which the liquid surface ruptures and ejects a small portion of the
liquid mass, which carries with it a very appreciable portion of the
original droplet charge. In studies using Millikan's oil-drop
apparatus, Doyle et al. (1964) elegantly verified this phenomenon by
experiment, and showed that approximately 30% of the droplet's original
charge departs on 5-10% of the mass.

An elementary analysis of the balance of forces acting within its
surface film, provides a mathematical relation to predict, as a function
of radius r_p and liquid surface tension Γ, the charge level causing the
onset of this instability for an evaporating droplet. Surface tension
creates a mechanical stress within the droplet's film of magnitude:

$$p_\Gamma = \frac{2\Gamma}{r_p} \qquad (29)$$

This stress tends to contract and bind the droplet into a spherical
mass. In contrast, a uniform surface charge creates a mechanical stress
in the film of magnitude:

$$p_q = \frac{q_p^2}{16\pi^2 \varepsilon_o r_p^4} \qquad (30)$$

which acts outward and tends to rupture and "explode" the droplet. A
limitation which applies to the maximum surface charge density on liquid
droplets, was first calculated by Lord Rayleigh (1896). A liquid
droplet will become unstable and break into smaller droplets, if an
attempt is made to charge it beyond that amount where the outward stress
due to the surface charge density just equals the inward stress of

surface tension. Thus, by equating the right sides of Equations (29) and (30), the Rayleigh limit on droplet charge, or the critical condition for the onset of hydrodynamic instability of a charged droplet, is given by:

$$\left[q_p \right]_{max.} = 8\pi \sqrt{\varepsilon_o \overline{\Gamma}} \; r_p^{3/2} \tag{31}$$

This Rayleigh limit on droplet charge is seen to depend upon the square root of the surface tension of the liquid; water sprays with surfactants (e.g. $\Gamma = 20$ mN m^{-1}) would, thus, be subject to droplet instability at a droplet-charge level only half that attainable on water sprays. Figure 7 plots the Rayleigh charge limit as a function of droplet size, for what are probably the extreme surface-tension values to be encountered in electrostatic spraying of aqueous-base pesticides. Also plotted, is the maximum charge-to-mass attainable on droplets, as governed by this hydrodynamic instability phenomenon.

Figure 7. Raleigh and other theoretical limits on the charge carried by air-borne pesticide droplets presented as functions of liquid surface tension and droplet radius.

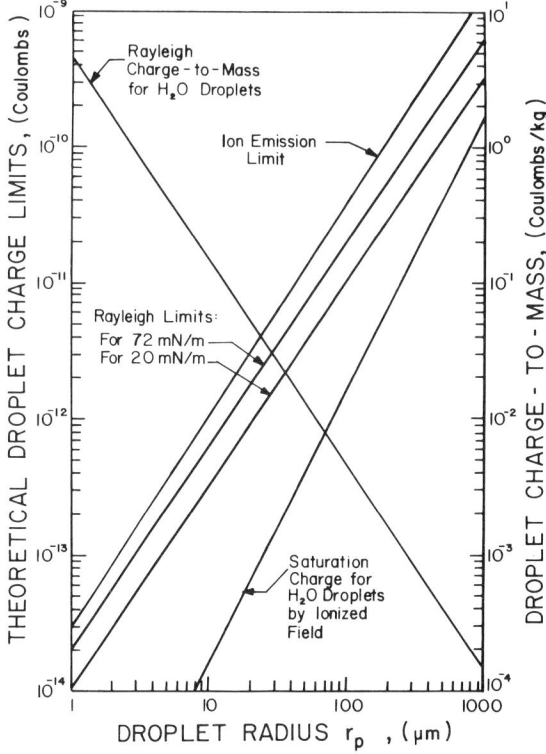

        Upon rupture the parent droplet again attains a stable state
with a reduced droplet charge. The rupture process is repetitive, so
long as evaporation continues, and the droplet charge is not lost by
other means. One such means of loss could be spontaneous dielectric
breakdown for the air surrounding a highly curved, charged droplet, and
subsequent gaseous discharge. Theoretical analysis verifies that,
throughout the size range relevant to pesticide spraying, the droplet-
charge level which causes hydrodynamic instability is reached before
discharge by ion emission can occur (see Figure 7). Therefore, the
rupture process continues for both the parent and the sibling droplets.

As compared with the parent droplet, Roth and Kelly (1983) have shown
the siblings to possess an increased charge-to-mass ratio. In the case
of a droplet evaporating to a non-volatile residue, the rupture process
should cease and the resulting residue should also possess a great
charge-to-mass ratio as compared with the parent droplet. In any
electric field, the motion of these highly charged siblings and residue
particles would be tremendously enhanced. The exact dynamics and
entomological consequences are experimentally unknown. However, it is
expected that with proper design, this droplet instability phenomenon
could have very favourable implications regarding space-charge
deposition within plant canopies. Incidentally, it also offers a novel
feature for charged-droplet spraying of crops from aircraft (Law, 1968).

Effect on evaporation rate. As a consequence of the surface-tension
pressure expressed by Equation (29), the equilibrium vapour pressure (p)
of a spherical droplet, is greater than the value (p_o) for a planar
surface of the same liquid subjected to similar conditions of
temperature and pressure. When the surface-charge stress from Equation
(30) is included, the net value for the vapour pressure (p) driving
droplet evaporation, can be calculated from the William Thomson (Lord
Kelvin) equation as:

$$\ln\left[\frac{p}{p_o}\right] = \frac{M}{\delta RT}\left[\frac{2\Gamma}{r_p} - \frac{q_p^2}{16\pi^2\varepsilon_o r_p^4}\right] \qquad (32)$$

Thus, by theory, surface charge is seen to reduce droplet evaporation
rate. Calculations of Equation (32) indicate $p \fallingdotseq p_o$ throughout the
pesticide-spray size range, and for droplets smaller than 0.1 μm the
effect of charge upon vapour pressure nullifies the surface tension
effect (Law and Bowen, 1975). It can therefore be concluded, on
theoretical grounds, that electrification has no appreciable effect upon
evaporation of agricultural pesticides.

Gaseous discharge of sprays
        From a single-droplet perspective, the Rayleigh instability
(described in the previous section) represents a charge-loss process;
however, for the entire spray cloud, charge is conserved. This is not
the case, unfortunately, for another charge-loss phenomenon acting on an
air-borne pesticide spray cloud - namely, partial neutralisation of the
approaching cloud of corona discharges induced to flow from sharp target

members. A brief description of this air-borne charge interaction, and
its introduction of a self-limiting feature into the electrostatic
crop-spraying process follows.

An earthed living plant, in the absence of nearby charged bodies or
space charge, has essentially zero net charge. However, as a charged
body or charged spray cloud approaches the plant, a charge transfer
tends to occur throughout the plant. In this manner, a charged state
opposite in polarity to that of the nearby spray cloud, is induced on to
the plant. As the charged spray cloud approaches more closely and
envelops the plant, a level of induced charge appropriate to maintain
the plant at ground electrical potential is naturally established. The
transient charge-transfer characteristic of living plants, presents no
limitation in this necessary plant-polarisation effect for charged
sprays applied at typical operational conditions, even for plants
drought-stressed beyond recovery (Lane and Law, 1982).

Electric field theory indicates that high levels of such induced charge
will be established upon those regions of a conducting body having small
radii of curvature. Thus preferentially, intense concentrations will
exist in the vicinity of sharp edges and points on a grounded charged-
spray target (e.g. leaf tips). For sufficiently intense levels of
induced surface charge, the electric field in the air just off the edges
and points, exceeds the air's dielectric strength. The resultant
ionisation and dielectric breakdown can provide a pathway for gaseous
conduction of charge between the grounded target points and the nearby
charged spray cloud. Greater values of cloud-space charge would be
expected to accentuate this gaseous exchange of charge. Neutralisation
of any appreciable portion of the droplet charge on an incoming spray
cloud by this ionisation-current effect, would probably result in a
corresponding and undesirable reduction in the electrostatic-deposition
benefit.

Ionisation currents, induced by space-charge fields to flow through
grounded vegetative points, have been confirmed for air-borne
charged-particulate clouds of greatly varying size. Polarisation
currents within trees, and ionisation currents from needle tips of
spruce forests, resulting from the overhead movement of thunderclouds,
have been widely studied (Ette, 1966; Stromberg, 1971). With smaller-
scale charged pesticide dust-clouds overhead, Bowen et al. (1964)
detected visible corona discharge at the tips of celery leaves.

These gaseous-discharge currents induced from target points have been
correlated (with a high level of statistical significance) with
degradation of the electrostatic-deposition process. Webb and Bowen
(1970) documented this using charged pesticide dusts. Law and Lane
(1982) report extensive charged-spray experiments, in which simultaneous
measurements of charge- and mass-transfer, showed that induced coronas
from target points reduced (to approximately half) the increase in
deposition achievable by the electrostatic effect. Of the total charge
which might possibly be exchanged between a target and an incoming

droplet cloud, greater than 80% is often dissipated via the gaseous-discharge component, and provides no enhancement in deposition. Figure 8 summarises the deposit-limiting effect caused by simply attaching a pointed 20 mm protrusion on to an originally smooth target surface. While the degradation is pronounced, it should be noted that, even for this severe corona condition, the electrostatic-deposition benefit is still 3.5-fold as compared with the application of similar, uncharged spray.

Figure 8. Effects upon electrostatic-deposition benefit caused by corona discharge induced from an earthed target point. (Reproduced with permission from Law and Lane (1982). Copyright, Institute of Electrical and Electronics Engineers.)

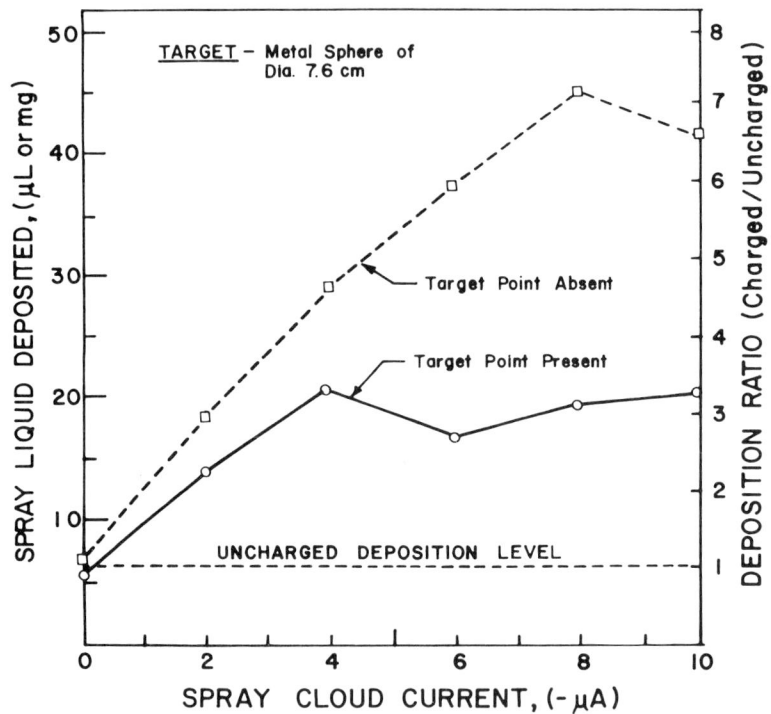

Corona repulsion of droplets
While the results in Figure 8 illustrate the effect which induced target corona has upon overall target deposition, additional studies were necessary to gain a fundamental understanding of the effect upon droplet motion on a more local scale. Laser Doppler anemometry experiments have now established the spatial extent to which the counterflow of corona ions from earthed target points interacts with the incoming charged-spray cloud and how trajectories of approaching charged droplets are consequently altered (Law and Bailey, 1983). Figure 9

depicts vertical droplet motion in the vicinity of a horizontal 6.4 mm diameter metal rod, which simulates a plant stem. A charged spray cloud ($\underline{c}$. -4 mC kg^{-1}) is being conveyed to the right in a 3 m sec^{-1} air-carrier stream, inclined 8° above the target centre-line. For this spray delivery, the 0 μA curve in Figure 9 indicates uncharged spray droplets were moving upwards, away from the target, at all seven of the laser-beam measurement locations along the target's length. For the same spray conditions, but with charging activated, the lower curve verifies vertical velocity reversal and attraction of the charged spray to the smooth cylindrical target along its entire length. Attachment of a sharp point on to the leading nose of this target is seen to totally negate the beneficial attraction of charged spray, at all measurement locations (see the curve -5 μA + point). At most of the locations, ionic-momentum transfer from the point to the droplets and the surrounding air, is seen to actually increase the departure speed (v_y) upward, away from the target, for the initially negatively-charged spray as compared even with uncharged spray. This increased departure speed probably results in part, to an electric gradient force, which would tend to pull droplets up into the cloud if the polarity of their charge shifts from negative to positive in passing through the point's ionic current.

In contrasting the deposition responses of Figure 8 and 9, it can be predicted that, for the complex plant morphologies encountered in electrostatic crop spraying, the overall plant deposition should benefit signficantly from spray charging even when induced target discharges are active. However, close attention should be paid to spray-charging conditions, and possible corona repulsion, whenever small specific plant structures comprise the primary spray target. Several bipolar spray-charging strategies to reduce the deleterious effects of target coronas are currently under development in the author's laboratory.

CONCLUSIONS

Electrical forces have been successfully incorporated into pesticide spray application to improve greatly the basic droplet-deposition process. Numerous mass-transfer studies now firmly document this improvement, and biological data support corresponding increases in pest-control efficacy.

Several distinct approaches to the engineering design of electrostatic spraying systems for agriculture have been carried through to the field-tested prototype stage by various researchers. These approaches, differ primarily in the combination of technical methods selected for liquid atomisation, droplet charging, transfer of the charged spray to the target region, and penetration therein. As there is no one universal pesticide chemical, there is likewise no single electrostatic spraying system which satisfies all of the diverse spray-application demands in crop production. Of the electrostatic systems currently created via the several distinct engineering research and development approaches, each is characterised by its unique benefits as well as limitations, when specific operational requirements are considered, such as: (1) electrically conductive versus non-conductive spray-tank mixes; (2) physical composition of spray liquid such as solutions versus concentrated wettable powders; (3) desired droplet size to be generated and charged; (4) solely electrostatic droplet transport versus supplementation by turbulent air entrainment; (5) exterior surface deposition versus charged-droplet penetration into the electrostatically-shielded plant canopy; and (6) the total electronic plus mechanical power input supporting the overall system. These currently developed electrostatic crop-spraying systems should be considered complementary in their efforts to enhance deposition efficiency throughout the broad spectrum of agricultural demands.

Figure 9. Droplet vertical velocity (v_y) at laser-measurement locations along a cylindrical target 1 cm above its centre-line for three spray conditions: (a) uncharged spray = 0 µA; (b) charged spray = -5 µA; and (c) charged spray on to target with point attached = -5 µA + point. (Reproduced with permission from Law and Bailey (1983). Copyright, Institute of Electrical and Electronics Engineers.)

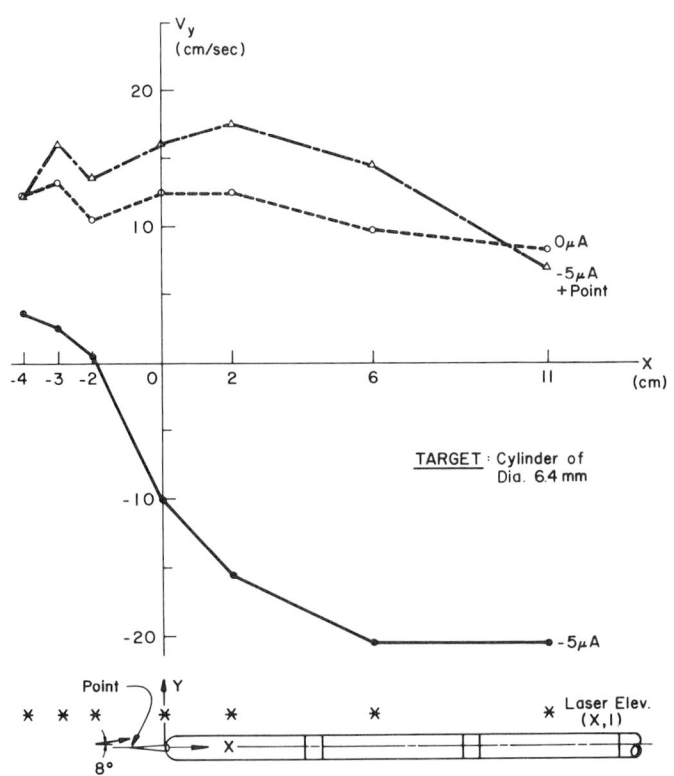

Common concerns of all electrostatic-spraying systems are
the exploitation of the most favourable force fields for achieving the
desired droplet motion, and the maintenance of droplet charge during the
in-flight trajectory of the electrified spray cloud. Thus, this paper
has established a theoretical foundation for comparing the relative
effectiveness of the various electric force fields which may be incorporated under realistic conditions, has discussed the primary charge-
loss phenomena which may become active during the air-borne phase of the
electrostatic-spraying process, and has outlined the implications
which these force fields and charge-interaction phenomena have upon
achieving reliable application of charged pesticide sprays on to crops.

It may be concluded with optimism that in the area of electrostatic
crop-spraying technology, sufficient fundamental understanding and
engineering-design know-how now exist internationally to support its
routine usage in the production of food and fibre for mankind.

REFERENCES
Anantheswaren, R.C. and Law, S.E. (1981). Electrostatic precipitation
    of pesticide sprays onto planar targets. Transactions of
    the American Society of Agricultural Engineers 24(2),
    273-276, 280.
Arnold, A.J. and Pye, B.J. (1981). Electrostatic spraying of crops with
    the APE 80. Proceedings of the British Crop Protection
    Council Conference 2, 661-666. British Crop Protection
    Council Publications.
Bowen, H.D., Hebblethwaite, P., and Carlton, W.M. (1952). Application
    of electrostatic charging to the deposition of insecticides
    and fungicides on plant surfaces. Agricultural Engineering
    Journal 33(6), 347-450.
Bowen, H.D., Splinter, W.E. and Carlton, W.M. (1964). Theoretical
    implications of electric fields on deposition of charged
    particles. Transactions of the American Society of
    Agricultural Engineers 7(1), 75-82.
Cobine, J.D. (1958). Gaseous Conductors: Theory and Engineering
    Applications. New York: Dover Publications.
Coffee, R.A. (1981). Electrodynamic crop spraying. Outlook on
    Agriculture 10, 350-356.
Doyle, A.D., Moffett, R. and Vonnegut, B. (1964). Behaviour of evapor-
    ating electrically charged droplets. Journal of Colloid
    Science 19, 136-143.
Ette, A.I.I. (1966). Estimation of displacement current in trees during
    point discharge. Journal of Atmospheric and Terrestrial
    Physics 28, 831-838.
Felici, N.J. (1965). Electrostatic engineering. Science Journal 1(9),
    32-38.
Giles, D.K. and Law, S.E. (1983). Space charge deposition of pesticide
    sprays onto cyindrical target arrays. American Society of
    Agricultural Engineers Paper No. 83-1501. St. Joseph,
    Michigan.
Griffiths, D.C., Arnold, A.J., Cayley, G.R., Ethridge, P., Phillips,

F.T., Pye, B.J. and Scott, G.C. (1981). Biological
effectiveness of spinning disc electrostatic sprayers.
Proceedings of the British Crop Protection Conference 2,
667-672. British Crop Protection Council Publications.
Herzog, G.A., Lambert, W.R., Law, S.E., Seigler, W.E. and Giles, D.K.
(1983). Evaluation of an electrostatic spray application
system for control of insect pests in cotton. Journal of
Economic Entomology 76(3), 637-640.
Hislop, E.C., Cooke, B.K. and Harman, J.P.M. (1983). Deposition and
biological efficacy of charged sprays of a fungicide in
cereal crops. Crop Protection 2, 305-316.
Inculet, I.I., Castle, G.S.P., Menzies, D.R. and Frank, R. (1981).
Deposition studies with a novel form of electrostatic crop
sprayer. Journal of Electrostatics 10, 65-72.
Johnson, J.C. (1954). Physical Meteorology. New York: John Wiley and
Sons.
Lane, M.D. and Law, S.E. (1982). Transient charge transfer in living
plants undergoing electrostatic spraying. Transactions of
the American Society of Agricultural Engineers 25(5),
1148-1153, 1159.
Law, S.E. (1968). Charge Loss Phenomena Active on Liquid Droplets.
Ph.D. Thesis, Department of Biology and Agricultural
Engineering, North Carolina State University, Raleigh.
University Microfilms No. 68-17, 566. Ann Arbor, Michigan.
Law, S.E. (1978). Embedded-electrode electrostatic-induction spray-
charged nozzle: theoretical and engineering design.
Transactions of the American Society of Agricultural
Engineers 21(6), 1096-1104.
Law, S.E. (1982). Spatial distribution of electrostatically deposited
sprays on living plants. Journal of Economic Entomology
75(3), 542-544.
Law, S.E. (1983). Electrostatic pesticide spraying: concepts and
practice. IEEE Transactions IA-19(2), 160-168.
Law, S.E. (1984). Physical properties determining chargeability of
pesticide sprays. In Advances in Pesticide Formulation
Technology, ed. H.B. Scher, Monograph Series 254, 219-230.
Washington, D.C.: American Chemical Society.
Law, S.E. and Bailey, A.G. (1983). Perturbations of charged droplet
trajectories caused by induced target corona: LDA analysis.
IEEE (IAS) Conference Record, 1042-1049.
Law, S.E. and Bowen, H.D. (1975). Theoretically predicted interactions
of surface charge and evaporation on air-borne pesticide
droplets. Transactions of the American Society of
Agricultural Engineers 18(1), 35-39, 45.
Law, S.E. and Bowen, H.D. (1984). Dual particle-specie concept for
improved electrostatic deposition through space-charge field
enhancement. IEEE (IAS) Conference Record.
Law, S.E. and Lane, M.D. (1981). Electrostatic deposition of pesticide
spray onto foliar targets of varying morphology.
Transactions of the American Society of Agricultural
Engineers 24(6), 1441-1445, 1448.

Law, S.E. and Lane, M.D. (1982). Electrostatic deposition of pest-
        icide sprays onto ionising targets: charge- and mass-
        transfer analysis. IEEE Transactions IA-18(6), 673-679.
Law, S.E. and Mills, H.A. (1980). Electrostatic application of low-
        volume microbial insecticide spray onto broccoli plants.
        Journal of the American Society of Horticultural Science
        105(6), 774-777.
Marchant, J.A. and Green, R. (1982). An electrostatic charging system
        for hydraulic spray nozzles. Journal of Agricultural
        Engineering Research 27, 309-319.
Miller, E.P. (1973). Electrostatic coating. In Electrostatics and
        Its Applications, ed. A.D. Moore, 250-280. New York: John
        Wiley and Sons.
Rayleigh, Lord (1896). The Theory of Sound, Vol. II. New York:
        Macmillan and Co.
Roth, D.G. and Kelly, A.J. (1983). Analysis of the disruption of
        evaporating charged droplets. IEEE Transactions IA-19(5),
        771-775.
Sherman, M.E. and Sullivan, J.G. (1983). Vehicle-mounted Electrodyn
        sprayer applications in cotton and soybeans. Proceedings
        of the 10th International Congress of Plant Protection 2,
        500.
Stromberg, I.M. (1971). Point discharge current measurements in a
        plantation of spruce trees using a new pulse technique.
        Journal of Atmospheric and Terrestrial Physics 33,
        485-495.
Webb, B.K. and Bowen, H.D. (1970). Electrostatic field breakdown
        phenomena in applying charged particles. Transactions of
        the American Society of Agricultural Engineers 3(4),
        455-458, 461.
White, H.J. (1963). Industrial Electrostatic Precipitation. Reading:
        Addison-Wesley.

9. THE EFFICIENT AERIAL APPLICATION OF SPRAYS

J.J. Spillman
Cranfield Institute of Technology, Cranfield,
Bedford, MK45 OAL, England

DEFINITION OF AN EFFICIENT APPLICATION
In aerial spraying, as in other forms of spray application,
several factors contribute to the overall efficiency of the operation.
An efficient application is one delivered:

1. on the scale dictated by the pest;

2. at the most effective time;

3. in the most effective place;

4. using the most effective formulation;

5. using the most effective chemicals; and

6. with the minimum wastage, and therefore of low cost
(both directly and indirectly associated with damage), and
causing minimal damage to the environment.

It is important to realise that each component of the spraying operation
is like a link of a chain - a weakness in any one will reduce the
strength of the whole. It is often worthwhile sacrificing the peak
efficiency that can be achieved in one aspect in order to improve
another of much lower efficiency. Thus, in order to control an insect
pest it may be worthwhile attacking the larval stage because the timing
is easier and the vulnerability greater, even though the eggs may be in
a more exposed position and therefore easier to reach.

It follows that an efficient application is one in which the whole
system is optimised rather than one aspect. This paper concentrates on
how to get aerial sprays to the most effective place and will assume
that the particular target position is known. However, efficiency of
deposition is also linked to timing, because atmospheric conditions
greatly affect the placement of sprays.

REACHING THE TARGET SITE
The target site will be a specific part of the crop plant,
such as the growing point, or the most probable site of the pest, and
not just the field in which the pest resides. Generally it is assumed

that the whole of the crop area has to be treated, but it would be
better to target selectively if the information were available. An
important consideration is whether the target surface is vertical or
horizontal and whether it is at the exposed edge of the canopy or hidden
within it since these factors determine to a major extent the droplet
size to be used.

Large droplets (>300 µm diameter) fall nearly vertically in all but
strong winds and therefore can be accurately aimed at a horizontal,
exposed target such as the ground or the top, horizontal leaves of a
plant. They do not change much in diameter due to evaporation unless
they comprise an extremely volatile formulation. However, it is almost
impossible for them to reach a vertical target, except by secondary
splashing; they are either caught by the upper horizontal surfaces of
the crop or penetrate right through it to the ground. Consequently,
they are of no use if the target surface is well within the canopy.
Even when big droplets are electrostatically charged they cannot be made
to deviate significantly from such behaviour. Their large size means
that relatively few droplets per litre are emitted, and this implies
either a relatively poor distribution over the field or else the use of
a large volume per hectare.

Small droplets (<100 µm diameter) have quite different characteristics.
Their fall direction is nearly horizontal in almost any wind-speed, and
as a consequence they are caught much more effectively by vertical
surfaces, such as the tops of conifers or cereals, provided the surfaces
are wide enough or rough enough to permit their impaction. The low
sedimentation speed of small droplets means that when they are caught in
a near windless zone, such as the wake of a tree or inside the canopy,
they will be distributed by turbulence and can reach the undersides of
leaves and the lower parts of the canopy, albeit in relatively small
numbers. Their paths can be significantly affected by charging them,
since the high charge: mass ratio which they can sustain, can create
attraction forces between them and a neutral surface of many times their
weight. However, this can be a disadvantage if deep penetration of a
crop canopy is required.

Because small droplets have low sedimentation speeds, they can be
carried some distance horizontally from the emission point and therefore
cannot be positioned accurately in the target. Atmospheric turbulence
influences their ultimate destination and they evaporate more rapidly.
The inability to predict the destination of a small droplet as
accurately as that of a large droplet is off-set by the fact that far
more droplets are generated per litre and hence there is a much higher
probability of sufficient of them reaching each target (Graham-Bryce,
1977).

It follows that there is an optimum droplet size associated with
reaching a particular target in a particular crop, under particular
atmospheric conditions. A change in target, crop or atmospheric
conditions might change the size of the droplet best suited for the job.
Figure 1 illustrates an example where a moderately large droplet is

required. The aircraft is required to spray a pesticide into a waterway. It is important that none of the chemical falls on the surrounding
fields. In this case the droplets must fall sufficiently vertically for
the pilot to aim them accurately into the water. With a cross-wind of
1 m s^{-1}, 300 µm diameter droplets will fall at an angle 50° to the
horizontal. If the wind-speed changed to 0.5 or 2 m s^{-1}, the angle
would change to 67° and 31°, respectively. The latter angles would be
achieved in a 1 m s^{-1} wind if the droplet diameters were 500 µm and
180 um respectively. It is obvious therefore that the tolerance in
accuracy allowed by the width of the bank borders of waterway is such
that whilst lower wind-speeds or larger droplets could be accepted,
perhaps by re-locating the flight path slightly relative to the waterway, higher wind-speeds or smaller droplets would make the accuracy
unacceptable. Thus, a wind-speed of less than 1.5 to 2 m s^{-1} is
required and droplets greater than about 225 µm diameter. The upper
limit on droplet size would be determined only by the allowable swath
width or aircraft offset from the waterway, associated with the pilot's
ability to see that he was doing.

Figure 1. Pesticide application on narrow waterways. Flight path shown
for 300 µm spray droplets applied at wind-speed of 1 m sec^{-1}. The
precision of application would be reduced if small droplets were used.

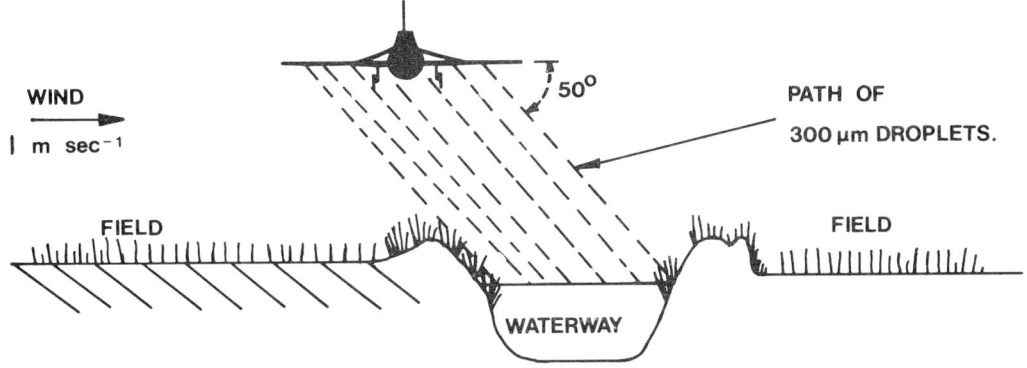

In contrast with the scenario just described, consider the problem
illustrated in Figure 2. Here the target, the grass which the range
caterpillar eats, consists of isolated tufts of grass which are dense in
themselves but sparsely dispersed over the ground. The narrow leaves
are primarily vertical. Clearly big drops (300 µm diameter) would fall
nearly vertically, most would miss the tussocks of grass and those that
did not would fall to the base and only be effective if the chemical was
systemic. However, if 100 µm droplets are used in a wind-speed of
1 m s^{-1} then their mean path is only 14° below the horizontal, and
viewed from this angle, it is clear that all the droplets are likely to
be caught. A stronger wind would reduce the angle and increase the
efficiency of capture. Clearly the limitations are associated with too

low a wind-speed or too large a droplet. The Plague Locust Commission of Australia applies spray droplets of about 80 μm - 100 μm volume median diameter (vmd) against both hoppers and swarms but only when the wind is 3 m s^{-1} or greater.

The efficiency of applying small, evenly distributed droplets, using the wind to carry the droplets into the canopy, is well documented. Joyce (1975) described how the technique was applied to cotton in the Sudan, whilst Holden and Bevan (1978) edited a report describing a wide-ranging investigation into the spraying of coniferous forests. Aerial spraying of flying insects requires droplets of 15 μm - 25 μm diameter in order to maximise their probability of being caught (Spillman, 1976; Hursey and Allsopp, 1983).

Figure 2. Control of range caterpillar on grass tussocks. An example of spray application problem where targets are discrete but irregularly spaced. Note that droplets are all caught by grass if small enough to have near horizontal path in the wind, but big enough to impact on grass blades.

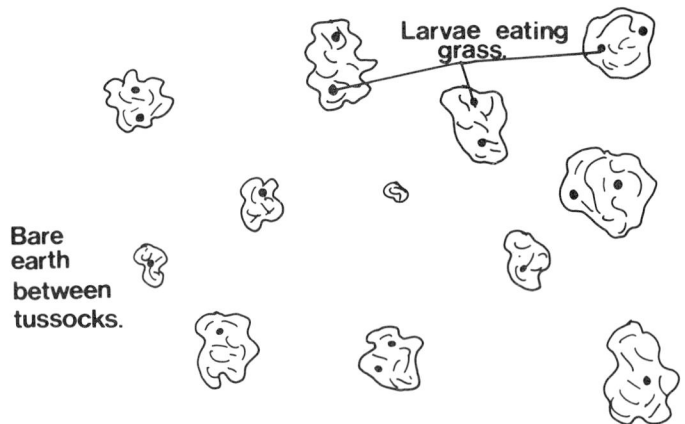

PLAN VIEW OF GRASS TUSSOCKS

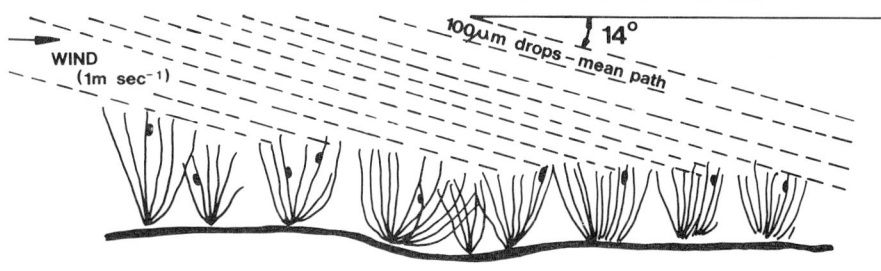

SIDE VIEW OF SAME TUSSOCKS

It follows that for most applications there is a most desirable droplet size, and experience suggests that droplets outside the range of $\pm25\%$ of that size are likely to be difficult either to direct or to deposit. Good atomisers should convert a high percentage of their throughput into droplet sizes within $\pm25\%$ of the required size. If this can be done there seems a high probability that application volumes can be brought down to levels of 1 to 5 l ha^{-1} for insecticides, 5 to 10 l ha^{-1} for fungicides and even 15 to 20 l ha^{-1} for herbicides. It should be remembered that 1 l ha^{-1} corresponds to 160 drops cm^{-2} of ground area for 50 μm diameter droplets and 16 l ha^{-1} to 40 drops cm^{-2} of ground area for 200 μm diameter droplets. Clearly the number of droplets cm^{-2} of leaf area depends upon the type of foliage target.

ATOMISATION

It is well known that the range of sizes in the spray emission of an atomiser varies tremendously depending upon the type of atomiser and the mode in which it operates. Thus, a vibrating needle or pulsed hole can emit a stream of almost mono-sized droplets whilst a hydraulic nozzle emits a very wide spectrum of sizes. The former devices generate droplets directly, but in the latter the liquid is made into a thin sheet, then into ligaments and eventually into droplets. The first method is unattractive for aerial applications because although the droplets are of uniform size the flow rate is far too small. The second method produces an adequate flow rate but in far too wide a range of sizes because of the random way the sheet disintegrates. Therefore, the "in-between" mode, that of forming ligaments which break into droplets, is likely to provide the required results, but only if the flow rate is even and the ligaments break up under steady, controlled conditions.

Spinning disc or cup type atomisers produce ligament flows and consequently near mono-sized droplets, but under normal circumstances the flow rates are severely limited by the flooding of the perimeter and the onset of unsteady flow behaviour. This characteristic can be overcome, except at much higher flow rates, by using a toothed perimeter and a radial airflow as shown by Spillman and Sanderson (1982, 1983). Atomisers of this sort produce sprays in which the ratio of volume median diameter to number median diameter (vmd/nmd) - a measure of droplet uniformity - is between 1 and 2.5, although still at flow rates less than 2 l min^{-1}.

Spinning cage or drum atomisers, such as the Micronair, Retomet or Becomist, produce flow rates up to 30 l min^{-1}, but because of their poorer control over the liquid disintegration, produce wider spectra of droplet sizes usually having vmd/nmd ratios of between 3 and 6. It is difficult to drive these devices satisfactorily using the forward speed of the aircraft below speeds of about 120 kph.

Normal fan and hollow cone hydraulic nozzles produce wide spectra of droplet sizes, with vmd/nmd ratios which can vary from about 5 to 30. Values are usually 8 to 10 for the better ones, with a vmd of about 150 μm, but increasing for ones giving a higher vmd. This is because in all cases a large number of very small droplets in 15 μm - 30 μm range are generated, the change in vmd being achieved by changing the size at the upper end.

Exceptions amongst hydraulic nozzles are those fitted with a 'raindrop extension', which produce fewer small droplets but some very big ones. Another system which produces very large droplets is the Microfoil, but these are so large that if clogging problems are to be avoided then there are few applications in which they can be used efficiently.

If small, charged droplets are required then they can be produced in almost uniform sizes by electrostatic means such as the ICI Electrodyn system. However, it remains to be shown that such a method can be used on an aircraft, or that emitting charged droplets from aircraft is desirable.

The use of formulation additives can influence the vmd and the vmd/nmd ratio of a spray, but is not really a universal alternative to using a good atomiser.

None of the atomisers commercially available for aerial applications gives sprays which are sufficiently near mono-sized to ensure very high application efficiencies. Considerably more effort is required to understand the physics of atomisation more precisely and to apply it to devise new and more efficient atomisers both for aerial and ground spraying.

However, it seems worthwhile spending some effort to improve the atomisers which are available. It is likely that the extremes of droplet sizes which are emitted by atomisers during aerial spraying can be reduced by selecting a smaller volume median diameter than one would normally use, and by catching the smaller droplets. In the oncoming airstream the smallest droplets will follow a trajectory almost directly backwards from the emitting points, whilst the larger ones will move along a more laterally positioned path. Thus, certain devices (cf. Fig. 3) can catch the smaller droplets, and accumulate the fluid to either re-emit it directly, re-cycle it or retain it. By utilising the dead air space behind the atomiser, the air-flow carrying the small droplets can be made to pass through a small mesh screen of wires which will not only catch the droplets and slow down the air-flow, but should also cause them to be concentrated into larger droplets which would be re-emitted. One might even reduce the overall drag of the atomiser! The geometry of the device relative to that of the atomiser would depend critically upon the airspeed of the aircraft and the drop sizes involved. However, since these should be standard for given jobs this does not seem to be a severe limitation on their use.

Figure 3. Screens for "catching" small droplets from three types of atomiser.

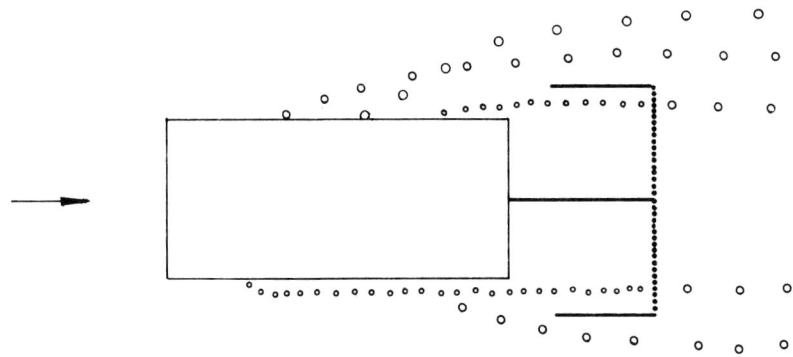

(a) SPINNING CAGE.

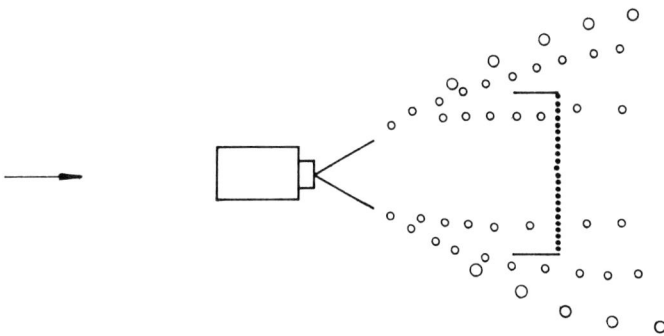

(b) HOLLOW CONE NOZZLE.

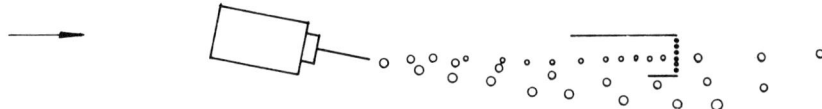

(c) FLAT FAN NOZZLE

It is apparent that an atomiser must be selected on the basis of its efficiency in producing droplets of the size desired, which must be a compromise between having a reasonable chance of the majority of the volume reaching the target sites, and a sufficiently small amount reaching non-target surfaces which might be damaged.

Clearly, herbicides are potentially dangerous where sensitive crops surround the target area and it is not surprising that drift is of great concern to all aerial operators. Also, any chemical which misses the true target is a form of pollution and the fall of excessively large droplets through the crop to the ground and their subsequent leaching into the streams and rivers can cause problems. Holden and Bevan (1978) comment upon the high risk to valuable fishing rivers which could arise from spraying insecticides unwisely in watershed areas, resulting in adverse changes to the whole food cycle.

EVAPORATION, CATCH EFFICIENCY AND LANE SEPARATION

Before the final decision on the desired droplet size is taken, the relationships between droplet size and evaporation, catch efficiency and the distance apart of the successive flight paths of the aircraft must be considered.

The rate of change of droplet diameter with fall time is constant for a given formulation, temperature and humidity for water droplets greater than 150 µm in diameter. Thus, big droplets with their much greater fall speed hardly change in size from atomisation to reaching the target. Whilst initially smaller droplets take longer to fall and will decrease in size, Spillman (1984) has shown that droplets below 150 µm diameter also evaporate more rapidly with time because of the different flow regime associated with the fall, and gives evidence that additives can change the evaporation characteristics very markedly, in some cases effectively stopping it.

It is the droplet size in the vicinity of the catching surface which is critical in determining whether it will be caught. Horizontal surfaces will catch all sizes of droplet efficiently, except perhaps the very smallest sizes which may not adhere well and get blown off again, or the biggest sizes which may bounce or shatter. However, the catch efficiency of vertical surfaces is quite different. Droplets carried by a wind are only caught well by smooth-surfaced vertical objects if they are large, or the wind-speed is high or the width of the catching surface small. It can be shown that a smooth, vertical object will catch only about half of the droplets in the oncoming air when the catching width in millimetres equals the product of the wind-speed (in metres per second) and the droplet fall speed (in centimetres per second). Fortunately, for hairy or crinkled natural surfaces, the catching width can be about ten times this size for the same catch efficiency.

It follows that small droplets in almost still air conditions are not easily caught and can be blown long distances, over long periods of

time, without being caught. The degree of natural turbulence in the air
is a major factor affecting the catch efficiency of droplets below 100
µm in diameter. Natural turbulence increases approximately linearly
with increase in wind-speed, until a wind-speed is reached which causes
the leaves to flutter and the branches to sway, when the turbulence
increases significantly more rapidly. The values naturally depend upon
the kind of ground cover, being quite small for smooth, short grass on
flat, open terrain and very large for hilly, forested areas.

Wind not only influences the collection efficiency of small droplets on
vertical surfaces, it also increases the horizontal component of their
mean flight path. Thus, by flying the aircraft several metres above the
crop in a cross-wind direction the pilot can spread the spray over a
much wider width. This technique, known as drift spraying, can increase
the distance between successive application runs by three or more times
without a significant reduction in the evenness of the cover. Indeed in
some spray operations in Canada, some separations of 100 metres or more
are used. In more usual applications, say over cotton, the evenness of
cover is maintained by the effects of the random spreading due to
turbulence. The wider distance between runs can significantly reduce
the cost of applications. However, spraying in a wind can lead to
problems with drift if proper care is not taken.

DRIFT
     It is often assumed that in a wind all the fine droplets of
a spray will be blown long distances downwind. This is a fallacy if the
spraying is over an extensive, rough canopy such as a forest or a cereal
crop. Most of the droplets will be brought down to the canopy, and
caught by it, sooner than they would have been in a very light wind.
This is because the turbulence induces extra vertical velocities to
packets of air containing the fine droplets, and whilst there is an even
possibility that a packet of air might at any instant move up or down,
if it goes down the droplets get caught. Each time this occurs most of
the droplets it contains are caught and the number of droplets still
air-borne is reduced. Under turbulent conditions such movements occur
continuously, the small frequent eddies making the greatest
contribution.

Thus, in a wind the turbulence brings most small droplets down quite
quickly, but of course a few will go upwards initially and may take a
long time and distance to reach the ground (De Almeida Texeida, 1983).
The variation of the deposit density of droplets in the canopy with
distance downward from the point of emission can be calculated. The
pattern depends upon the size of the droplet and the degree of
turbulence as well as on the wind-speed and the height of the emission
point above the canopy. Figure 4 shows the variation of deposit with
distance downward for a typical spray in which the smallest droplets
were about 15 µm diameter and the largest 900 µm. The dotted line in
Figure 4 shows the distribution which would occur if there was no
turbulence. It is clear that turbulence reduces the amount of
deposition well downwind of the emission point. In this example, the

reduction occurs beyond ten emission heights downwind. Figure 5 shows
the distribution for three groups of sizes of this spray. The first
size range is 15 µm to 100 µm, i.e. the smallest 30% of the emitted
volume; the second size range, 100 µm to 250 µm, comprises another 30%
of the volume; whilst the remaining 40% is in the droplet size range of
250 µm to 900 µm. For the 250-900 µm spray, there was no measureable
deposit beyond 250 m whilst even at 100 m downwind it was only 10% of
that resulting from the same amount of emission in the 100 µm to 250 µm
range, and only about 4% of the full 15 - 900 µm spray (Fig. 5).

Figure 4. Downwind spray deposit distribution, with and without
turbulence (taken from Spillman, 1983). Conditions: height of emission,
5 m; wind-speed 5 m sec^{-1}; turbulence velocity, 0.5 m sec^{-1}; emission
10 1 min^{-1} from single Micronair AU3000. - - - - - - , no turbulence;
——————— with turbulence.

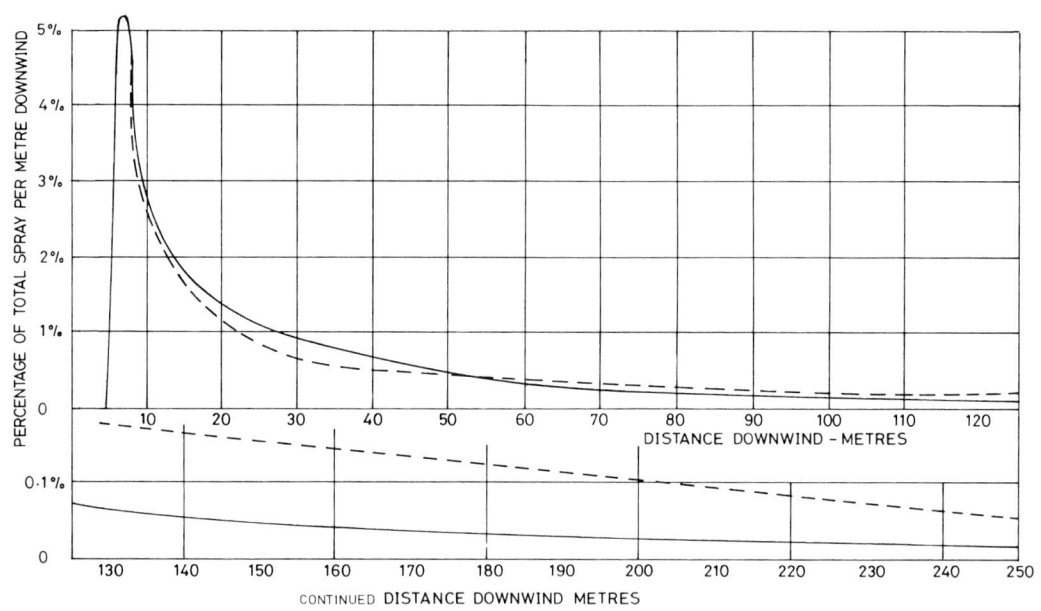

If this deposit density at 100 m of about one twenty-thousandth part of
the initial emission per metre flown, spread over one square metre, is
sufficiently low that it will not cause damage then it means that the
aircraft can spray to within 100 metres upwind of the sensitive crop and
not do significant damage. This is illustrated in Figure 6; the
dimensions are appropriate to the conditions shown in Figures 4 and 5.

Since the downwind distances are directly proportional to the height of
the aircraft above the canopy, then by halving the flying height and the
emission rate per metre flown the aircraft could spray to within 50 m
upwind of the start of the sensitive crop. Obviously, there is a

Figure 5. Downwind deposition of different spray droplet size groups – conditions as for Figure 4. Droplet size groups: 0-100 µm (30% of spray volume), x-----x; 100-250 µm (30% of spray volume) 0 —— 0; 250-900 µm (40% of spray volume), Δ ——— Δ .

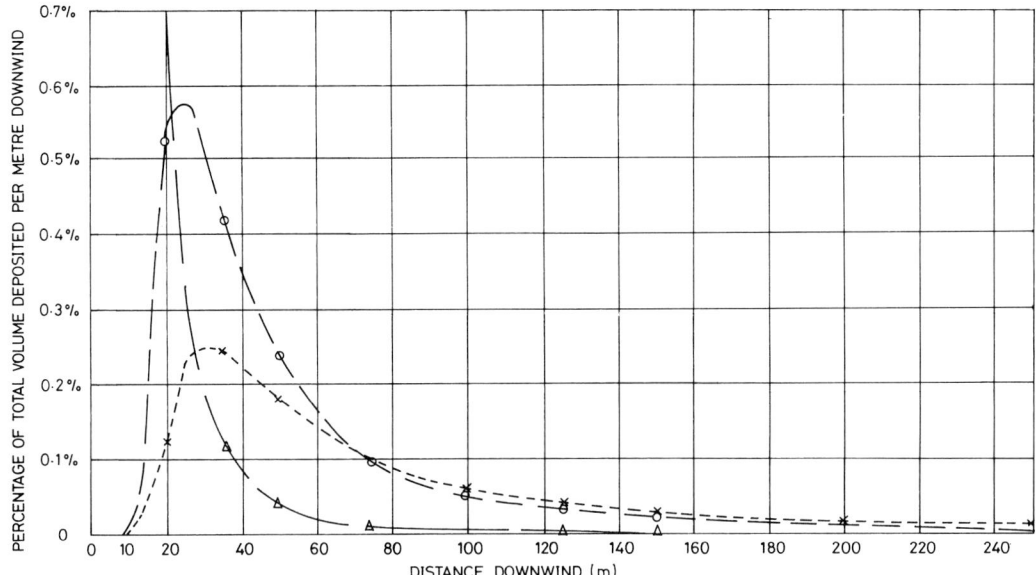

Figure 6. Crosswind spray technique.

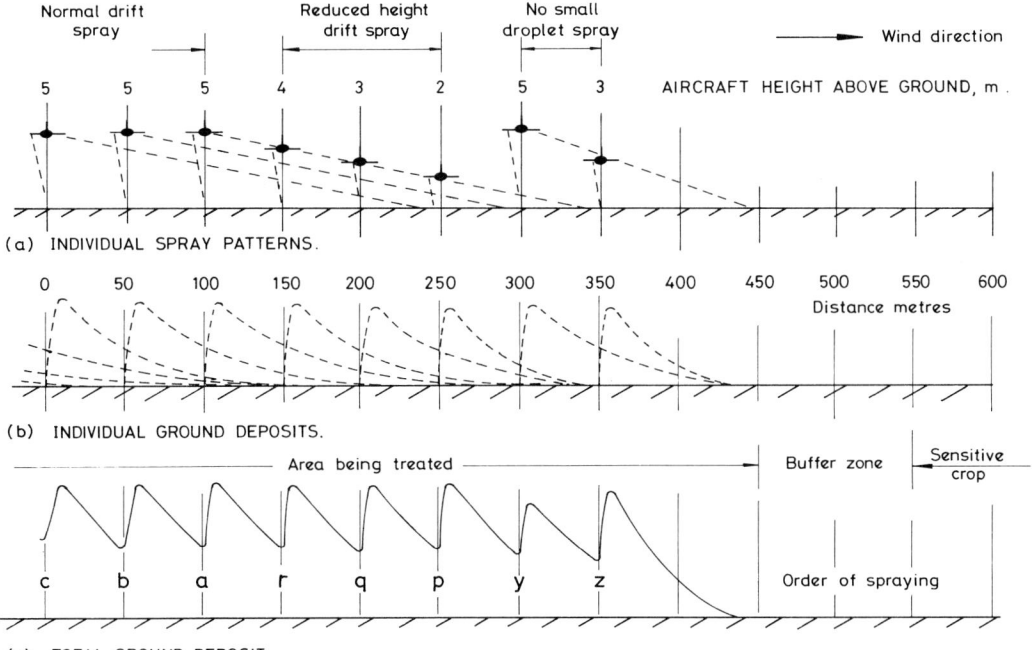

minimum flying height above the crop, and consequently there is a need
for a barrier zone between the most downwind path of the aircraft and
the start of the sensitive crop.

The extent of the barrier zone depends upon the atmospheric conditions
and the sensitivity of the crop downwind to the chemical. It is
unreasonable to expect the pilot to calculate these distances before
flying because of the complexity of the calculation, and the lack of a
simple way of determining the level of turbulence. Watt (1980) has
suggested that for most herbicide applications a barrier of 300 m should
be adequate.

It seems likely that in the future, ground markers will have simple
instruments to measure wind-speed and turbulence which will enable the
pilot to decide which routine spray procedure he should adopt. These
must be planned and practiced before the spraying season starts. Often
off-target sprays result from switching on the spray too soon or
switching off too late, usually when descending or climbing, when
wing-tip vortices are stronger. Practice in switching on and off whilst
in the steady straight and level spray condition is essential, followed
by special runs to spray the headlands.

CONCLUSIONS
        Maximum spray efficiency can be achieved only by ensuring
that the scale, timing, target, formulation and chemical are selected
together to make the best, compatible overall system, rather than being
individually most effective. This requires co-operation between all
those involved.

The target site is seldom the whole field but usually specific parts of
the crop.

There is a droplet size which is optimum for a given target, in a given
crop, under given meteorological conditions. Droplets which differ in
size from this optimum by more than $\pm25\%$ can be considered inefficient.
If the bulk of the spray emitted can be kept within this range, it is
likely that spray application rates can be reduced with savings in cost
and pollution damage.

Hydraulic nozzles emit wide spectra of droplet sizes because of the
random break-up of the sheets of liquid produced. Adequate control of
droplet sizes can only be achieved by carefully controlling the
atomisation. Spinning disc and cup devices do this well and are being
developed to produce spray in adequate quantities.

There is a possibility that the range of sizes released by an atomiser
can be decreased by catching the smaller-sized droplets close to the
atomiser.

The efficiency with which small droplets are caught by a crop depends
upon the size of the droplet, the local wind-speed and the width of the

catching surface. Evaporation of the droplet can significantly reduce
its diameter before reaching the canopy unless suitable formulations are
selected. The effective local wind-speed must include the effects of
turbulence.

The lane separation of aircraft can be increased significantly by flying
cross-wind and allowing it to carry the small diameter spray downwind.
Such a process reduces the costs and the probability of missing the
target sites.

The turbulence created by a wind over a rough canopy actually decreases
the downwind distance to which the bulk of a spray is carried. However,
a small part of the spray will be carried well downwind and could cause
damage to a sensitive crop. The extent of this depends upon the flying
height, the type of crops and the meteorological conditions.

The risk of downwind damage by small droplets can be minimised by
eliminating the smaller-sized droplets in the spray, by reducing the
flying height of the aircraft, and by using a buffer zone between the
spray zone and any sensitive downwind crops.

It is essential that operators of spraying aircraft have routine and
practiced procedures which will ensure that the amount of off-target
spray is minimised. To allow them to decide which procedure is
required, better information on crop and meteorological conditions than
is available currently is required.

REFERENCES
De Almeida Teixeira, M.E.F. (1983). Measurements of downwind ULV droplet
        profiles under various meteorological conditions. European
        and Mediterranean Plant Protection Organisation Bulletin
        13(3), 433-437.
Graham-Bryce, I.J. (1977). Crop protection: a consideration of the
        effectiveness and disadvantages of current methods and of
        the scope for improvement. Philosophical Transactions of
        the Royal Society, Series B, 281, 163-179.
Holden, A.V. and Bevan, D. (eds) (1978). Control of Pine Beauty Moths
        by Fenitrothion in Scotland, 1978. Edinburgh: Forestry
        Commission.
Hursey, B.S. and Allsopp, R. (1983). Sequential applications of low
        dosage aerosols from fixed-wing aircraft as a means of
        eradicating Tsetse Flies (Glossina spp.) from rugged terrain
        in Zimbabwe. Tsetse and Trypanosomiasis Control Branch,
        Department of Veterinary Services, Zimbabwe.
Joyce, R.J.V. (1975). Sequential aerial spraying of cotton at VLV rates
        as a contribution to synchronised chemical application over
        the area occupied by the pest population. Proceedings of the
        5th International Agricultural Aviation Congress,
        Stoneleigh, England.

Spillman, J.J. (1976). Optimum droplet sizes for spraying against flying
        insects. Agricultural Aviation 17 (1/4).
Spillman, J.J. (1983). A rapid method of calculating the downwind
        distributions from aerial atomisers. European and
        Mediterranean Plant Protection Organisation Bulletin 13 (3),
        425-431.
Spillman, J.J. (1984). Evaporation from freely falling droplets. The
        Aeronautical Journal (May/June), 181-185.
Spillman, J.J. and Sanderson, R. (1982). A disc-windmill atomiser for
        the aerial application of pesticides. Proceedings of the
        2nd International Conference on Liquid Atomisation and Spray
        Systems, Madison, Wisconsin, USA.
Spillman, J.J. and Sanderson, R. (1983). Design and development of a
        disc-windmill atomiser for aerial applications. European
        and Mediterranean Plant Protection Organisation Bulletin
        13 (3), 365-370.
Watt, J. (1980). The importance of correct aerial spraying. The
        Australia Cotton Grower 28, (Jan.29).

10. MATHEMATICAL MODELLING IN SPRAY ENGINEERING RESEARCH

J. A. Marchant
National Institute of Agricultural Engineering, Wrest Park,
Silsoe, Bedfordshire, MK45 4HS, England

INTRODUCTION
Models used in engineering

Modelling is widely used in engineering and whilst this paper is primarily concerned with mathematical modelling it is important to realise that this is only one of a number of modelling tools available to the engineer. Hence, before discussing mathematical modelling in detail, modelling will be considered briefly in more general terms.

In a review article, O'Dogherty (1981) covers those aspects of modelling especially relevant to agricultural engineering. He classifies models into three broad categories:

1. Iconic Models. This type of model looks like reality. Examples are a scale model of a civil engineering structure, or a mock-up of the control panel of a motor car. The main use of this type of model is in communicating ideas amongst a group of designers, or between designer and customer.

2. Analogues. An analogue consists of a model system which operates on the same principles as the real one. The advantage of experimenting with the analogue rather than the real system, is that it is easier to control the variables and to visualise the results. An example is an electrical resistance analogue used to model heat flows. The real system contains variables such as temperatures, thermal resistances, and heat flows. These are represented in the analogue by voltage, electrical resistances, and electrical currents which are much easier to vary and to measure.

3. Mathematical Models. In this type of model, the observed behaviour of a real system or a component of a system, is described by a mathematical equation. The equation or set of equations is then solved to predict the behaviour of the system. An example of a simple mathematical model is Stokes' law of viscosity, and an example of a more complex model is that used to predict the trajectory of a lunar rocket.

Mathematical modelling

This is often regarded as a new technique, but in fact it
has been used from the beginning of quantitative science. An example of
a very early model is the Archimedes principle, which equates the up-
thrust on an immersed body with the weight of water it displaces. The
modern connotations associated with the technique have probably been
brought about by the advent of the digital computer, which has significantly increased the power and scope of modelling.

An example relevant to spraying research is the calculation of spray
drop trajectories. Because of their small size and relatively low
speed, spray drops can be treated as spherical objects when calculating
air drag forces (Berry, 1974). If spraying were carried out in a
vacuum, the trajectory of a drop could be calculated from its initial
velocity, using only the simple tools of pencil and paper. The relevant
mathematical description of the system would be obtained from Newton's
laws of motion (Meriam, 1959). The model could be refined, by including
the viscous drag exerted on the drop when moving through air. If the
drop was very small and moving very slowly (Spillman, 1983), Stokes' law
could be used to calculate the air drag. In this case, pencil and paper
would still be adequate, but a calculator would speed up the process
considerably. To make the model representative of a wider range of drop
sizes and speeds, it would be necessary to represent the non-linear
nature of the air drag force. Drag coefficients drawn from established
experimental work (Streeter, 1962) rather than Stokes' law would need to
be used to calculate the drag force. In this case, pencil and paper
would be found to be completely inadequate, and the calculations would
have to be done using a computer (see Marchant, 1976).

Although the resulting model would be rudimentary - for example, it
would probably assume the air stream in which the drop was moving to be
perfectly smooth flowing - it would still be useful. For instance, it
would show the extremely sensitive nature of the downstream drift
distance to the drop diameter. Consequently, it may help to determine a
drop diameter above which drift could be assumed to be negligible.

The model just described is deterministic. That is, for any set of
input conditions the computer always calculates the same answer.
Recently, models which have a stochastic component have been constructed
to describe the transport of spray drops (Thompson and Ley, 1983). In
this type of model the random nature of the wind can be simulated using
measured parameters of the wind statistics. It is possible to predict
the diffusion pattern of drifting spray, and also to identify the
important variables affecting drift. This is the first step towards
suggesting a means of controlling drift, and in my opinion this approach
has exciting prospects for the future.

The foregoing discussion has shown how mathematical modelling might be
used in research. That is, in helping the investigator to understand
the process that he is dealing with, identifying which variables are
important and which have little effect on the process, and suggesting
hypotheses that can be tested by experiment. The advantage of using a

model over a purely experimental approach, is that by the time
experimental work is started, the investigator has a reasonably clear
understanding of how the process works. The modelling phase can thus be
used to rule out likely unproductive and "shot in the dark" experiments.
Because the experimental phase is generally relatively time consuming
and labour intensive, the modelling work increases the overall cost
effectiveness of the investigation.

Mathematical modelling can also be used in design, where the end point
to the investigation is specified beforehand, in contrast to research
where the investigation tends to be open ended. In this case, the model
needs to be much more accurate to be of real value. Models used for
design tend to be based on physical laws which are known to apply more
or less exactly to the problem in hand, and which have been used
successfully in similar design problems over a number of years.

The rest of this paper will be devoted to two examples, where the author
has used mathematical modelling in spray engineering at the National
Institute of Agricultural Engineering (NIAE). The first example is from
the area of research, and the second one from the area of design.

STUDIES OF CHARGE LEVELS IN AN ELECTROSTATIC SPRAYING SYSTEM

Charging spray from hydraulic nozzles
The NIAE has been working for some time on the engineering
problems, and the physical aspects, of charging the spray from hydraulic
nozzles. Information was required on the charge levels likely to be
achieved, and the way that variables such as operating voltage, nozzle
size and pressure, and electrode spacing, affect the charge level. The
analysis and experimental confirmation is reported by Marchant et al.
(1985a,b). A system for charging spray from a hydraulic flat fan nozzle
is fully described by Marchant and Green (1982), but the basic principle
of operation is shown in Figure 1. An electric field at the surface of
the liquid induces a charge on it, and the spray drops depart from the
liquid carrying this charge.

A model for liquid break-up
In order to construct a mathematical model for this process,
an analysis of the way drops are formed is required. Dombrowski and
Johns (1963) show that the liquid forms into a fan shaped sheet on
emerging from the orifice. By analysing the dynamics of the sheet they
show that waves form on it due to the air drag at the sheet surface.
The waves grow in amplitude until the sheet eventually breaks into
drops. The real process can be clearly seen in Figure 2. Dombrowski
and Johns' model the real process by an extremely simplified representation, where the sheet first breaks into cylindrical ligaments parallel
to the sheet end, and then the ligaments break into drops of equal size.
Dombrowski and Johns derive several equations and a table, from which
properties such as the length of the sheet, the thickness and wavelength
at break-up and the drop diameter, can be calculated from the nozzle and
liquid variables (i.e. orifice size, viscosity, pressure etc.). The

single drop diameter of the model is taken to represent the Sauter mean diameter of the real spray.

Figure 1.   Charging spray from a hydraulic nozzle.

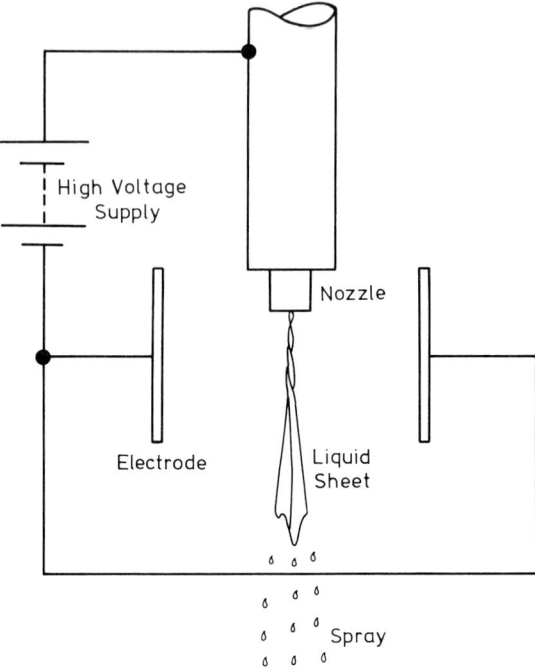

Figure 2. Liquid break-up from a flat fan nozzle.

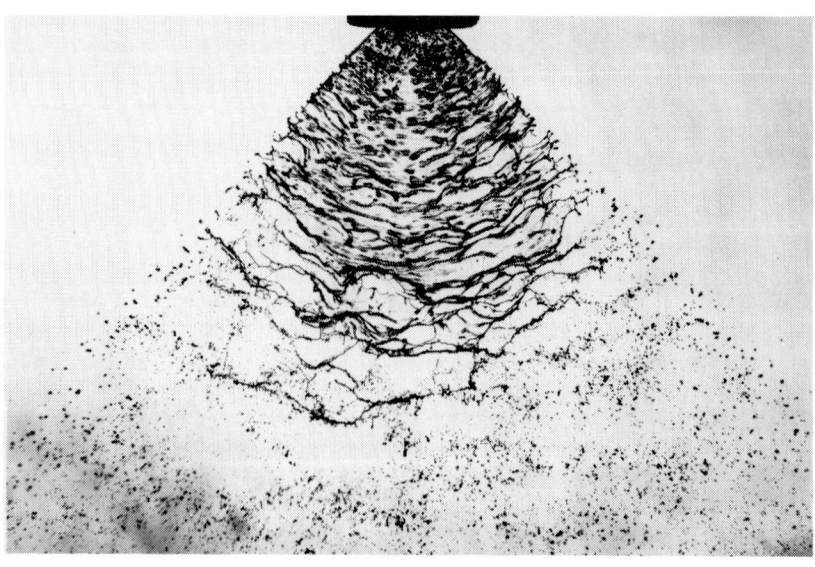

Marchant et al. (1985a,b) have taken Dombrowski and Johns'
equations and simplified them by using liquid properties relevant to
spraying water-based pesticides, whereupon some of the terms become
negligible. This results in a formula for the Sauter mean diameter:

$$d = 1.68 \left(\frac{A}{p \sin \left(\frac{\alpha}{2} + 0.14 \right)} \right)^{1/3} \tag{1}$$

where p is the nozzle pressure (Pa), α is the nominal nozzle angle
(radians) and A is the orifice area (m^2).

Figure 3 shows results from the practical testing of a range of nozzle
sizes at different pressures, and the line corresponding to Equation 1.
Considering the scatter of the experimental results, Equation 1 gives a
good representation of the real drop diameters, despite the fact that
Dombrowski and Johns' model of drop formation is an extreme idealisation
of reality.

Figure 3. Measured (•) and predicted drop size from a
flat-fan nozzle.

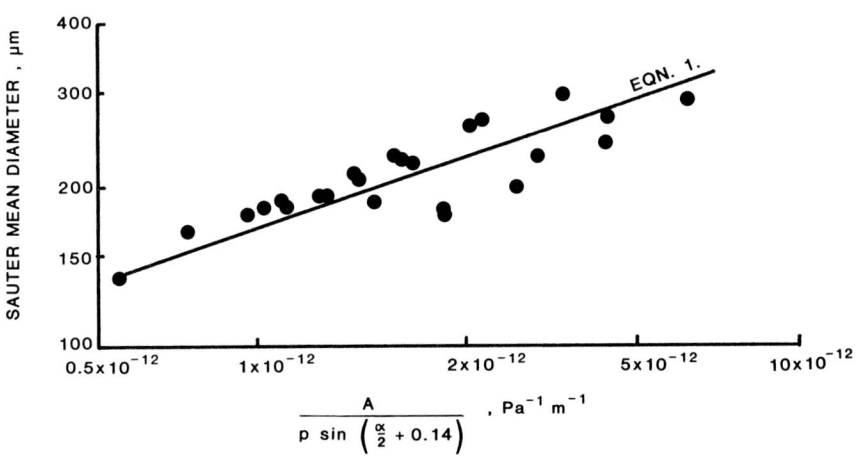

A disadvantage of using the model rather than testing is that only the
average diameter is calculated. The model does not give any information
on the drop size distribution from a given nozzle. However, it does
lead to an understanding of how the physical variables affect the nature
of the spray, and the relative degree of importance of each variable on
the drop size.

A model for charging

Marchant et al. (1985a,b) combined their simplified version of Dombrowski and Johns' model with the analysis of the electric field at the liquid sheet, to give a formula for the charge level on the spray:

$$q = 8.8 \times 10^{-12} \frac{\alpha_m}{\alpha_a} \left[\frac{1}{p} \left(\frac{\sin \left(\frac{\alpha + 0.14}{2} \right)}{A} \right)^2 \right]^{1/3} \frac{V}{Y} \tag{2}$$

where q is the charge-to-mass ratio (C/kg), V is operating voltage (V), and Y is the electrode/liquid sheet spacing (m). α_m and α_a are given by

$$\alpha_m = \alpha + 0.27 - 1.11 (0.67)^{p \times 10^{-5}} \tag{3}$$

$$\alpha_a = \alpha + 0.27 \tag{4}$$

p, α, and A have been defined previously, and once again α is in radians.

An important point when using mathematical models, is to be aware of their limits of applicability. In this particular model, there is no mechanism to describe electrical breakdown of the air between the electrodes and the liquid sheet. The model is therefore only valid for an operating voltage below the limit for air ionisation. Also, Equation 2 gives the charge-to-mass ratio for the whole of the spray, including the fraction that is attracted to the electrodes. In practice, the charge-to-mass ratio of the spray that escapes the influence of the electrodes will be less than that given by Equation 2. A discussion of the magnitude of these effects appears in Marchant et al. (1985a,b).

Figure 4 shows experimental results together with the line derived from Equation 2. Once again the experimental results are very scattered, but overall they are reasonably close to the calculated values. The theoretical line is however outside the 95% confidence limits of the experimental results for part of its course, suggesting that there may be a real difference between the way the model behaves and reality.

Whether Equation 2 (which from Figure 4 would give charge-to-mass ratios up to 34% different from the expected values) is useful in practice depends on the accuracy of alternative methods of estimating the expected charge-to-mass ratio. The obvious alternative is to test a number of nozzles of the same type, and take a mean of the results. Marchant et al. (1985a,b) show that, due to the scatter of the experimental results, up to 14 nozzles would need to be tested to be reasonably confident that the experimental result is as close to the expected value as the theoretical one. This would only establish one point on the graph of Figure 4.

Figure 4. Measured (●) and predicted charge-to-mass ratio. Broken lines indicate 95% confidence limits for measured values.

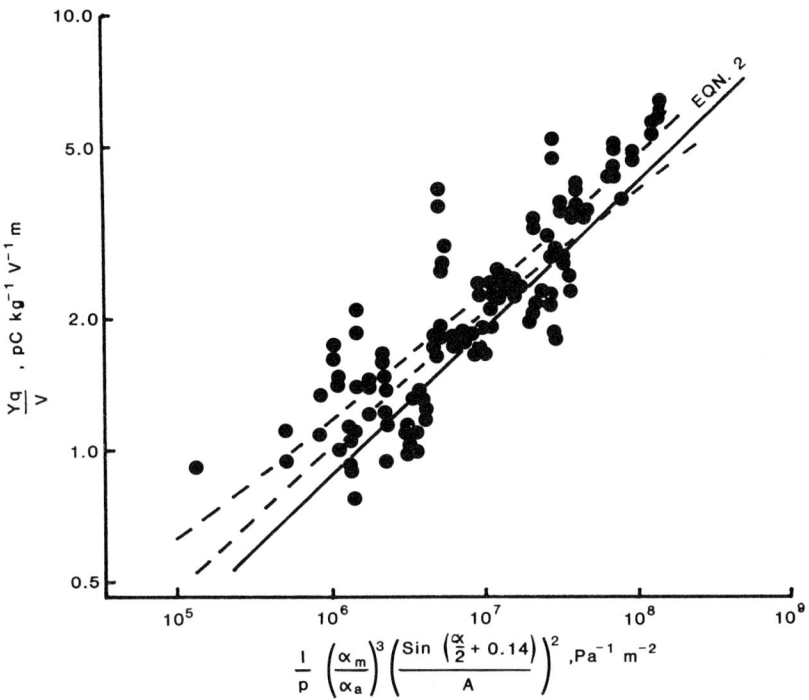

As well as being more convenient, using the model shows how the variables combine to give the charge-to-mass ratio. For example, the charge-to-mass ratio is shown to be proportional to the operating voltage, and inversely proportional to the electrode/liquid sheet spacing. Also, q is reduced when A increases, so that larger nozzles would be expected to give lower charge-to-mass ratios than small ones, for the same operating pressure.

The foregoing predicted behaviour would probably be rather obvious to the intelligent investigator. However, less obvious behaviour can also be predicted from Equation 2. For instance, pressure affects the charge-to-mass ratio directly via Equation 2, and indirectly via α_m in Equation 3. The direct effect of increasing the pressure is to reduce the charge-to-mass ratio (Equation 2). However, increasing the pressure also increases α_m (Equation 3) and hence indirectly increases the charge-to-mass ratio. The net result, is that the effect of pressure on charge-to-mass ratio is fairly small, and whether the charge-to-mass ratio is increased or decreased by an increase in pressure, depends on the nozzle angle. The model predicts that a pressure rise leads to a rise in charge-to-mass ratio for angles less than 66°, and a fall for angles above this. Practical tests (Marchant et al., 1985a,b) have confirmed this to be precisely the case.

DESIGN OF AN ACTIVE SPRAY BOOM STABILISATION SYSTEM

A problem when spraying from long booms, is that the irregularity of the ground surface is transmitted via the tractor to the boom. This causes unwanted roll and yaw movements, resulting in uneven spray deposit and wastage, and also gives rise to increased drift (Nation, 1978). The problem has been partially overcome by using passive suspension systems, and one particular type designed by Nation (1978) is shown in Figure 5. Note that the simplified system, as drawn, will stabilise in roll only; the complete system as specified by Nation also stabilises in the yaw plane. This system works well in removing relatively high frequency inputs from the uneven ground. Also, it gives a boom attitude that is parallel to the tractor frame in the steady state. This means that if the tractor frame is parallel to the ground, the boom is also parallel to the ground, which confers the ability to work on a hillside. There are still however some problems to be solved. First, the tractor frame is not always parallel to the ground — differential tyre sag which may occur on a hillside and cause a significant height difference at each end of a long boom. Also, low frequency inputs can still be transmitted to the boom, and those around the resonant frequency of the suspension/boom combination may be amplified. This can result in damped, low frequency oscillations of the boom after the tractor goes over a sudden bump.

Figure 5. Passive spray boom suspension system.

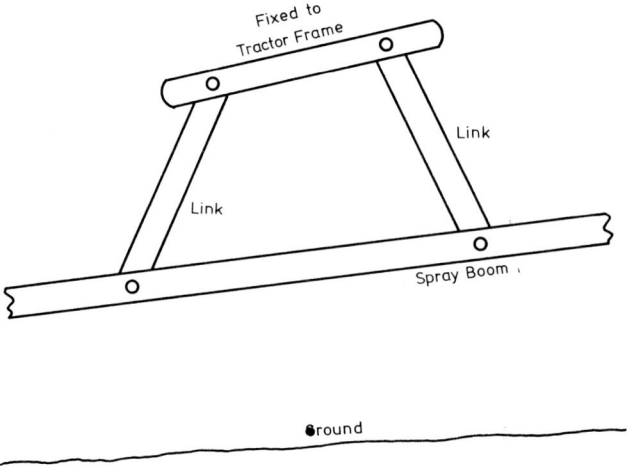

Frost (1985) has suggested that one of the links in the suspension (Fig. 5) should be variable in length. Height sensors, mounted at each end of the boom, could then be used to provide a roll error signal (i.e. the difference between the ground angle and the boom angle) which could be used to change the length of the link, and to reduce the roll error to zero. A properly designed control system would ensure that the resulting closed loop system would be stable (i.e. would not oscillate wildly), capable of maintaining the roll error within reasonable bounds, and able to respond quickly.

Frost (1985) has combined Newton's laws of motion (Meriam, 1959) with the laws of geometry, to produce a mathematical model of the suspension shown in Figure 5. Given the properties of the suspension (lengths of links, positions of pivots, damping constants, etc.), it can calculate the response of the system for any input. Figure 6 shows typical responses for a passive system, i.e. where the actuator is replaced by a fixed length link, when the ground angle changes suddenly by 0.1 radians. The boom angle starts off at zero (horizontal) and rises in response to the change of ground angle. The boom then overshoots the target angle of 0.1 radians, and then settles down to meet it. In this case, the designer's job might be to choose a value for the damping, μ, to limit the overshoot, and enable the designer to calculate a value and to choose an actual component to do the job.

Figure 6. Boom angle response for a passive system with two values of damping (μ).

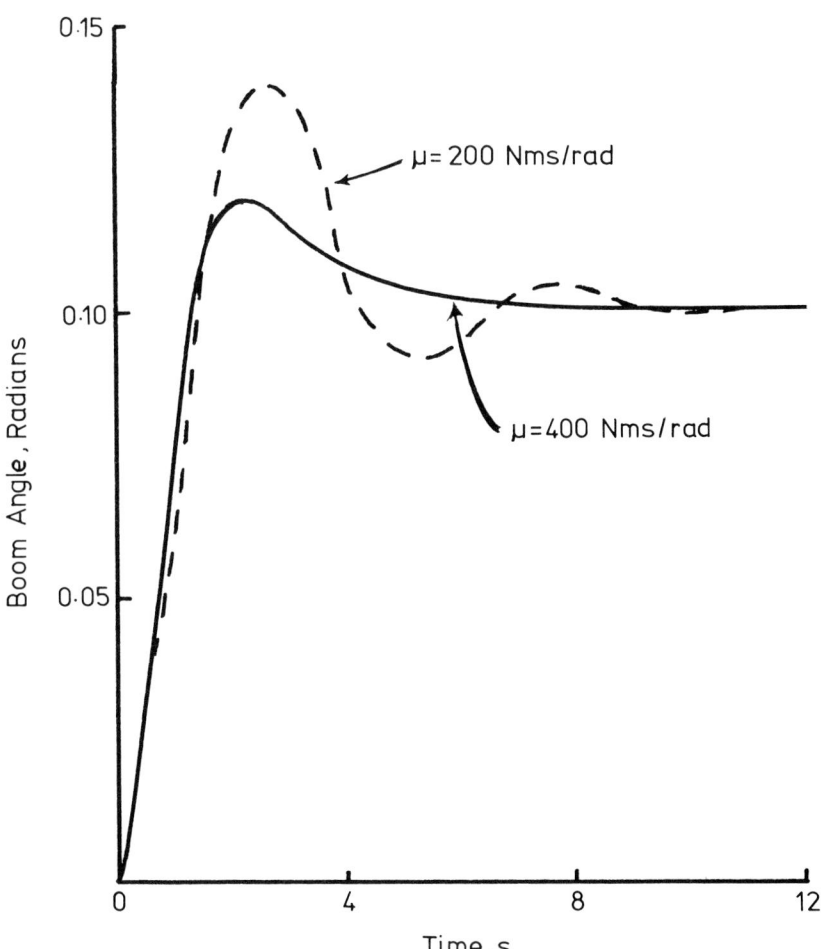

In order to design an active system, the control engineer would first draw a block diagram of his proposed system (Fig. 7). Here, the existing passive part is shown as a solid line. The tractor frame angle is determined by adding a "noise" component to the ground angle. The noise is caused by tyre sag, wheels running in ruts etc. The resulting frame angle, acting on the boom dynamics, eventually gives rise to a boom angle. The broken lines indicate the extra components required to activate the system. The difference between the boom angle and the ground angle is the roll error, and is measured with the height sensors. The actuator is changed in length in response to the roll error. The length change can be considered as equivalent to a perturbation of the frame angle, which acts through the boom dynamics to change the boom angle, until the roll error is zero.

Figure 7. Block diagram of an active boom stabilisation system.

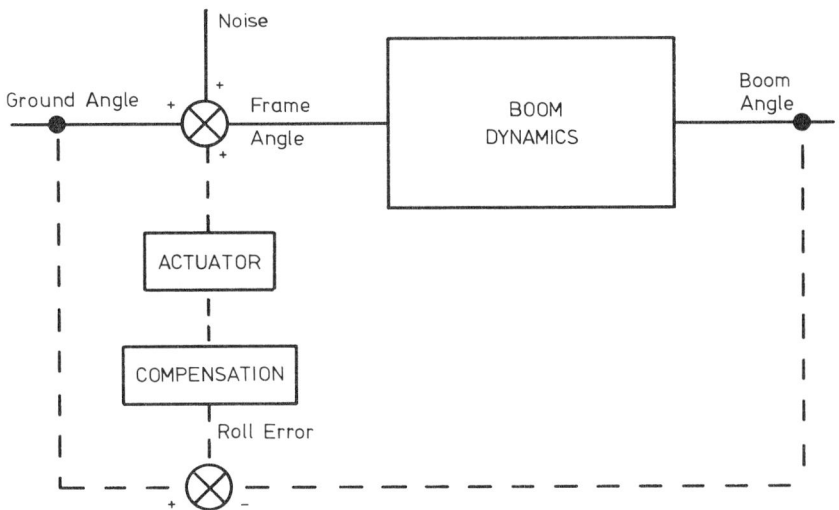

Note that a block labelled "compensation" appears on Figure 7. The purpose of the compensation is to modify the roll error signal so that the response of the system meets certain design criteria. The control engineer may design the compensator, for example, to improve the stability while maintaining a given level of accuracy of control. Alternatively, he may wish to make the boom respond more quickly to a change in ground angle. To do this, the engineer needs a description of the boom dynamics in mathematical terms. Unless a mathematical model is available, the engineer will have no idea of how to process the roll error signal, and a poor control system will result. This situation will lead to a much lengthened prototype-testing phase, and increase the overall cost of the development. Added to this, there will be no certainty of success, because it will not be clear what the engineer needs to change in order to achieve the desired result.

As part of the research programme at the NIAE, simple
descriptors of the boom dynamics have been derived by using Frost's
mathematical model. These descriptors are the passive gain (i.e. the
ratio between boom angle and frame angle in the steady state), the
natural frequency, and the damping tractor. Using these parameters,
experimental control systems have been designed, and Figure 8 shows the
predicted behaviour of two systems. The first is a poorly-designed
system which is in fact unstable. The second has been designed by
combining the mathematical model with proper control system design
techniques. The next step in the programme is to test a spray boom
fitted with this prototype system.

Figure 8.   Predicted boom angle response for two active
stabilisation systems.

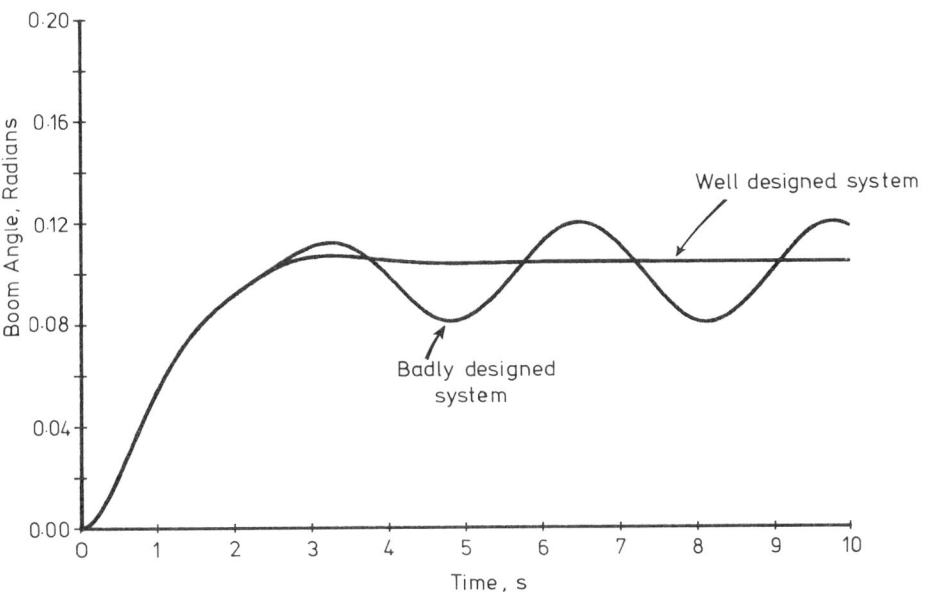

DISCUSSION AND CONCLUSIONS

This paper has shown how mathematical modelling can be
useful in the engineering of crop protection systems. As part of
research, it can help the investigator to understand how a process
operates, and show how each variable contributes to the final result.
This understanding can generate hypotheses, which can then be tested by
using classical experimental procedures. As part of design, modelling
can help the engineer to chose values for parameters to satisfy
predetermined design criteria. This can cut down the time devoted to
prototype testing, which is generally a resource and labour intensive
exercise. By this means, modelling can increase the cost-effectiveness
of a design project.

It must be remembered that modelling is not a substitute for either experimentation or prototype testing. Rather it is an aid to enable the investigator to carry out these activities in a more intelligent fashion. Experimentation and testing are of course vital to a project programme.

As its title suggests, this paper has considered only the engineering aspects of crop protection, where the requirements can be expressed in physical terms. For example, given a requirement for the positioning of spray deposit to achieve a particular biological effect, the charge levels in an electrostatic spray could be chosen to achieve a suitable trajectory (Dix and Marchant, 1984). In another example, the movement of a spray boom could be controlled so that the variation in spray deposit does not exceed the level where poor control results. The ultimate requirements in crop protection are biological, environmental and economic, but there is a gap in our knowledge between the biological effect on the one hand, and the physical causes of this effect on the other. Unless this gap is filled with quantitative information, the final link in the chain between good engineering design and good crop protection cannot be made. It is unlikely that the necessary skills will be found in one group of people, and hence the growing co-operation between engineers and biologists needs to be encouraged, and further strengthened, to ensure the future success of crop protection research.

REFERENCES

Berry, E. X. (1974). Equations for calculating the terminal velocities of water drops. Journal of Applied Meteorology 13, 108-110.
Dix, A. J. and Marchant, J. A. (1984). A mathematical model of the transport and deposition of charged spray drops. Journal of Agricultural Engineering Research 30, 91-100.
Dombrowski, N and Johns, W. R. (1963). The aerodynamic instability of viscous liquid sheets. Chemical Engineering Science 18, 203-214.
Frost, A. R. (1985). Simulation of an active spray boom suspension. Journal of Agricultural Engineering Research 30, 313-325.
Marchant, J. A. (1976). Calculation of spray droplet trajectory in a moving airstream. Journal of Agricultural Engineering Research 22, 93-96.
Marchant, J. A. and Green, R. (1982). An electrostatic charging system for hydaulic spray nozzles. Journal of Agricultural Engineering Research 27, 309-319.
Marchant, J. A., Dix, A. J. and Wilson, J. M. (1985a). The electrostatic charging of spray produced by hydraulic nozzles, Part I. Theoretical Analysis. Journal of Agricultural Engineering Research 31, 329-344.
Marchant, J. A., Dix, A. J. and Wilson, J. M. (1985b). The electrostatic charging of spray produced by hydraulic nozzles, Part II. Measurements. Journal of Agricultural Engineering Research 31, 345-360.
Meriam, J. L. (1959). Mechanics, Part II: Dynamics. New York: Wiley and Sons Inc.

Nation, H. J. (1978). Developments in spray boom design. Proceedings
          of British Crop Protection Conference - Weeds, 649-656.
          British Crop Protection Council Publications.
O'Dogherty, M. J. (1981). Modelling for engineering research and
          development. Agricultural Engineer 36, 81-90.
Spillman, J. J. (1983). Notes for short course on the aerial
          application of pesticides. Cranfield Institute of
          Technology, Bedford, UK.
Streeter, V. C. (1962). Fluid Mechanics. New York: McGraw-Hill.
Thompson, N. and Ley, A. J. (1983). Estimating spray drift using a
          random-walk model of evaporating drops. Journal of
          Agricultural Engineering Research 28, 419-436.

Section 3. RESISTANCE

11. FUNGICIDE RESISTANCE IN CROPS - ITS PRACTICAL SIGNIFICANCE AND MANAGEMENT

K.J. Brent
Department of Agricultural Sciences, University of Bristol,
Long Ashton Research Station, Long Ashton,
Bristol, BS18 9AF, England

INTRODUCTION
The emergence of fungicide-resistant strains of target
pathogens has caused major problems in many countries and many types of
crop over the last fifteen years. In the UK and other temperate
countries it now creates much greater difficulties overall for
agriculture than insecticide resistance, despite the longer history of
the latter problem, and it dwarfs the more recent occurrence of
herbicide resistance. This contribution considers the incidence of
fungicide resistance amongst crop pathogens, its impact on agriculture
and related industries, and current approaches to its prevention or
containment. Professor Dekker, in the next chapter, discusses in some
depth the biological, chemical and other factors that determine the
accumulation and persistence of the resistant forms, and hence confer
the risk of resistance on particular fungicides and fungicide uses.

I shall use the term 'resistance' freely, since it has been recommended
by the International Society of Plant Pathology as the preferred term
for an inherited loss in sensitivity to fungicides. It is also one
which has been in use for many years to denote loss of effectiveness of
antibiotic medicines, insecticides and rodenticides. Some workers
prefer the expressions 'insensitivity' or 'tolerance' - mainly because
they consider that fungicide 'resistance' may be confused with natural
disease 'resistance' in plant varieties. Also in some difficult
commercial situations 'insensitivity' or 'tolerance' have been preferred
because they sound less alarming than 'resistance'! Many persons,
including the author, regard all these terms as synonyms which can be
used inter-changeably and according to personal preference.

OCCURRENCE
Until the late 1960s, the appearance of fungicide-resistant
strains of plant pathogens was a rare phenomenon. Observations of
resistance to biphenyl and sodium orthophenylphenate in storage moulds
of citrus fruit (Penicillium spp.), to hexachlorbenzene in the wheat
bunt pathogen (Tilletia foetida) and to organomercurials in leaf-stripe
pathogen of oats (Pyrenophora graminea) were documented (Duran and
Norman, 1961; Harding, 1962; Kuiper, 1965; Noble et al., 1966).
Failures of disease control were associated with these occurrences, but
they remained of local importance and happened only after many years of
use of the fungicides concerned. After 1970, however, and coincidental

with the almost simultaneous introduction of four major groups of
systemic fungicides - benzimidazoles, pyrimidines, morpholines and
carboxanilides - many more resistance problems arose. Some occurred
within a year or two of the first commercial use of the fungicide
concerned, whereas others took several years to develop (Table 1).

Table 1. Occurrence of fungicide resistance problems in crops.

Date (approx.)	Fungicide or fungicide class	Years of commercial use prior to resistance (approx.)	Main crops, diseases and pathogens affected
1960	Aromatic hydrocarbons	20	Citrus storage rots: Penicillium spp.
1964	Organo-mercurials	40	Barley leaf-stripe: Pyrenophora graminea
1969	Dodine	10	Apple scab: Venturia inaequalis
1970	Benzimidazoles	2	Many target crop diseases and pathogens
1971	2-Amino-pyrimidines	2	Cucumber powdery mildew: Sphaerotheca fuliginea
1972	Kasugamycin	5	Rice blast: Pyricularia oryzae
1976	Phosphorothiolates	9	Rice blast: Pyricularia oryzae
1980	Acylalanines	2	Potato blight: Phytophthora infestans
1982	Dicarboximides	5	Grape-vine: Botrytis cinerea
1982	Sterol demethylation inhibitors	7	Cucurbit and barley powdery mildews: Sphaerotheca fuliginea and Erysiphe graminis

The considerable variation between the periods taken for resistance to be observed makes an interesting subject for study, bearing in mind the factors considered to determine rates of build-up of resistance (see Dekker, Chapter 12). In some instances, differences probably exist in the intrinsic capacity of the target pathogen to mutate to produce forms resistant to different fungicides. This is probably the reason why loss of sensitivity of barley powdery mildew (Erysiphe graminis) to the 2-amino-pyrimidine sterol demethylation inhibitor fungicides has been detected many times, whereas shifts in response to the equally widely used morpholine fungicides have not been recorded. In other cases, differences in the degree of exposure to the fungicide in space and time may have occurred. Thus the more rapid and severe occurrence of resistance of cucumber powdery mildew to dimethirimol, in comparison with the relatively slow development of a less marked resistance of barley powdery mildew to the closely related ethirimol, probably reflected both the longer periods of exposure to the fungicide and the virtual absence of refugia of 'wild-type' forms which were associated with the case of the cucumber mildew (Brent, 1982). Resistance to the benzimidazole fungicides has arisen more rapidly in pathogens of horticultural crops, such as Venturia inaequalis (which causes apple scab), and Botrytis cinerea (grey mould in grapes), than in cereal pathogens, again probably due mainly to the more frequent applications that are made to the former. Surprisingly, clear instances of resistance of apple powdery mildew to fungicides applied repeatedly for its control (including single-site inhibitors such as bupirimate and triadimefon) have not yet been reported; possibly this reflects the practical difficulties of testing changes in response to fungicides of this particular pathogen. The availability and use of a rather large number of different types of fungicides to control apple mildew may also be relevant.

FUNGICIDES AFFECTED
Relatively few of the surface ('protectant') fungicides have encountered resistance problems, and then only after many years of use; the first three examples listed in Table 1 are of this type. However, amongst the newer, systemic fungicides resistance problems appear to be the general rule, and only the morpholine fungicides (e.g. tridemorph, fenpropimorph) and fosetyl-Al have escaped such problems so far. Most commonly, fungicide resistance has resulted from an altered sensitivity at the primary site of fungitoxic action, so that the greater vulner-ability of systemic fungicides with their highly specific, single sites of action is readily understandable. Systemicity per se is probably not a major factor; indeed, resistance of Botrytis cinerea to the dicarb-oximide fungicides procymidone and vinclozolin has arisen readily even though these fungicides show little or no systemic movement. However, the greater persistence of action of systemic fungicides and the more uniform redistribution they achieve may also add to the risk.

If mutation to resistance to one fungicide automatically confers resistance to a second fungicide the fungus is said to show cross-resistance to both fungicides. In practice, resistant mutants are

generally resistant to all fungicides which share a common biochemical
mechanism of action. For example, strains resistant to benomyl are
usually resistant also to the related fungicides carbendazim,
thiophanate-methyl and thiabendazole. However, the degree of resistance
may vary according to the particular fungicide which is selected for
test, and occasionally one or more fungicides within a class may remain
effective. Cross-resistance between fungicides known to have different
mechanisms of action is rare or absent. One strain of a fungus may in
fact be resistant to two or more biochemically unrelated fungicides, but
such diverse shifts in sensitivity have been found to result fortuitouly
from independent mutations and are termed multiple resistance. Amongst
fungicides whose mechanism of action are unknown, unexpected cross-
resistance patterns have been encountered; for example, fungal strains
which have become resistant to aromatic hydrocarbon fungicides such as
quintozene and dicloran appear always to be resistant also to the
dicarboximide fungicides.

Negative cross-resistance is the term applied when loss of sensitivity
to one fungicide confers automatically an increase of sensitivity to
another. There are several experimental examples of this phenomenon
(see De Waard, 1984) and in practice there is increasing evidence that
an inverse correlation exists between the sensitivities of different
barley powdery mildew isolates to ethirimol and triadimenol (Butters et
al., 1984; Hunter et al., 1984); ethirimol-resistant forms tend to be
particularly sensitive to triadimenol, and vice versa. The biochemical
basis for this example of negative cross-resistance is hard to
understand, since ethirimol and triadimenol, as far as we know, are
unrelated in their sites of action or in mechanisms of acquired
resistance.

ADAPTATION OF THE PATHOGEN
        Two main types of fungicide resistance can be distinguished.
Discrete resistance ('black-and-white' resistance) denotes the emergence
of variants which are much less sensitive to the fungicide concerned
than are members of the normal 'wild-type' population. Continuous
resistance ('shades of grey' resistance) involves the appearance of
forms with many different degrees of resistance, some of which may
differ little in response from the more resistant members of the normal
population. Determination of which of these two types of resistance
operates appear to depend more on the class of fungicides involved than
on the particular type of pathogen. Thus, resistance to the benzimidazole and acylalanine fungicides is discrete, whereas resistance to the
2-amino-pyrimidines and to the sterol demethylation inhibitors (DMIs) is
continuous. Discrete resistance is generally associated with a rapid
and semi-permanent build- up of highly resistant forms in the field,
whereas continuous resistance favours the more gradual and reversible
build-up of moderately resistant forms. There is now growing evidence
that discrete resistance results from major single-gene mutations,
whereas continuous resistance results from the mutation of one or more
of a number of genes each of which confirm only a proportion of the
highest degree of resistance achievable. Both types of resistance are

of great agricultural and economic importance, but the more explosive, severe and stable nature of discrete resistance tend to make it the more difficult type of problem to control.

Fungicide resistance has arisen in all the major classes of fungal plant pathogens. However, it has built up more rapidly in pathogens such as the powdery and downy mildews which reproduce frequently and produce many spores, but relatively slowly in pathogens such as loose smut of barley (Ustilago nuda) or the wheat eyespot fungus (Pseudocercosporella herpotrichoides) which only have one generation per year.

PRACTICAL IMPACT
Whilst it is fairly easy to detect differences in fungicide sensitivity between different fungal isolates in laboratory bioassays, it is much harder to demonstrate that such differences have caused losses of disease control in the field. Indeed much confusion has arisen through the unwarranted extrapolation of laboratory findings to field situations, such that the Fungicide Resistance Action Committee (FRAC) - described by Ruscoe, see Chapter 15 - proposed that the term 'laboratory resistance' should be used to describe the results of laboratory tests for response and 'field resistance' to describe shifts in the response of field populations sufficient to cause problems of disease control. In the author's opinion the use of these terms would increase rather than dispel confusion, since 'laboratory resistance' revealed under controlled conditions must always be used to demonstrate 'field resistance', and since it is wrong to classify variants found in the field which are insufficiently resistant, frequent or pathogenic to cause loss of control as 'laboratory resistant' together with mutants produced artificially in the laboratory. Use of the term 'practical resistance' to denote resistance causing loss of disease control in crops is more acceptable. When absolute terms such as 'resistant' and 'sensitive' are used however, it remains vitally necessary to define their precise meanings and to assess carefully their practical significance.

Whilst the detection of resistant strains in laboratory cultures or even in field populations does not necessarily indicate that practical problems of disease control have arisen or are likely to arise, equally it must be emphasised that many crop protection problems of considerable economic importance have been shown beyond all reasonable doubt to result from fungicide resistance. The extent to which such problems affect the grower varies greatly. If a disease affects directly the quality and marketability of the produce of a crop, then a sudden breakdown in its control may be disastrous. For example, in the mid-1970s the unexpected appearance of scab lesions on fruit from apple orchards in several countries which had been treated with benzimidazole fungicides according to the manufacturers' recommendations caused considerable financial loss to the growers concerned. Similarly, the sudden failure of the acylalanine fungicide, metalaxyl, to control late blight of potatoes after a very short period of use resulted in severe losses of yield of saleable produce in Ireland and Holland. In each of these examples the growers even sued the manufacturers for damages attributable to misleading claims for effective disease control. The subsequent inclusion of warnings on labels that resistant strains of the target pathogens may occur and interfere with the normal action of the production now gives the manufacturer some measure of protection against claims for compensation.

In other situations the effects of resistance have developed more slowly, and the precise financial impact at farm level is harder to identify. For example, in barley powdery mildew in Europe, the build-up of the resistance to the DMI and 2-amino-pyrimidine fungicides has been both gradual and partial; the resulting increase in amounts of disease in treated crops will have reduced yield to relatively small extents (10%), and has not affected greatly the quality of the grain. In such cases the onset of a considerable resistance problem can even go unnoticed by some growers or their advisers.

To the manufacturers and suppliers of fungicides, resistance has brought many serious problems additional to the claims for financial compensation already mentioned. Resistance has not yet forced the complete removal of a product world-wide, but it has necessitated the withdrawal of product recommendations for specific uses in certain countries. For example, dimethirimol was withdrawn by the manufacturers for use against cucumber powdery mildew in glasshouses in Holland in 1971, following a rapid and widespread breakdown in disease control associated with the presence of dimethirimol-resistant mildew populations. More recently, recommendations for the application of metalaxyl against potato blight in Holland and Ireland were suspended by the manufacturers.

Partial restrictions in use imposed by the manufacturers themselves or by official advisory bodies, or a lessening of use by the growers in the light of their own experience are more common consequences, and again can lead to substantial losses of fungicide sales. Because plant pathologists are unable to predict precisely the rate of build-up of resistance and of its practical effects at farm level, it is equally hard for marketing departments of agrochemical companies to assess the relative financial risks they incur either by restricting product use and/or by adding companion compounds at effective concentrations in order to extend the effective life of a product, or alternatively of encouraging more extensive use in order to recoup research and development costs and to concentrate sales within its patent life. The appearance of resistance to another competitive product can of course boost sales, provided that there is no cross-resistance and especially if the first product is considered to carry a lower risk of encountering resistance.

In industrial research, the resistance phenomenon has mixed influences. It can discourage further exploratory research on the particular class of fungicides concerned; for example, efforts to find further benzimidazole fungicides declined rapidly once the widespread resistance to benomyl and thiophanate-methyl was confirmed. It can also weaken confidence more generally in research to find new systemic fungicides with their selective and hence 'vulnerable' mechanisms of action. Conversely, however, it can open up fresh opportunities for the development and marketing of new types of fungicide, to replace compounds which have lost their activity or to use alongside them in the avoidance strategies discussed below. Hence, resistance can sustain and stimulate biological and chemical research efforts. Partly for this reason, but also because of the increased awareness of more general market opportunities, fungicide research in industry has tended to increase in recent years.

PREDICTION
The ability to predict the build-up of fungicide resistance would clearly be valuable in guiding research priorities and establishing strategies of marketing and application. Up to now, attempts at prediction have not been entirely successful. For example, the potato blight fungus, Phytophthora infestans, did not adapt readily to acylalanine fungicides in laboratory 'training experiments' and it was inferred that the risk of resistance was low (Staub et al., 1979); in practice, severe resistance problems developed rapidly in the field (Davidse, 1982). Moreover, most 'experts' would have predicted that acylalanine resistance in Bremia lactuca, the causal pathogen of downy mildew in lettuces, where several repeated applications per season were made in a closed environment, would have arisen earlier than in potato blight, but in fact resistance in this pathogen built up relatively slowly. Predictions of resistance to the morpholine fungicides have so far proved unfounded. The combination of their specific mode of action coupled with their extensive use against cereal powdery mildews might appear to be a recipe for resistance; moreover, the ability to mutate to morpholine resistance was detected in another fungus, Ustilago maydis, in laboratory experiments (Barug and Kerkenaar, 1984). In practice, no resistance has been detected after about 15 years of use, although of course the possibility of future problems cannot be ruled out.

However, the combination of experience and experiment results which has now built up has increased greatly our capacity reliably to assess risks. If current 'know-how' had been available earlier, it does seem likely that all instances of resistance so far encountered could have been predicted by considering the mode of action of the fungicide, its degree of use, the fecundity of the target pathogen and the ease of generating pathogenic, resistant forms of the pathogen in laboratory mutagenesis tests. Equally all cases of fungicide durability, except that of the morpholines mentioned above, also would have been predicted from such knowledge. Thus, the lack of resistance over many years' use to copper, sulphur, dithiocarbamate and phthalimide fungicides is entirely predictable. Whilst 'low', 'medium' and 'high' levels of risk can be indicated, and at least can give some guidance in establishing technical and commercial strategies for fungicide use, it must be said that quantitative rates and degrees of build-up cannot be predicted with any degree of precision.

A number of mathematical models have been proposed which aim to predict the rates of proportional increase in resistant forms when an initially predominantly sensitive pathogen population is exposed to single fungicides, mixtures or rotations of fungicides and single ('major gene') mutations towards resistance occur (Kable and Jeffery, 1980; Skylakakis, 1981; Levy et al., 1983). There are considerable differences in detail between these models in inputs, assumptions and outputs; in general they all predict delays in build-up if at-risk fungicides are mixed or rotated with non-risk fungicides (perhaps not surprisingly), but the relative advantages of mixture versus rotation vary between the models and estimates of rates of build-up also differ. At present there is little experimental evidence to validate the models (Skylakakis, 1984), and validation is hard to achieve because they depend greatly on factors such as the initial frequency of mutation of the pathogen, and the degree of cover and the persistence of applied fungicides, which are difficult if not impossible to assess with sufficient precision and also may vary considerably between local crop situations. Nevertheless, as experimental data accumulates concerning epidemiological aspects of fungicide resistance, and as further refinements in models are developed for both single-step and multiple-step resistance, our ability to predict the onset and speed of resistance seems likely to improve.

MONITORING
This term is used commonly but also loosely in discussions on resistance problems. Generally it seems to take samples of pathogen populations in crops and test them for their degree of sensitivity to one or more fungicides. However, such an exercise can at one extreme be continued surveillance from year to year over a large area and at the other an intensive 'one-off' investigation of the incidence of resistant forms at one site where difficulty in disease control has arisen. Considered in this broad-ranging sense, monitoring serves several important purposes and much can be learnt from it.

If done prior to the widespread use of a product, it can provide valuable initial ('base-line') data revealing the normal variation in response of a pathogen population which exists before selection of resistant forms occurs. Such initial data become badly needed for comparison if problems of disease control by the product and the need for further testing do arise later. They are also useful at the time they are obtained, since taken together with results of laboratory mutation tests they can indicate the degree of genetic variability in fungal response and assist the assessment of risk of resistance and the establishment of an initial guidance strategy for the application of the particular treatment. Monitoring at a later stage can then indicate whether, when and where reistance is building up, and also how quickly and to what degree. This knowledge in turn can indicate how far avoidance strategies already adopted are working, and indicate any need for their modifiction. In a few situations (e.g. the use of benomyl to control Sigatoka leaf-spot of bananas, caused by Mycosphaerella spp. in

Central America, and the use of benomyl, thiabendazole and sec-
butylamine to control citrus storage rots caused by Penicillium spp. in
California), when the incidence of fungicide-resistant forms is known to
fluctuate considerably, on-site monitoring of spore populations for
percentage resistance by simple agar-plate colony counts can give local
guidance from month to month on the selection of appropriate fungicides.

There are many pitfalls in fungicide resistance monitoring. Probably
the commonest error is to assume that variation in sensitivity between
samples observed in laboratory tests necessarily reflects similar
variation in field response, whereas often this is not the case. For
example, variation in sensitivity detected in laboratory tests can arise
from the presence or absence of a minor proportion of resistant spores
which may be of no significance in the field, or the variation may be
too small to cause practical difficulties. Thus, a correlation between
laboratory responses and disease control in the field must be discern-
ible before firm conclusions can be reached regarding any possible
failure of action at the crop level. To achieve this, sustained and
systematic surveillance of the field performance of fungicide
applications is essential; often is is done either skimpily or not at
all.

It is also important that methods of sampling and testing be as
'realistic' or close to the field situation as possible. For example,
in vivo sensitivity assays are often more reliable than in vitro
methods, and tests are best done on samples which are fresh from the
crop and have not been sub-cultured under laboratory conditions.
Internationally accepted methods should be used wherever possible to
permit comparisons between different surveys, and suitable 'standard'
methods have been published for a number of major crop diseases (Anon.,
1982).

COUNTERMEASURES
    First we will consider strategies which aim to avoid or
delay the build-up of resistance, and then discuss what can be done in
situations where a major resistance problem has already been allowed to
develop. The surest way to avoid resistance is, of course, not to use
the particular type of fungicide concerned. Indeed, the likelihood of
resistance arising certainly has influenced manufacturers' decisions not
to market certain experimental fungicides. Assuming, however, that the
product will be used, in order to profit the suppliers from its sale and
the farmers from its effect, then one somewhat less stringent policy is
to restrict use to times when the special properties of the compound are
critically necessary for crop disease control. At other times either no
fungicide or an alternative one is used. For example, ethirimol was
used in the UK as a systemic seed treatment for both autumn-sown
('winter') and spring-sown ('spring') barley. When the threat of
build-up of resistant strains was detected through monitoring, the
recommendation for use in winter barley was withdrawn by the manufacturer (Shephard et al., 1975). It was considered that ethirimol-
treated winter barley formed a selective bridge on which less sensitive

forms could persist and multiply between successive crops of spring
barley. Moreover, at that time winter barley was a much less important
crop than spring barley, and the value of using ethirimol was less well
established. With later knowledge that the more resistant forms tended
to be slightly less fit than more sensitive ones, with the increasing
use in barley of a range of alternative fungicides of different types,
and with a growing importance of the winter barley crop in the UK,
ethirimol seed treatment was re-introduced into winter barley about six
years later.

The timing of fungicide applications may also affect the appearance of
resistant strains. For example, is it more risky to apply a systemic
fungicide (such as ethirimol or triadimenol) as a seed treatment or as a
spray, in order to control foliage diseases of cereals? Such seed
treatments are applied routinely as 'insurance' to protect against
disease that may arise; the fungicide is slowly taken up by the plant
and therefore is present at fungitoxic levels for many weeks. Sprays
also may be applied for insurance, but generally are used when the first
visible symptoms of disease are observed in the crop, i.e. only when
they are really needed. They enter plants more quickly, but have a less
persistent action. It has been argued that the seed treatment would
give longer and sometimes unnecessary selection pressure and that short-
duration spray treatments applied at need may be 'safer'. Equally,
however, it can be argued that sporulating lesions allowed to appear in
the crop would provide a large population of individual genetic units
which would be very vulnerable to selection by a potent spray
application; a seed treatment should keep the pathogen population (if
present at all) very low, and when disease ultimately arises it is
because the fungicide has declined to low levels in the tissue and is no
longer exerting a significant selection pressure. Similar conflicting
arguments can apply to the different timings of sprays - 'little and
often' prophylactic applications versus the curative ('fire brigade')
approach. There is virtually no experimental evidence relevant to this
question, and current judgements rely mainly on theory. No doubt many
quantitative factors such as the frequency and timing of epidemics and
of sprays, the fecundity of the pathogen, and the persistence of the
fungicide must form part of the question.

Altering fungicide dose (amount applied on each occasion) could also
affect the selection of resistant strains. It remains an open question,
however, as to whether it is more risky to lower or to raise the dose.
Lowering it would tend to decrease fungicidal activity and hence
selection pressure that favours highly resistant forms. However,
lowering the application rate might also encourage the survival and
spread of moderately or weakly resistant strains, or permit the stepwise
growth of resistance if it is polygenically determined. Conversely,
increasing the rate might increase the selection of highly resistant
forms, but decrease more gradual build-up of resistance. Experimental
data bearing on this question are at present surprisingly sparse and
inconclusive. Experience suggests that resistance tends to arise more
readily when disease control is very efficient, and that lowering the
rate of application to the minimum necessary to give an acceptable

economic benefit is more likely to decrease than increase the risk of
resistance in most situations. However, more research on this aspect is
badly needed.

The sustained exposure of pathogen populations to a simple 'high-risk'
fungicide can be avoided in a different way - by the use of a companion
fungicide of a different type, applied either mixed with the former
fungicide or rotated with it. The companion compound must be one which
is known to control strains of fungi that are resistant to the at-risk
fungicide, i.e. there must be no 'cross-resistance' to the two
fungicides. With present experience there appears little to choose
between mixing and alternating treatments. On theoretical grounds, the
former seems the more effective method of stopping the selection of
resistant forms, provided the companion compound is potent enough and
present in sufficiently high concentrations to exert its own independent
action on the pathogen. Either multi-site or specific-site fungicides
can be used as companion compounds, but if suitable multi-site compounds
are available these are preferred because they themselves carry a very
low risk of acquiring resistance. Thus, protectant fungicides such as
mancozeb and copper have found renewed use as partners to the more
recent systemic fungicides. Proprietary formulated mixtures
increasingly are being applied; one example is 'Fubol' (Ciba-Geigy), a
mixture of metalaxyl and mancozeb for use against potato blight.

Combining two site-specific compounds may also be effective, since it is
very unlikely that the pathogen will mutate to both components
simultaneously. If a 'fully' systemic action is required, for example
as in seed treatment to control foliar diseases, then any companion
compound must itself be systemic in order to reach the target pathogen.
If exposure to both components is not entirely simultaneous, because of
differences in their rates of uptake or degradation, the possibility of
successive mutations giving double resistance will arise. However, the
risk would still be relatively small compared with the chances of
mutation in response towards continued use of a single compound, and
also the double mutants may well be less fit as pathogens than the wild-
type. Several mixtures of site-specific fungicides have been introduced
recently; one example is 'Sportak Alpha' (Schering), a spray formulation for cereals containing prochloraz and carbendazim.

Application of a combination of two fungicides showing negative-cross-
resistance could be particularly effective, since a fungus 'switching
off' against one fungicide partner will automatically 'switch on'
against the other. The cereal seed treatment 'Ferrax', recently
introduced by ICI (Northwood et al., 1984) includes a 2-amino-pyrimidine
fungicide (ethirimol) and a triazole (flutriafol), and is designed to
exploit the negative cross-resistance of powdery mildew which is known
to exist between these two classes of fungicide (Hunter et al., 1984;
Butters et al. 1984). The experimental fungicide MDPC (methyl N-(3,5-
dichlorophenyl) carbamate) described by Kato et al. (1984) exhibits
marked negative cross-resistance toward existing benzimidazole
fungicides; it acts only against benomyl- or carbendazim-resistant
strains of several plant pathogens. Since these strains are widespread,

the use of mixtures of MDPC or closely related compounds either with established benzimidazole fungicides or with other types of fungicide could give both effective and durable disease control.

Formulated mixtures have the considerable advantage that the site-specific component or components cannot be applied alone. It is hard to 'police' the use of other strategies such as tank-mixing or rotation of fungicides, which involve decision-making by the grower and whose implementation cannot be guaranteed. A broader spectrum of action against different pathogens is often obtained by using mixtures, and this can be an additional advantage.

Once resistance is well-established, either on a local or a national scale, possibilities for taking counter-measures are much more limited. Increasing the dose generally raises problems of crop, human or animal safety, and is seldom effective. The application of fungicide mixtures containing the 'failed' fungicide is also ineffective at this stage, unless the other component shows negative cross-resistance. Often the only practical strategy is to stop use of the fungicide concerned, use an alternative fungicide of a different type, and monitor to determine whether sensitivity is regained. Re-entry with the original fungicide either mixed or rotated with a companion compound can then be attempted, although there is still a considerable risk of a 'relapse'. Considerable variation occurs in the degree of 'recovery' of different pathogen populations to different fungicides and factors that determine this are discussed by Dekker (see Chapter 12).

INTEGRATED RESISTANCE MANAGEMENT

This concept, which we may designate IRM by analogy with and as a part of IPM (integrated pest management), implies the bringing together of different chemical and biological measures, so as to combat resistance but sustain the required standard of disease or pest control. It combines several approaches which are summarised below:

Components of Integrated Resistance Management

1. Predict risks for chemical.

2. Obtain 'base-line data' on response of field samples.

3. Monitor performance.

4. Monitor sensitivity.

5. Introduce avoidance strategies (e.g. minimising use, applying mixed formulations or programmes).

6. Use complementary non-chemical methods.

7. Establish co-operation between personnel and organisations in the public and private sectors.

The initial assessment of risk and establishment of an appropriate strategy of use should be regarded as an essential part of any pesticide development programme. Monitoring, first for base-line data and subsequently for product performance in the field and for shifts of sensitivity as revealed by laboratory tests, is a key activity. Above all, it is the integrated use of stringently selected and timed chemical treatments together with non-chemical methods such as growing disease-resistant cultivars (if available) and following hygienic cultural practices that will best delay or avoid the build-up of resistant strains.

To foster IRM, co-operation between public-sector research organisations, agrochemical companies, advisory bodies and growers must increase. Companies which market fungicides of the same class (with regard to cross-resistance) need to collaborate in producing and operating joint strategies of resistance management. It is useless for one company to restrict frequency of application of a fungicide, or to insist it is used only in conjunction with a companion compound, whilst other companies with closely related products are not pursuing such policies. Thus, FRAC (see Ruscoe, Chapter 15) has already proved its worth by establishing joint monitoring programmes, and by producing some agreed strategies of use to which member companies have adhered.

The co-operation and commitment of national pesticide regulatory bodies is vital to the success of IRM. Whilst some organisations have given every assistance, it must be said that others have raised difficulties. In particular, registration officials of certain countries have objected to the introduction of mixtures of fungicides, although the evidence that their use can either avoid or delay resistance continues to grow. Willingness to register 'alternative' fungicides of different classes to those already in use, even though their effectiveness against normal pathogen populations may merely equal or may be slightly inferior to those already in use, has also been lacking in certain countries.

It is very important that as great a diversity as possible of fungicides should be available for crop disease control, and we must hope that the considerable efforts of industry to discover new fungicides will be sustained. An impressive series of major new classes of fungicide appeared in the late 1960s and mid-70s, but no further new classes of fungicide have been announced since 1977. In some situations, for example the control of eyespot disease of cereals (caused by Pseudocercosporella herpotrichoides) and of grey mould diseases in several different crops (caused by Botrytis cinerea) there is now very little choice of fungicides which work well, and effective new materials are badly needed.

REFERENCES
Anon. (1982). Recommended methods for the detection and measurement of resistance of agricultural pests to pesticides. FAO Plant Protection Bulletin 30, 36-71, 141-143.

Barug, D. and Kerkenaar, A. (1979). Cross-resistance of uv-induced
    mutants of Ustilago maydis to various fungicides which
    interfere with ergosterol biosynthesis. Mededelingen van
    der Faculteit der Landbouwwettenschappen Rijksuniversiteit
    Gent 41, 421-427.
Brent, K.J. (1982). Case Study 4: Powdery mildews of barley and
    cucumber. In Fungicide Resistance in Crop Protection, eds
    J. Dekker and S.C. Georgopoulos, 219-230. Wageningen:
    Centre for Agricultural Publishing and Documentation.
Butters, J.A., Clark, J. and Hollomon, D.W. (1984). Resistance to
    inhibitors of sterol biosynthesis in barley powdery mildew.
    Mededelingen van der Faculteit der Landbouwwettenschappen
    Rinksuniversiteit Gent 49, 143-151.
Davidse, L.C. (1982). Acylalanines: Resistance in downy mildews,
    Pythium and Phytophthora spp. In Fungicide Resistance in
    Crop Protection, eds J. Dekker and S.G. Georgopoulos,
    118-127. Wageningen: Centre for Agricultural Publishing
    and Documentation.
De Waard, M.A. (1984). Negatively correlated cross-resistance and
    synergism as strategies in coping with fungicide resistance.
    Proceedings of the 1984 British Crop Protection Conference -
    Pests and Diseases 2, 573-584. British Crop Protection
    Council Publications.
Duran, R. and Norman, S.M. (1961). Differential sensitivity to biphenyl
    amongst strains of Penicillium digitatum Sacc. Plant
    Disease Reporter 45, 475-480.
Harding, D.R. (1962). Differential sensitivity to sodium orthophenyl
    phenate by biphenyl-sensitive and biphenyl-resistant strains
    of Penicillium digitatum. Plant Disease Reporter 46,
    100-104.
Hunter, T., Brent, K.J. and Carter, G.A. (1984). Effects of fungicide
    regimes on sensitivity and control of barley mildew.
    Proceedings of the 1984 British Crop Protection Conference -
    Pests and Diseases 2, 471-476. British Crop Protection
    Council Publications.
Kable, P.F. and Jeffery, H. (1980). Selection for tolerance in
    organisms exposed to sprays of biocidal mixtures: a
    theoretical model. Phytopathology 70, 8-12.
Kato, T., Suzuki, K., Takahashi, J. and Makoshita, K. (1984).
    Negatively correlated cross-resistance between benzimidazole
    fungicides and methyl N-(3,5-dichlorophenyl) carbamate.
    Journal of Pesticide Science 9, 489-495.
Kuiper, K. (1965). Failure of hexachlorobenzene to control common bunt
    of wheat. Nature, Lond. 206, 1219.
Levy, Y., Levi, R. and Cohen, Y. (1983). Build-up of a pathogen sub-
    population resistant to a systemic fungicide under various
    control strategies: A flexible simulation model.
    Phytopathology 73, 1475-1480.
Noble, M., Maggarvie, Q.D., Hams, A.F. and Leafe, L.L. (1966).
    Resistance to mercury of Pyrenophora avenae in Scottish seed
    oats. Plant Pathology 15, 23-28.
Northwood, P.J., Paul, J.A. and Gibbard, M. (1984). FF4050 seed

treatment - a new approach to control barley diseases. Proceedings of the 1984 British Crop Protection Conference - Pests and Diseases 1, 47-52. British Crop Protection Council Publications.

Shephard, M.C., Bent, K.J., Woolner, M. and Cole, A.M. (1975). Sensitivity to ethirimol of powdery mildew from UK barley crops. Proceedings of the 8th British Insecticide and Fungicide Conference 1, 59-66. British Crop Protection Council Publications.

Skylakakis, G. (1981). Effects of alternating and mixing pesticides on the build-up of fungal resistance. Phytopathology 71, 1119-1121.

Skylakakis, G. (1984). Quantitative evaluation of strategies to delay fungicide resistance. Proceedings of the 1984 British Crop Protection Conference - Pests and Diseases 2, 565-572. British Crop Protection Council Publications.

Staub, T., Dahmen, H., Urech, P. and Schwinn, F.J. (1979). Failure to select for in vivo resistance in Phytophthora infestans to acylalanine fungicides. Plant Disease Reporter 63, 385-389.

12. BUILD-UP AND PERSISTENCE OF FUNGICIDE RESISTANCE

J. Dekker
Department of Phytopathology, Agricultural University,
Wageningen, The Netherlands

INTRODUCTION
In a population of a fungal plant pathogen, which is
sensitive to a particular fungicide, cells may emerge by mutation or
other genetic change, which are significantly less sensitive. As long
as the proportion of such resistant cells is very low, the fungicide
will remain effective. However, under selection pressure by the
fungicide, a major part of the pathogen population may become resistant
and result in failure of disease control. In some cases build-up of
resistance, leading to failure of disease control, has occurred very
rapidly, even in the first season in which the fungicide was applied,
but in others it has happened only after a number of years. This
difference is important in practice. In the case of slow development of
resistance over a number of years, monitoring will make it possible to
detect build-up of a resistant pathogen population in an early phase, so
that timely counter measures can be taken, and crop losses avoided.

When fungicide resistance leads to failure of disease control in a
particular area, it is useless to continue application of the fungicide
involved. When application is stopped, the proportion of resistant
cells in the pathogen population may decrease rapidly, slowly or not at
all, depending on different factors. The persistence of fungicide
resistance varies greatly. The question arises whether, when the
proportion of resistant cells has dropped to a low level, re-use of the
chemical, either alone or in combination or alternation with other
fungicides, should be considered.

The factors, which influence the build-up of a resistant pathogen
population and the possible failure of disease control, and those which
determine the persistence of resistance will be discussed in this paper.
The accent will be on the principles involved, illustrated by a few case
studies, with special attention on the fitness of resistant strains.

FITNESS OF RESISTANT STRAINS
General
Fitness is a comparative concept: one strain may be more or
less fit than another under certain environmental conditions. Fitness
is also a complex property, with many parameters involved. In vitro
fitness parameters are the degree of spore germination, mycelium growth
on artificial medium and sporulation on agar. Parameters of fitness of

a pathogen on the host plant are the chance of infection, the speed of colonisation of the plant tissue and the degree of sporulation. Strains with greater fitness will be more competitive. To compare fitness, competition tests between two or more strains may be carried out, either in the presence or absence of the fungicide.

It has been observed with several fungicides that the fitness of resistant strains is lower than that of wild-type strains. In such cases, the changes in fungal metabolism that are responsible for resistance apparently are somewhat disadvantageous under normal conditions, i.e. in the absence of the fungicide. For some fungicides there even seems to be a direct biochemical link between increased resistance and reduced fitness, as has been illustrated in studies with the antifungal antibiotic, pimaricin. This compound complexes with ergosterol in the fungal membrane, causing leakage and cell death. In pimaricin-resistant mutants of certain yeasts, ergosterol appears to be replaced by a precursor, which has two effects. Firstly, there is less affinity between the precursor and the antibiotic, resulting in resistance, and secondly the precursor functions less well than ergosterol in the membrane, which means reduced fitness (Dekker and Gielink, 1979a).

Often it has been reported for a particular fungicide that many, but not all resistant strains show a reduced fitness. In such cases there is, apparently, no absolute link between resistance and reduction in fitness. Only when such a link exists are all resistant strains necessarily less fit. When there is, as with pimaricin, an inverse relation between the level of resistance and the degree of fitness, it has important consequences with respect to the risk that disease control failure will occur in practice. Also for this reason studies on the mechanism of resistance to various fungicides, and its possible relation with fitness, are important.

Fitness and mechanism of resistance
The fitness of resistant strains will be reviewed below for those compounds of which the mechanism of action and/or the mechanism of resistance has been elucidated. More extensive literature is reviewed by Dekker (1984).

Carbendazim and related benzimidazoles inhibit mitosis in fungi by attachment to the tubulin subunits of the spindle. Changes in these tubulins, brought about by mutation, result in reduced affinity to the fungicide and consequently resistance (Davidse and Flach, 1977). These changes are sometimes detrimental to the fungal cell, reducing its fitness, and sometimes not.

Carboxanilides interfere with respiration in fungi at a site which lies somewhere between succinate and coenzyme Q in the complex II region. Resistance may be brought about by mutation which results in a slight change at the target site. Such a change however, did not appear to be linked to reduced fitness of carboxin-resistant strains of Ustilago maydis (Georgopoulos et al., 1975), or oxycarboxin-resistant strains of Puccinia chrysanthemi (author's unpublished results).

Acylalanines interfere with RNA synthesis in Oomycete fungi, by inhibition of one or more RNA polymerases. In studies with Phytophthora megasperma f. sp. medicaginis Davidse et al. (1983b) obtained evidence that resistance is due to a change at the target site. This change, apparently, is not necessarily detrimental to the pathogen, as several of the resistant strains were as virulent as the wild-type fungus.

Antibiotics for control of plant diseases are primarily used in Japan. Polyoxin B interferes with chitin synthesis, by inhibition of the enzyme chitin synthase. Resistance in Alternaria kikuchiana was not due to a change at the target site, as chitin synthesis of resistant and sensitive strains was equally inhibited by the antibiotic in cell-free extracts. However, it was correlated with a strongly reduced uptake of the antibiotic. In competition tests the resistant inoculum was somewhat inferior to that of the wild type (Ito and Yamaguchi, 1979). Perhaps a change in membrane permeability, resulting in some degree of resistance, was slightly disadvantageous to the pathogen, at least in absence of the fungicide. Kasugamycin inhibits protein synthesis in Pyricularia oryzae, by preventing the binding of aminoacyl-tRNA to the ribosome. Resistance is caused by mutation leading to a change at the target site in the ribosome, which decreases its affinity to the antibiotic (Siegel, 1977). In competition tests, the resistant strains gave slightly fewer lesions than the sensitive ones, indicating a small disadvantage in the former in the absence of the fungicide (Ito and Yamaguchi, 1979).

A number of structurally unrelated groups of fungicides interfere with biosynthesis of ergosterols (sterol biosynthesis inhibitors, s.b.i.'s) as shown by inhibition either of ^{14}C-demethylation (pyrimidine, imidazole, triazole, piperazine and pyridine fungicides) or $\Delta^8 - \Delta^7$ isomerisation (morpholine fungicides). In studies with the pyrimidine fenarimol and mutants of Aspergillus nidulans, resistance appeared due to a constitutive, energy-dependent efflux mechanism which decreased the accumulation of the fungicide in the fungal cell (De Waard and Van Nistelrooy, 1979).

Fenarimol-resistant mutants of A. nidulans and the citrus pathogen Penicillium italicum showed a partially decreased fitness (De Waard and Van Nistelrooy, 1982). The presence of a constitutive efflux mechanism might be an extra burden for the cell metabolism, which is slightly disadvantageous to the cell in absence of the fungicide. A reduced fitness of resistant strains has also been reported for the piperazine triforine (Fuchs et al., 1977), and there are indications that it might hold also for other s.b.i.'s. It is not yet known, however, whether other mechanisms of resistance to s.b.i.'s occur, for example modification of the target site, and what consequences for fitness there might be in such cases.

There is little to say about a possible relation between fitness and the mechanism of resistance if this mechanism is not yet known. This is the case with dicarboximides, 2-amino-pyrimidines, organophosphates, dodine and various other groups. Strains of Botrytis cinerea with a relatively high degree of resistance to dicarboximides were reported to be osmotically sensitive, which may be a cause for the reduced fitness of such forms observed in the field (Beever and Byrde, 1982). Evidence for decreased fitness of ethirimol-resistant strains of Erysiphe graminis (Shephard et al., 1975) and of pyrazophos-resistant strains of Sphaerotheca fuliginea (Dekker and Gielink, 1979b) has been reported. Dodine-resistant mutants of Venturia inaequalis, however, seem to be as fit as the wild-type fungus (Gilpatrick, 1982).

Conclusion
The fitness of fungicide-resistant strains depends on the type of fungicide and the mechanism by which a pathogen becomes resistant to this fungicide. Some metabolic changes, responsible for resistance, cause lower fitness in the absence of the fungicide, others do not. On this basis, fungicides may be classified as carrying a high, moderate or low risk with respect to the build-up of a resistant pathogen population and the breakdown of disease control.

Whether the latter will be realised in practice, depends, however, also on several important factors other than fitness.

BUILD-UP OF RESISTANCE AND FAILURE OF DISEASE CONTROL
General
There are several cases, where development of fungicide-resistance, followed by failure of disease control, occurred rather rapidly after the introduction of a new fungicide. Examples are benomyl resistance in cucumber powdery mildew (Schroeder and Provvidenti, 1969) and metalaxyl resistance in Phytophthora infestans (Davidse et al., 1981). On the other hand, it took several years before control of rice blast by kasugamycin failed due to development of resistance (Uesugi, 1982), and more than ten years of continued application of dodine, before problems with control of apple scab arose (Gilpatrick, 1982).

In addition to the fitness of fungicide-resistant strains, discussed above, factors which affect the rate of build-up of a resistant pathogen population include the type of disease and the life cycle of the pathogen, the selection pressure exerted by the fungicide (dose, frequency of application, effectiveness, persistence in the crop), environmental factors which influence the disease pressure and the influx of sensitive strains from untreated fields.

It should be mentioned, however, that build-up of resistance will not always, without exception, result in failure of disease control. When the level of resistance is low in relation to the concentration of the fungicide in the crop, control may remain satisfactory.

The influence of the factors mentioned above will be illustrated by considering some selected cases of high-risk, moderate-risk or low-risk compounds.

Compounds with high risk

When changes in the fungal cell, responsible for resistance to a particular fungicide, occur without reduction of fitness, this fungicide can be considered as highly risky with respect to development of resistance problems in practice. Examples are benzimidazole fungicides and acylalanines. However, it would be wrong to conclude that the use of such compounds will invariably lead to failure of disease control. This will depend on several other factors, as discussed earlier.

Development of resistance to benzimidazoles has occurred rapidly with powdery mildews, fast-spreading diseases, which need frequent fungicide applications for control, but it happened only after a number of years in Pseudocercosporella herpotrichoides, the cause of eyespot in wheat (Horsten and Fehrmann, 1980; Obst, 1984). In eyespot disease, the nature of the soil-borne, slow-spreading pathogen with only one reproductive cycle per year, and the low selection pressure by the fungicide (normally only one application per season against eyespot) apparently do not lead to rapid build-up of resistance. The influence of the life-cycle of the pathogen is clearly shown by another example. In benomyl-treated apple orchards, build-up of resistance did occur in Venturia inaequalis, the cause of apple scab, but not in Gymnosporangium juniperi-virginianae, the cause of apple rust. In contrast to the former pathogen, G. juniperi-virginianae, does not have a repetitive summer cycle on the main host, favouring a rapid build-up of resistance, since it forms on apple only spermagonia and aecidia (Gilpatrick, 1982).

A very rapid change took place in the response of Phytophthora infestans to metalaxyl, soon after its introduction for control of potato blight (Davidse et al., 1981). The explosive character of epidemiology of this disease, and the sustained high selection pressure needed for control, favour a rapid build-up of resistance. Metalaxyl, however, has remained effective against some other Phytophthora spp. which cause root diseases.

Samoucha and Cohen (1984) made the interesting observation that synergy exists between metalaxyl-sensitive and metalaxyl-resistant strains in Pseudoperonospora cubensis. The sensitive strains stimulated release of zoospores, and also, probably enzymatically, penetration by resistant strains, a phenomenon which may contribute to a rapid build-up of resistance.

Compounds with moderate or low risk

When changes in the fungal cell, responsible for fungicide resistance, at the same time cause reduced fitness of the pathogen, this will hamper the build-up of resistance to the fungicide concerned. Usually they will not cause problems during the first years after introduction, but only after a more prolonged period of high selection pressure by the fungicide, depending on the degree of fitness reduction.

There are several examples of moderately risky compounds. Failure of rice blast control by kasugamycin occurred only after a number of years

and especially in certain areas of Japan, where this antibiotic had been
used almost exclusively, and resistance to Kitazin P developed in
practice only after 10 years of intensive use (Uesugi, 1982). Other
examples are the relatively slow development of polyoxin resistance in
Alternaria kikuchiana, the causal agent of black spot on Japanese pear
(Nishimura et al., 1976) and pyrazophos resistance in Sphaerotheca
fuliginea on cucumber (Dekker and Gielink, 1979b).

Of interest is the situation with respect to the amino-pyrimidine
fungicides, ethirimol and dimethirimol. Although samples of Erysiphe
graminis with decreased ethirimol sensitivity were obtained from many
treated fields in the UK over several years, no control failure has yet
occurred. Most isolates showed a relatively low level of resistance,
and the percentage of strains with a relatively high level of resistance
did not increase under selection pressure by the fungicide, probably
because reduced fitness was inherent to high resistance (Shephard et
al., 1975). Decreased selection pressure due to the withdrawal by the
manufacturer of a recommendation for use in autumn-sown barley, and a
relatively rapid breakdown of the fungicide in the plant may also have
counteracted the build-up of resistance. In contrast, application of
the closely related compound dimethirimol for control of powdery mildew
on glasshouse cucumbers rapidly led to resistance problems (Bent et al.,
1971). In this case the pathogen was subjected to continuous selection
pressure, by repeated soil applications throughout the growing season.
Moreover, the glasshouse provides a more or less closed environment so
that after elimination of the sensitive strains by the fungicide, the
resistant strains hardly encountered competition by wild-type strains
from outside. It is of interest that the use of dimethirimol on outdoor
cucumber crops in the mediterranean region has not led to resistance
problems.

Dicarboximide fungicides are important for the control of Botrytis and
Monilinia disease, especially B. cinerea on grapes. Although isolates
of this pathogen with decreased fungicide sensitivity were obtained from
vineyards, control of grey mould remained satisfactory in most cases.
The probable explanation is that most strains show only a moderate
resistance level, and that strains with higher resistance appeared to be
less virulent (Beever, 1983). Recently, however, failures in the
control of B. cinerea have been encountered in some areas of France.
Davis and Dennis (1981) found poor sporulation of dicarboximide-
resistant strains to be the single most important factor affecting their
development on the strawberry crop. Frequent use of these fungicides
during the season causes a shift in the pathogen population towards
lower fungicide sensitivity, but this increases again in between the
growing seasons (Lorenz and Eichhorn, 1982). It is not yet known
whether the fitness of resistant strains improves during continuous
selection pressure by the fungicide. This possibility should be taken
into consideration in planning management strategies.

In view of the rapidly increasing use of fungicides which inhibit
biosynthesis of sterols (s.b.i.'s) there is much interest and concern
about the possibility of development of resistance. Several authors
consider s.b.i.'s to be low risk compounds, but the situation is not yet

clear. The use of the triazole, triadimefon, against barley powdery
mildew has led to an increase in the frequency of less sensitive strains
in the UK (Heaney et al., 1984; Wolfe et al., 1984). Following
complaints about less satisfactory control of wheat powdery mildew by
triadimefon in The Netherlands, De Waard at our laboratory investigated
whether this could be due to development of triadimefon-resistant
strains. Isolates were collected from treated and untreated fields in
various parts of The Netherlands, and tested for sensitivity to
triadimefon in foliar spray tests in comparison with that of an isolate
of Erysiphe graminis f. sp. tritici, which had never been exposed to
s.b.i. fungicides. The majority of the isolates collected at the end of
the growing season of 1982 from treated fields in the province of
Limburg showed an EC_{50} between 3 and 6 µg ml^{-1}, while the EC_{50} isolates
from untreated fields in Noord-Brabant and Gelderland was lower than 2
µg ml^{-1}, and comparable with the reference isolate (Fig. 1). In the
following spring, isolates with decreased triadimefon sensitivity had
almost disappeared, which indicates a reduced fitness. However,
isolates collected in June and July 1983, after one or two fungicide
treatments, showed again a significantly reduced triadimefon
sensitivity, comparable to that observed in 1982. The level of
resistance was not high enough to explain failure of disease control at
the rate the fungicide is used in practice, but it might have
contributed to it under conditions favouring disease development (De
Waard, pers. commun.).

During the last decade several s.b.i.'s have been introduced for control
of cucumber powdery mildew in The Netherlands, namely triforine,
imazalil, fenarimol and bitertanol. In the autumn of 1981, H.T.A.M.
Schepers started research at our laboratory to investigate whether
build-up of resistance to these fungicides was occurring in the causal
pathogen, Sphaerotheca fuliginea. In leaf disc and foliar spray tests,
the fungicide sensitivity of glasshouse isolates was compared with that
of a wild-type isolate, which had never been exposed to fungicides.
Schepers (1983a) observed a significant shift towards lower fungicide
sensitivity for all s.b.i.'s used, but not for dinocap (Table 1). The
high level of resistance to triforine was remarkable, because this
fungicide had been used successfully in the period 1972-77 without
resistance problems. The general absence of resistance problems with
this fungicide has been attributed to lower fitness of resistant
strains, as observed by Fuchs et al. (1977) in Cladosporium cucumerinum,
and to the short half-life of this compound which is unfavourable for a
constantly high selection pressure. The use of the more persistent
compounds imazalil and fenarimol may, in view of cross resistance, have
contributed significantly to the build-up of triforine resistance
(Schepers, 1983a). Apparently this resistance may increase to a much
higher level than originally thought, without crucial loss of virulence.

Schepers (1983b) noticed also an increase of strains with lower
fenarimol sensitivity during each of two successive growing seasons. In
the autumn of 1982 the proportion of isolates with increased EC_{50} was
significantly higher than in the spring of that year, presumably due to

selection pressure by frequent fungicide application. Early in 1983,
s.b.i. sensitivity had increased, as compared with the autumn of 1982,
indicating a somewhat lower fitness for resistant strains in the
fungicide-free winter period. Later in 1983 the resistance level rose
above that of 1982.

Figure 1. Shift towards lower triadimefon sensitivity in
isolates of <u>Erysiphe</u> <u>graminis</u> f. sp. <u>tritici</u> from fungicide
treated fields in Limburg, as compared with normal
sensitivity of isolates from untreated fields in
Noord-Brabant and Gelderland. Isolates collected in 1982;
n = number of isolates (data from De Waard, pers. commun.).

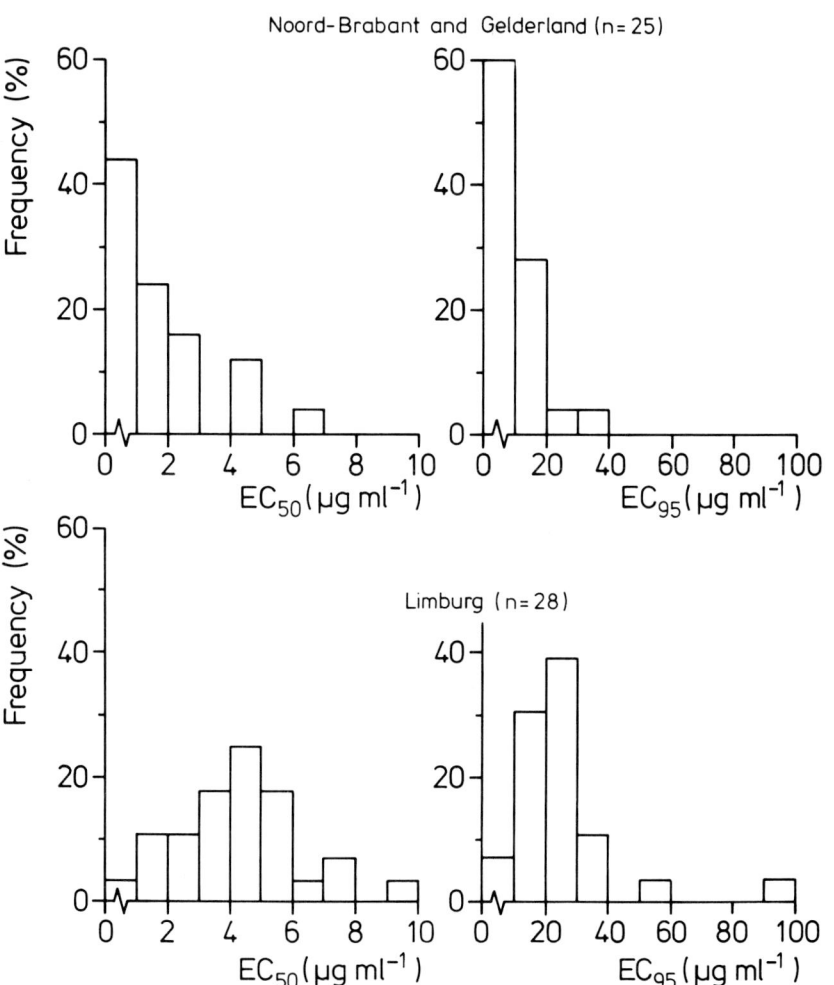

It is an open question, whether the level of resistance to fenarimol, imazalil and other s.b.i.'s may increase from year to year to such a level that efficient control will become impossible. Failure of powdery mildew control by fenarimol has already been observed in outdoor cucumber crops in Israel, but not yet in glasshouse crops in The Netherlands, although the level of resistance observed in isolates appeared to be about the same (Table 1).

Table 1. Sensitivity to ergosterol biosynthesis inhibitors of isolates of Sphaerotheca fuliginea, obtained from cucumber in The Netherlands (glasshouse) and Israel (open air). Results given as EC_{50} in mg 1^{-1} determined in a foliar spray test.

Fungicide and dosage	Sensitive Reference isolate	D-17 (Neth.)	Isr.-1 (Israel)
Bitertanol (300)+	1.5	7.0	9.0
Fenarimol (24)	0.2	3.0	2.5
Imazalil (50)	0.5	3.5	15.0
Triforine (200)	6.0	600	600
Dinocap (135)	13.5	13.5	-

+ Recommended dosage in practice in mg a.i. 1^{-1}. After Schepers (1983a,b).

Discussion
The build-up of fungicide resistance in practice is influenced by various factors. With intelligent management of fungicide application, even very risky compounds may continue to be used for control of some diseases. Manipulation of disease control with compounds of moderate risk will give fewer problems. This will hold even more for low-risk compounds, although too little information on these is available to predict the risks in the long run.

When a fungicide which has earlier given effective disease control performs below expectation, the farmer will consider increasing the concentration. On the other hand it has been argued that a high selection pressure by the fungicide will, by elimination of all sensitive or moderately resistant strains, favour the build-up of a highly resistant pathogen population. The latter may be true in those cases, where mutants with a high degree of resistance emerge, such as with benzimidazole fungicides and acylalanines. If, however, only low-

level resistance to a particular fungicide occurs, and the development
of greater resistance seems unlikely in view of an inverse relation
between resistance and fitness, increasing the dosage might prevent
failure of disease control.

PERSISTENCE OF RESISTANCE
General
The persistence of resistance in the field after application
of the fungicide has been discontinued, varies considerably between
different fungicides. In some cases the percentage of resistant strains
in the pathogen population drops to a low level in one or a few years,
but in others no decrease can be observed over ten years or more.

Because, after withdrawal of the fungicide, resistant strains have to
compete with sensitive wild-type strains which have survived, or which
arrive from other untreated fields, the fitness of resistant strains
seems of prime importance for persistence. If it equals that of wild-
type strains, the pathogen population might stay resistant for an
indefinite period. In the long run, however, sensitive strains may
again appear by back mutations.

When resistant strains have a much reduced fitness, resistance is
unlikely to persist very long. If, however, resistant strains show an
only slightly reduced fitness, persistence of resistance may depend on
several other factors, as discussed below.

Persistence in practice, and re-use of fungicide
There are several reports about continued persistence of
resistance after application of the fungicide has stopped. Benomyl
resistance of Cercospora beticola, the cause of leaf spot of sugar beet,
which developed after only two years of intensive and almost exclusive
use of benomyl in Greece, has not declined since the withdrawal of the
fungicide in that country in 1972 (Georgopoulos, 1982). In the
commercial glasshouses in The Netherlands, where benomyl was introduced
to control powdery mildew in 1972 and abandoned in 1973 after the
development of resistance, all strains isolated ten years later by
Schepers (1984) appeared to be highly resistant. In such cases it is
not sensible to re-use the fungicide.

After the dramatic failure of Phytophthora control in Dutch potato
fields in 1980, due to the sudden development of resistance to metalaxyl
(Ridomil), the fungicide resistance was monitored. Most isolates from
potato fields in the next year appeared again to be metalaxyl-sensitive
- probably because the seed tubers used in 1981 had been harvested
before the epidemic struck in 1980, so that incidentally infected tubers
carried mainly metalaxyl-sensitive inoculum. On old cull piles of
potatoes, however, mainly resistant isolates were found, and in at least
two cases metalaxyl-resistant strains had overwintered in the planted
seed tuber (Davidse et al., 1983a), which indicates the potential
persistence of resistant strains. Although the current pathogen

population became again predominantly sensitive, renewed use of Ridomil
in the same way as in 1980, has not been advised, because rapid build-up
of resistance has to be expected with this type of disease and
fungicide. Also in Israel, resistant strains of Phytophthora persisted
in fields where metalaxyl was not used any more (Cohen and Reuveni,
1983).

In the use of oxycarboxin to control chrysanthemum rust, caused by
Puccinia chrysanthemi, resistance problems arose in various countries
(Dirkse et al., 1982). Isolates of this pathogen were obtained from
glasshouses in The Netherlands with unsatisfactory disease control, and
investigated in laboratory tests. They were able to infect leaf discs,
placed on filter paper containing a 64 ug ml^{-1} solution of the
fungicide, whereas the wild-type fungus was inhibited at 8 ug ml^{-1}. The
resistant strains appeared to be as pathogenic as sensitive strains.
One year after use of oxycarboxin was stopped, isolates were found with
the same degree of resistance (author's unpublished results). Carboxanilides may therefore be considered as high risk compounds, with a good
chance that resistance will persist. Resistance problems have been slow
to arise in the control of smuts by carboxin, probably because of the
relatively low frequency of reproduction of this type of pathogen, which
has only one generation per year and hence does not favour rapid
build-up of resistance.

After discontinuation of the use of dimethirimol for control of cucumber
powdery mildew in 1970, the incidence of less sensitive strains remained
high in the next year, but decreased gradually in following years.
There was some renewed application of the fungicide in 1977 and 1978,
but again failure of disease control soon occurred, and its use was
abandoned (Brent, 1982). After a number of years, in 1981, isolates
from these glasshouses appeared, according to Schepers (1984), to have
retained their level of resistance.

Unsatisfactory control of powdery mildew on cucumber by pyrazophos was
observed only after several years of use of this compound. Isolates
then appeared to be less sensitive to pyrazophos, although they showed a
somewhat reduced fitness in the absence of the fungicide (Dekker and
Gielink, 1979b). When isolates from these glasshouses were tested again
by Schepers (1984), they had retained their pyrazophos resistance; they
were resistant also to other types of fungicide.

In 1972 the application of kasugamycin was stopped in a district in
Japan, where it failed to control rice blast. The proportion of
resistance strains then dropped gradually from 100% to less than 10% in
the following five years. Re-use of the fungicide as a single disease
control agent would certainly result again in build-up of resistance.
It was therefore decided to use kasugamycin in mixtures with
tetrachlorophthalide (Rabcide) and alternately with organophosphorus
fungicides, and this approach seems to have been successful (Ito and
Yamaguchi, 1977).

Four years after the introduction of polyoxin B for control of black spot in Japanese pear, caused by <u>Alternaria</u> <u>kikuchiana</u>, disease control failed in some orchards. Upon discontinuation of application of the antibiotic, the proportion of highly fungicide-resistant strains declined, becoming undetectable after ten years. At the same time there was an increase in the percentage of moderately resistant strains, and to lesser extent of sensitive strains. The decrease of resistant strains carrying unnecessary genes after removal of directional selection pressure, and the increase of moderately resistant strains was considered as stabilising selection (Udagawa <u>et al</u>., 1982). Polyoxin is still in use for control of pear black spot, but now in combination or rotation with structurally unrelated fungicides.

Discussion
After withdrawal of a fungicide, which has encountered resistance problems, resistance may persist not only in cases where resistant strains are as fit as sensitive strains, but sometimes also when resistance is linked to some reduction in fitness. This may happen, for example, when all wild-type strains have been eliminated by prolonged and effective application of the fungicide, and when sensitive strains from untreated sites are not available or do not have access. Such a situation may exist in glasshouse crops, or when all field crops in a given area have been thoroughly treated with the same fungicide. In such cases re-use of the fungicide after a period of interruption cannot of course be recommended.

Presumably the occurrence of strains which have acquired multiple resistance may contribute to persistence of resistance. In such cases even alternation of chemicals with different mechanism of action may not counteract resistance. If fungal cells are resistant to unrelated compounds A, B and C, application of A will not only select for resistance to A, but also for resistance to B and C.

When, after withdrawal of a fungicide, the proportion of the resistant pathogen drops significantly, the possibility of its re-use can be considered, in mixtures or in alternation with unrelated fungicides, but re-use of the compound alone should be avoided. In the design of strategies in such cases, theoretical models may be helpful (see Brent, Chapter 11).

GENERAL CONCLUSIONS
The potential of a fungal cell to become resistant to a particular fungicide, depends primarily on the mechanism of action of this fungicide. Virtually all fungicides with specific-site action appear to be vulnerable to resistance. The development of resistance problems in practice, however, depends crucially upon whether substantial build-up of a resistant pathogen population does take place. An important factor in this respect is the fitness of resistant strains, which depends largely on the mechanism of resistance. On the basis of site-specificity and of the fitness of resistant forms, fungicides may be classified as high-, moderate- or low-risk compounds with respect to

development of resistance in field populations of pathogens. Whether build-up of resistance will in fact be realised, depends on various other factors, namely the selection pressure exerted by the fungicide, the type of disease and life-cycle of the pathogen and environmental conditions.

There is much variation in the degree of persistence of resistance in fungal populations after the withdrawal of the fungicide involved. Factors which determine persistence include the fitness of resistant strains in absence of the fungicide, the occurrence of multiple resistance, and the presence of wild-type strains or access of such strains from outside.

When resistance persists, re-use of the fungicide is senseless, even in combination or alternation with other fungicides. If, however, the proportion of resistant fungal cells has dropped to a low level, re-use of the fungicide may be considered together or in rotation with other fungicides. These should preferably be conventional fungicides with multi-site action, to avoid the danger of development of multiple resistance.

REFERENCES

Beever, R.E. (1983). Osmotic sensitivity of fungal variants resistant to dicarboximide fungicides. Transactions of the British Mycological Society 80, 327-331.

Beever, R.E. and Byrde, R.J.W. (1982). Resistance to dicarboximide fungicides. In Fungicide Resistance in Crop Protection, eds J. Dekker and S.G. Georgopoulos, 101-107. Wageningen: Pudoc, Centre for Agricultural Publishing and Documentation.

Bent, K.J., Cole, A.M., Turner, J.A.W. and Woolner, M. (1971). Resistance of cucumber powdery mildew to dimethirimol. Proceedings of the British Insecticide and Fungicide Conference, Brighton, 3, 274-282. British Crop Protection Council Publications.

Brent, K.J., (1982). Powdery mildews of barley and cucumber. In Fungicide Resistance in Crop Protection, eds J. Dekker and S.G. Georgopoulos, 219-230. Wageningen: Pudoc, Centre for Agricultural Publishing and Documentation.

Cohen, Y. and Reuveni, M. (1983). Occurrence of metalaxyl-resistant isolates of Phytophthora infestans in potato fields in Israel. Phytopathology 73, 925-927.

Davidse, L.C. and Flach, W. (1977). Differential binding of methylbenzimidazol-2-yl carbamate to fungal tubulin as a mechanism of resistance to this antimitotic agent in mutant strains of Aspergillus nidulans. Journal of Cell Biology 72, 174-193.

Davidse, L.C., Danial, D.L. and van Westen, C.J. (1983a). Resistance to metalaxyl in Phytophthora infestans in The Netherlands. Netherlands Journal of Plant Pathology 89, 1-20.

Davidse, L.C., Hofman, A.E. and Velthuis, G.C.M. (1983b). Specific interference of metalaxyl with endogenous RNA polymerase

activity in isolated nuclei from Phytophthora megasperma
f.sp. medicaginis. Experimental Mycology 7, 344-361.

Davidse, L.C., Looyen, D., Turkensteen, L.J. and van der Wal, D. (1981).
Occurrence of metalaxyl-resistant strains of Phytophthora
infestans in Dutch potato fields. Netherlands Journal of
Plant Pathology 87, 65-68.

Davis, R.P. and Dennis, C. (1981). Studies on the survival and
infective ability of dicarboximide-resistant strains of
Botrytis cinerea. Annals of Applied Biology 98, 395-402.

Dekker, J. (1984). The development of resistance to fungicides.
Progress in Pesticide Biochemistry and Toxicology 4,
165-218.

Dekker, J. and Gielink, A.J. (1979a). Acquired resistance to primaricin
in Cladosporium cucumerinum and Fusarium oxysporum f.sp.
narcissi associated with decreased virulence. Netherlands
Journal of Plant Pathology 85, 67-73.

Dekker, J. and Gielink, A.J. (1979b). Decreased sensitivity to
pyrazophos of cucumber and gherkin powdery mildew.
Netherlands Journal of Plant Pathology 85, 137-142.

De Waard, M.A. and van Nistelrooy, J.G.M. (1979). Mechanism of
resistance to fenarimol in Aspergillus nidulans. Pesticide
Biochemistry and Physiology 10, 219-229.

De Waard, M.A., Greeneweg, H. and van Nistelrooy, J.G.M. (1982).
Laboratory resistance to fungicides which inhibit ergosterol
biosynthesis in Penicillium italicum. Netherlands Journal
of Plant Pathology 88, 99-112.

Dirkse, F.B., Dil, M., Linders, M. and Rietstra, I. (1982). Resistance
in white rust (Puccinia horiana) of chrysanthemum to oxycarboxin and benodanil in The Netherlands. Mededelingen
Faculteit Landbouwwetenschappen Rijksuniversiteit Gent 47,
793-800.

Fuchs, A., de Ruig, S.P., van Tuyl, J.M. and de Vries, F.W. (1977).
Resistance to triforine: a nonexistent problem? Netherlands
Journal of Plant Pathology 83, Suppl. 1, 189-205.

Georgopoulos, S.G. (1982). Cercospora beticula of sugar beets. In
Fungicide Resistance in Crop Protection, eds J. Dekker and
S.G. Georgopoulos, 187-194. Wageningen: Pudoc, Centre for
Agricultural Publishing and Documentation.

Georgopoulos, S.G., Chrysayi, M. and White, G.A. (1975). Carboxin
resistance in the haploid, the heterozygous diploid and the
plant parasitic dicaryotic phase of Ustilago maydis.
Pesticide Biochemistry and Physiology 5, 543-551.

Gilpatrick, J.D. (1982). Venturia of pome fruits and Monilinia of stone
fruits. In Fungicide Resistance in Crop Protection, eds J.
Dekker and S.G. Georgopoulos, 195-206. Wageningen: Pudoc,
Centre for Agricultural Publishing and Documentation.

Heaney, S.P., Humphreys, G.J., Hutt, R., Montiel, P. and Jegerings,
P.M.F.E. (1984). Sensitivity of barley powdery mildew to
systemic fungicides in the UK. Proceedings of the British
Crop Protection Conference - Pests and Diseases, 3, 549-464.
British Crop Protection Council Publications.

Horsten, J.A.H.M. and Fehrmann, H. (1980). Fungicide resistance of

Septoria *nodorum* and *Pseudocercosporella* *herpotrichoides*.
I. Effect of fungicide application on the frequency of
resistant spores in the field. Zeitschrift für Pflanzen-
krankheiten und Pflanzenschutz 87, 439–453.

Ito, I. and Yamaguchi, T. (1977). Occurrence of kasugamycin-resistant
rice blast fungus influenced by the application of
fungicides. Annals of the Phytopathological Society of
Japan 43, 301–303.

Ito, I. and Yamaguchi, T. (1979). Competition between sensitive and
resistant strains of *Piricularia* *oryzae*, Cav. against
kasugamycin. Annals of the Phytopathological Society of
Japan 40, 40–46.

Lorenz, D.H. and Eichhorn, K.W. (1982). *Botrytis* *cinerea* and its
resistance to dicarboximide fungicides. EPPO Bulletin 12,
125–129.

Nishimura, S., Kohomoto, K., and Udagawa, H. (1976). Tolerance to
polyoxin *Alternaria* *kikuchiana* Tanaka, causing black spot
disease of Japanese pear. Review of Plant Protection
Research 9, 97–57.

Obst, A. (1984). Schwietigkeitlen mit neuen fungiziden bei der
Kranheitsbekämpfung im Getreidebau. Gesunde Pflanzen 36,
143–148.

Samoucha, Y. and Cohen, Y. (1984). Synergy between metalaxyl-sensitive
and metalaxyl-resistant strains of *Pseudoperonospora*
cubensis. Phytopathology 74, 376–378.

Schepers, H.T.A.M. (1983a). Decreased sensitivity of *Sphaerotheca*
fuliginea to fungicides which inhibit ergosterol
biosynthesis. Netherlands Journal of Plant Pathology 89,
185–187.

Schepers, H.T.A.M. (1983b). Decreased sensitivity of *Sphaerotheca*
fuliginea to ergosterol biosynthesis inhibitors and other
fungicides in cucumber glasshouses in the Netherlands.
Proceedings of an International Symposium on Systemic
Fungicides, 20–26 May at Reinhards-Brunn, DDR.

Schepers, H.T.A.M. (1984). Persistence of resistance to fungicides in
Spaerotheca *fuliginea*. Netherlands Journal of Plant
Pathology 90, 165–171.

Schroeder, W.T. and Provvidenti, R. (1969). Resistance to benomyl in
powdery mildew of cucurbits. Plant Disease Reporter 53,
271–275.

Shephard, M.C., Bent, K.J., Woolner, M. and Cole, A.M. (1975).
Sensitivity to ethirimol of powdery mildew from UK barley
crops. Proceedings of the British Insecticide and Fungicide
Conference, Brighton, 3, 59–65. British Crop Protection
Council Publications.

Siegel, M.R. (1977). Effect of fungicides on protein synthesis. In
Antifungal Compounds, eds M.R. Siegel and H.D. Sisler, Vol.
II, 399–438. New York and Basel: Marcel Dekker.

Udagawa, H., Kohguchi, T., Otani, H., Kohmoto, K. and Nishimura, S.
(1982). A decade of transition of polyoxin-tolerant strains
of *Alternaria* *alternata* Japanese pear type in the field
ecosystem. Journal of the Faculty of Agriculture, Tottori

University, XVIII, 9-17.
Uesugi, Y. (1982). *Pyricularia oryzae* of rice. In Fungicide Resistance
in Crop Protection, eds J. Dekker and S.G. Georgopoulos,
207-218. Wageningen: Pudoc, Centre for Agricultural
Publishing and Documentation.
Wolfe, M.S., Minchin, P.N. and Slater, S.E. (1984). Dynamics of
triazole sensitivity in barley mildew, nationally and
locally. Proceedings of the British Crop Protection
Conference - Pests and Diseases, 3, 465-470. British Crop
Protection Council Publications.

13. RESISTANCE AND HORMOLIGOSIS AS DRIVING FORCES BEHIND PEST OUTBREAKS

V. Dittrich
Agrochemicals Division, Ciba-Geigy Ltd, Basel, Switzerland

INTRODUCTION

Primary pests appear in each season in most important crops and require human interference to avoid intolerable damage to the crop. Secondary pests may also make their appearance from time to time, but their control may not always be necessary from an economic point of view. This paper considers why secondary pests in intensive crops may switch roles with the established, regular targets of control efforts and become primary pests themselves. To ask this question is not only of academic interest. In 1961, severe mite problems developed in Egyptian cotton following a Spodoptera outbreak in that year. Subsequently, the mites assumed a primary pest status and had to be controlled by specific acaricides, applied either in mixtures with leafworm or bollworm sprays or separately. A parallel situation, more serious in its effects, occurred in Sudanese cotton and elsewhere, when the whitefly (Bemisia tabaci) superseded the bollworm (Heliothis armigera) as the primary pest and assumed the new status so efficiently that all established control methods to check H. armigera were rendered superfluous. What were the driving forces behind such developments with their critical economic consequences? This discussion of the case-history of these two examples will provide an insight into the causes of the puzzling and frustrating experience of insect outbreaks, and lead to suggestions for avoiding such critical situations in the future.

MITES IN EGYPT

Following an emergency programme of carbaryl applications, which were done to check bollworms (Spodoptera littoralis) which had escaped control by the usual toxaphene applications in 1961, mites became increasingly damaging in Egyptian cotton. Both Tetranychus urticae (syn. T. arabicus) and T. cinnabarinus (syn. T. cucurbitacearum) had been present before, but never attained pest status in cotton. This changed in the early 1960s when special acaricides, such as chlorfenson, tetradifon and dicofol, had to be used besides the leafworm and bollworm insecticides. This situation still prevailed when carbaryl was succeeded by Torbidan, a mixture of toxaphene (40%), DDT (20%), and methylparathion (7.5%). Spider mites remained a problem and had to be controlled by spraying special acaricides. However, with the development and large-scale use of a new leafworm and bollworm insecticide, monocrotophos, mites were reduced to what they had been before: a secondary pest and a minor problem.

What were the reasons for the mite problem? Followers of the beneficial insect hypothesis are probably inclined to lay the blame for the surging mite population on the elimination of parasites and predators by the sprays aimed at the leafworms and bollworms. However, both toxaphene which was used before 1961 and monocrotophos introduced in the early 1970s were very potent general insecticides and not liable selectively to spare the beneficials. If, therefore, the beneficial insect hypothesis cannot serve as an explanation for the mite surge, could some property intrinsic to carbaryl and DDT but missing in toxaphene and monocrotophos be responsible for the mite problems?

A demonstration of the effects of carbaryl and Torbidan with its DDT content on mites is given by field experiments performed in upper Egypt by Ghobrial (1972). In Assuit, Behera and Dakhlia he encountered the same situation: treated plots showed a marked increase in mite numbers over untreated after application of both insecticides (Tables 1 and 2). In extreme cases the mite counts on treated plots were more than 100% higher than on untreated. This stimulation was accompanied by resistance development towards the supposedly acaricidal component of Torbidan, methylparathion. Within three seasons, T. cinnabarinus had resistance ratios (LC_{50} for resistant form: LC_{50} for sensitive form) increasing up to 90-fold which demonstrated the efficiency of selection by methylparathion (Fig. 1). Thus, both population increase and resistance development occurred in mites, resulting in their acquiring primary pest status, as a consequence of control efforts directed against the targets S. littoralis and Pectinophora gossypiella.

Following these field experiments, we investigated the nature of the acceleration effect on mite populations of the insecticides carbaryl and DDT in the laboratory. Were plant stimulation and an altered host capacity the reason, or was it a direct influence on the mites by contact with chemical residues on the plant? For various reasons we investigated the latter question (Dittrich et al., 1974) taking as a working hypothesis the concept of hormoligosis described by Luckey (1968). This suggests that exposure to small doses of toxic chemicals can stimulate reproduction, and enhance the population growth, of certain pest organisms.

Both DDT and carbaryl treatments increased egg numbers, in closely controlled experiments featuring single females maintained over their life-time on residues of both these insecticides, which are almost non-toxic to mites (Figs. 2 and 3). Apart from a change in egg-laying patterns the treated mites also showed a shift in the female/male ratio of progeny in favour of the females (Table 3).

This effect could be further analysed due to a specific arrangement in our tests. Earlier investigators had shown that repellency of DDT residues occurs only on freshly prepared surfaces, but not when DDT residues have aged (Putman, 1963). Therefore, we shifted the mites to fresh bean leaves every 2-4 days and reared the offspring on each isolated leaf to adulthood. Thus, the ratio of females to males could be assayed as the mother grew older, and we found a decline of the

Table 1. Stimulatory effect on mites (<u>Tetranychus cucurbitacearum</u>) by carbaryl (Sevin) in an acaricide field test, Assuit 1970. (Source: A. Ghobrial, PhD thesis, Faculty of Science, University of Cairo, 1972.)

No.	Pretreatment (12.8.70) Acaricide	Control (%)	1st evaluation (19.8.70) Acaricide	Control (%)	2nd evaluation (26.8.70) Acaricide	Control (%)	3rd evaluation (2.9.70) Acaricide	Control (%)	Table of 3 evaluations Acaricide	Control (%)
1	Zolone	-10.01[a]	Sevin	-47.04[a]	Sevin	-54.46[a]	Sevin	-22.86[a]	Sevin	-56.76[a]
2	Sevin	7.35[a]	Untreated	0.00[b]	Untreated	0.00[b]	Untreated	0.00[a]	Untreated	0.00[b]
3	Galecron	-4.94[a]	Zolone	47.83[c]	Kelthane	67.27[c]	Zolone	74.84[b]	Zolone	64.77[c]
4	Kelthane	-1.52[a]	Kelthane	61.81[c]	Zolone	72.76[c]	Kelthane	81.49[b]	Kelthane	66.11[c]
5	Untreated	0.00[a]	Galecron	83.85[a]	Galecron	94.73[c]	Galecron	97.29[b]	Galecron	89.10[c]

Concentrations used (in g.a.i. ha^{-1}) were: Zolone, 380; Sevin, 3030; Galecron, 590; Kelthane, 540.

Treatments followed by the same superscript letter were not significantly different at each evaluation.

Table 2. Stimulatory effect on mites (Tetranychus cucurbitacearum) by Torbidan in an acaricide field test, Assuit 1970. (Source: A. Ghobrial, PhD thesis, Faculty of Science, University of Cairo, 1972.)

No.	Pretreatment (12.8.70)	Control (%)	1st evaluation (19.8.70)	Control (%)	2nd evaluation (26.8.70)	Control (%)	3rd evaluation (2.9.70)	Control (%)	Table of 3 evaluations	Control (%)
	Acaricide		Acaricide		Acaricide		Acaricide		Acaricide	
1	Galecron	-4.46^a	Untreated	0.00^a	Torbidan	-62.38^a	Torbidan	-12.47^a	Torbidan	-22.89^a
2	Torak	-4.09^a	Torbidan	14.38^a	Untreated	0.00^b	Untreated	0.00^a	Untreated	0.00^b
3	Zolone	-3.34^a	Rogor	45.04^b	Thiocron	57.24^c	Nuvacron	71.73^b	Nuvacron	56.62^c
4	Rogor	-2.35^a	Nuvacron	50.06^{bcd}	Nuvacron	61.68^c	Rogor	72.51^b	Rogor	58.24^{cd}
5	Kelthane	-0.86^a	Thiocron	57.80^{bcd}	Rogor	63.31^c	Kelthane	78.36^b	Thiocron	67.92^{cde}
6	Folimate	-0.37^a	Folimate	71.37^{cde}	Torak	65.88^{cd}	Zolone	79.72^b	Zolone	73.27^{cde}
7	Untreated	0.00^a	Zolone	73.27^{de}	Kelthane	70.09^{cd}	Kilval	82.26^b	Torak	74.00^{de}
8	Tartan	0.49^a	Kilval	75.16^{de}	Folimate	72.19^{cd}	Tartan	82.84^b	Tartan	74.72^{def}
9	Kilval	1.36^a	Tartan	75.98^{de}	Zolone	73.83^{cd}	Torak	84.01^b	Kelthane	75.65^{ef}
10	Torbidan	1.61^a	Kelthane	78.69^{de}	Kilval	76.40^{cd}	Thiocron	84.79^b	Kilval	75.87^{ef}
11	Nuvacron	2.97^a	Torak	80.73^e	Tartan	77.33^{cd}	Folimate	88.10^b	Folimate	77.18^{ef}
12	Thiocron	4.59^a	Galecron	88.87^e	Galecron	91.58^d	Galecron	91.61^b	Galecron	90.18^f

Concentrations used (in g.a.i ha^{-1}) were: Galecron, 595; Folimate, 666; Torak, 576; Tartan, 476; Zolone, 440; Kilval, 476; Rogor, 714; Torbidan, 714; Kelthane, 6380; Nuvacron, 238; Thiocron, 357;

Treatments followed by the same superscript letter were not significantly different at each evaluation.

Figure 1.   Development of resistance in Tetranychus cucurbitacearum to methylparathion, a component in Torbidan. (Source: A. Ghobrial, PhD thesis, Faculty of Science, University of Cairo, 1972.)

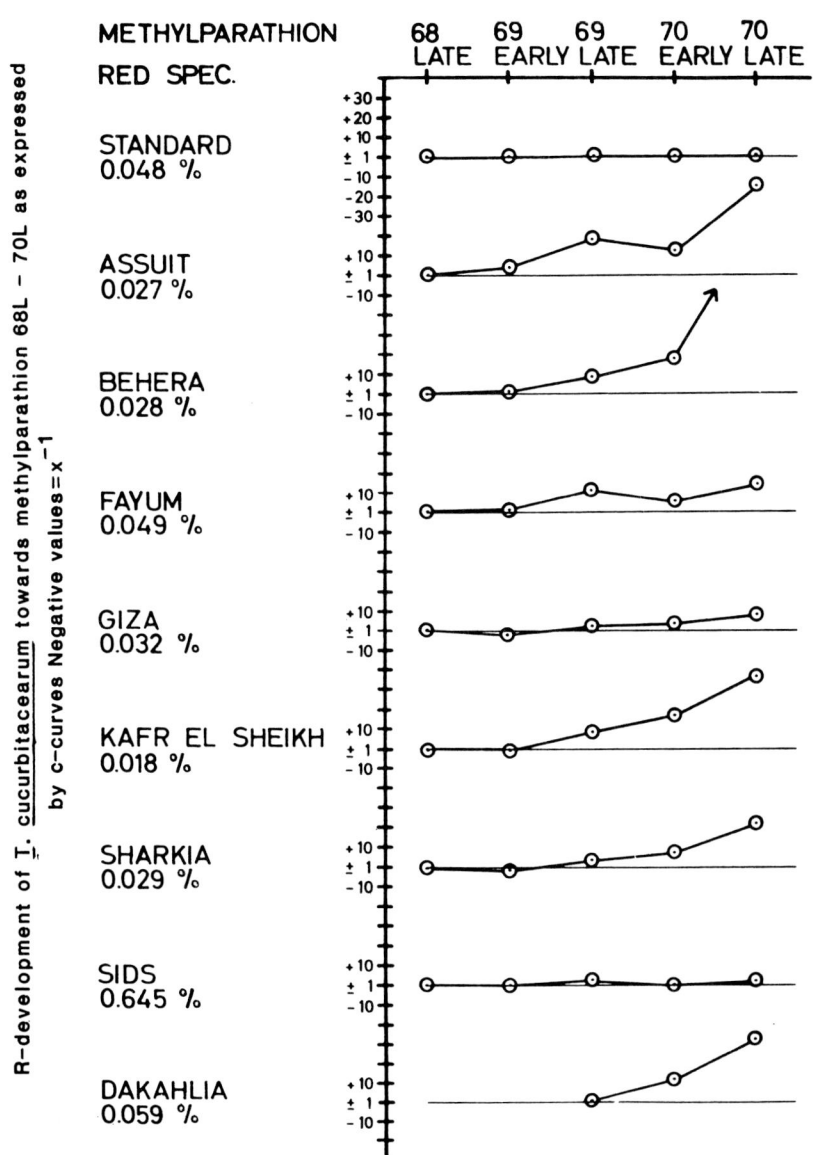

Figure 2. Effect of DDT residues on cotton leaves, on egg laying of female mites, _Tetranychus urticae_. (Reproduced by courtesy of _Environmental Entomology._)

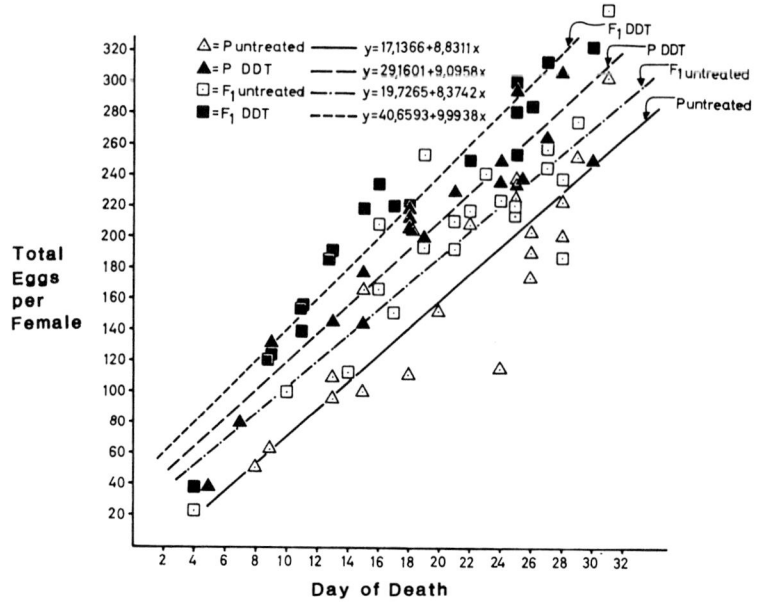

Figure 3. Effect of carbaryl residues on cotton leaves, on egg laying of female mites, _Tetranychus urticae_. (Reproduced by courtesy of _Environmental Entomology._)

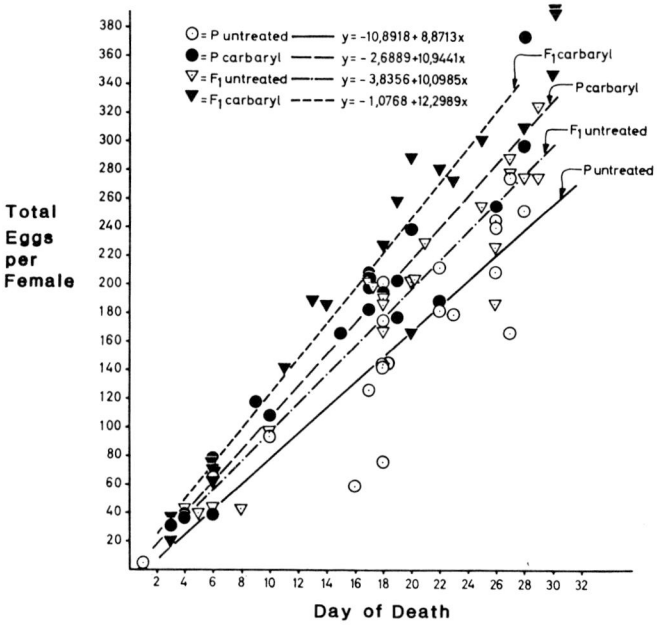

number of female offspring in favour of the males as time went on (Fig. 4). Mites and whiteflies are arrhenotokous insects, males develop from unfertilised eggs and are haploid, females are diploid and originate from fertilised eggs. In 90% of the cases observed, female mites mated only once. Therefore, a declining provision of sperm may have accounted for the declining number of females. On DDT, however, or even more pronounced, on carbaryl, such a decline was not observed. This is attributable to a higher sperm production by the males when exposed to the stimulatory effects of the residues. Thus, stimulation of both the male and female reproductive system would be responsible for the greater number of females which determines the reproductive capacity. Resistance towards parathion would add to the problems encountered with control of mites in the field, as long as DDT and carbaryl are part of the regime of treatments used to control the primary pests.

Table 3. Female/male ratio in a mite generation (Tetranychus cucurbitacearum) raised on insecticide residues to which parent mites were also exposed. (Reproduced by courtesy of Environmental Entomology).

Group	No. of female parents	F_1 progeny no. Female	Male	Female (%)	χ^2
Untreated	10	954	549	63.5	73.7**
Carbaryl	10	1131	321	77.9	
Untreated	17	1498	846	63.9	173.0**
DDT	17	2047	744	73.3	
Untreated	17	1784	1021	63.5	0.02
Dioxacarb	18	2002	1154	63.4	
Untreated (carbaryl) Untreated (DDT) Untreated (dioxacarb)					0.09

** Statistically significant at P = 0.01.

All this changed when the shift was made to monocrotophos for controlling leafworm and bollworm. Monocrotophos is a straight chain organophosphorus (OP) insecticide, and highly effective against mites. It controls methylparathion-resistant types very well and in this respect resembles vamidothion, another straight-chain OP which retained its efficacy despite extreme resistance towards methylparathion in the Egyptian mite races. Thus, I conclude from this evidence that mite problems in Egyptian cotton had little to do with the presence of predators and parasites, but rather were directly related to the type of leafworm and bollworm insecticides used during the period of the mite problems.

Figure 4. Effect of exposure to carbaryl residues on the female/male ratio of off-spring of single females of Tetranychus urticae. (Reproduced by courtesy of Environmental Entomology.)

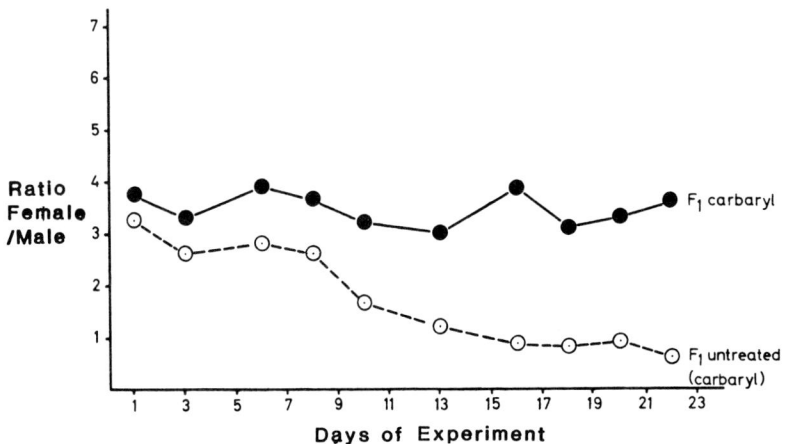

WHITEFLIES IN THE SUDAN

In Sudanese cotton, the whitefly (Bemisia tabaci), normally plays a secondary role, as mites did in Egypt, but it's economic importance made its shift from a secondary pest to the dominating primary one more spectacular. Bemisia tabaci has always been present on Sudanese cotton, just as mites were prior to 1961 in Egypt. First reports on a possible stimulatory influence of DDT sprays applied to control the bollworm, Heliothis armigera, were made by Joyce (1955) and Van der Laan (1961). Nothing resulted from this early work and for more than 20 years the mixture of DDT and dimethoate was the standard insecticide against bollworms, jassids, and whiteflies. Later on, through the 1970s, monocrotophos was used in increasing quantities but did not supplant DDT in the regime of treatments. In the second half of the decade, whiteflies became an increasing nuisance, and finally assumed the role of the primary pest of cotton and reduced H. armigera to a secondary pest.

In 1981, research was started on the whitefly problem in the cotton growing area of the Gezira near Wad Medani. A scan of reports and published literature had indicated three possible subjects which could be investigated in greater detail:

> 1. The cotton host and its changes in terms of cultivar and cultivation-methods;
>
> 2. The presence of stable insecticides and their influence on whitefly proliferation;
>
> 3. The possibility of resistance in whiteflies being due to intensive selection pressure over the years.

After developing a suitable technique, resistance was measured in field populations from different parts of the country. This was continued over three seasons, to analyse the dynamics of resistance under the impact of current insecticide treatments. By the 1980/81 season, both monocrotophos and DDT/dimethoate had been abandoned as bollworm insecticides, but dimethoate continued as a selectant in deltamethrin/ dimethoate and endosulfan/dimethoate mixtures to control H. armigera and secondary pests such as jassids and aphids.

Our resistance surveys, according to expectation, showed high resistance to be present in the whiteflies (Table 4). Resistance ratios were particularly high for dimethoate and monocrotophos, but not so, however, for profenofos. This confirms that cross-resistance between this mixed ester and other OP compounds was only partial. Contrasting with this was the high cross-resistance to OP compounds encountered in tests with carbofuran. This is notable, since neither carbofuran nor other carbamates have been used to any significant extent in the Sudan. A decline of resistance-ratios due to cessation of monocrotophos selection pressure was obvious; by contrast, ratios for dimethoate continued to increase in consequence of its continued application (Table 4). Resistance thus could partly account for the whitefly surge, but a further facet was added by our investigation of the influence of DDT on the fertility of the whitefly.

When a generation of whiteflies during their immature development was exposed to different amounts of DDT on cotton leaves, and this was continued throughout the adult life also, the response was similar to that of the mites as seen earlier (Table 5). Egg production in both a sensitive and a resistant field strain increased due to increasing DDT quantities on leaves, when measured as egg total or as eggs per female per day. The average life-time was shortened as DDT residues increased from 1 to 10 ppm in the applied solutions indicating the toxic action of the higher concentrations. Yet, with a total life-time shortened by 25%, the total egg production was higher. This means a shortened generation cycle at equal or higher fertility, compared with untreated whiteflies, and hence an increase in propagation potential by 25%.

Table 4. Resistance development in adult whiteflies (Bemisia tabaci) under standard insecticide applications; Wad Medani, Sudan, 1981 to 1983 seasons.

Insecticides	Reference strain 1981 LC_{50} (ppm)	Field strain: resistance ratio		
		1981	1982	1983
Monocrotophos	6.5	300	160	150
Dimethoate	12.2	242	290	311
Profenofos	4.9	9	13	13
Endosulfan	1.6	10	9	8
DDT	11.5	21	8	14
Cypermethrin	2.9	3	10	14
Decamethrin	0.2	–	25	13
Carbofuran	2.4	190	–	300

CONCLUSIONS

Both mites and whiteflies were treated with insecticide mixtures containing DDT and an OP insecticide. In both cases, sprays were mainly directed against lepidopterous primary targets. DDT is practically non-toxic to mites and has an intermediate toxicity for adult whiteflies. As residues on the crop decline, certain low amounts can be expected which cause stimulation upon contact. In both cases an accompanying OP served for selection among the secondary pests, methylparathion in one, dimethoate in the other. The difference in the speed by which both species reacted with resistance development is remarkable. Three seasons of selection with Torbidan produced a 90-fold increase in resistance ratio of methylparathion in T. cinnabarinus whereas dimethoate selection required about 20 years to produce an equivalent effect in B. tabaci. With parameters such as reproductive potential and number of generations per season being similar, the difference reflects the quality of both insecticides as resistance selectants. It appears that the gene conferring dimethoate resistance is much less frequent in whiteflies than the gene for parathion resistance in mites. There is historic evidence that these observations are paralleled in various pest species. As regards control of the new primary pests, the situation in Egyptian cotton proved to be less critical than that in the Sudan. Effective acaricides were available to control the mites, which were accessible by ground application. In the Sudan, by contrast, specific new whitefly control agents were not available, the most efficient insecticides being devalued by resistance. Further, a thorough coverage of plant parts below the canopy was not possible due to the necessity of aerial application.

Table 5. Acceleration of fertility of two whitefly (Bemisia tabaci) strains caused by exposure of immature stages and adults to DDT residues on cotton leaf discs. (Source: S.O. Hassan, MSc thesis, University of Reading, 1982.)

Insecticide treatment	Number of females	Period observed (days)	DDT concn. (ppm)	Mean life-time of females (days)	Total eggs/ female	Total eggs/ female/day
Sensitive strain						
None	25	38	0	29.4 ± 8.8	309.0 ± 115.2	10.6 ± 2.2
DDT	25	38	1	28.6 ± 7.0	344.8 ± 119.8	12.0 ± 2.4
Resistant strain						
None	7	34 – 42	0	37.6 ± 3.4	257.3 ± 172.9	6.9 ± 4.7
DDT	19	41 – 44	5	30.7 ± 8.8	261.9 ± 93.1	8.5 ± 2.1
DDT	12	40	10	26.7 ± 8.5	286.3 ± 97.7	10.7 ± 2.2

Soon after monocrotophos supplanted DDT and carbaryl in Egypt, the mites were reduced from their primary pest status and became a minor problem again, due to the excellent acaricidal properties of monocrotophos. Cessation of DDT in the Sudan by 1981/82 season coincided with three seasons of low whitefly pressure. This may be due to the absence of the accelerator DDT, but other reasons such as favourable weather conditions are also quoted as a possible cause for the change in whitefly dynamics.

From this it follows that if a secondary pest is to become a primary one, it must be de-stabilised in its natural equilibrium. The elimination of beneficials through chemicals is most frequently held responsible for such dis-equilibration, but for both the Egyptian and the Sudanese cases this explanation is not convincing. Rather, direct de-stabilisation of the population equilibrium by certain pesticides applied to control primary pests is indicated. DDT and carbaryl are effective de-stabilisers, their most obvious mode of action being enhancement of fertility. If, in addition to the effect of de-stabilising chemicals an insect population is released from check by an insecticide through its becoming resistant, nothing will restrain it from accelerated growth. In both Egyptian mites and Sudanese whiteflies this happened, methylparathion-resistance and dimethoate-resistance frustrating any control effort with these insecticides.

The consequences were less critical in Egypt than in the Sudan because the acaricides required for mite control differ in structure and biochemical effect from insecticides, such that cross-resistance to methylparathion did not exist.

Less fortunate circumstances prevailed when the insect B. tabaci had to be controlled by insecticides available in the Sudan. Here most of the pesticides were of little use because of the cross-resistance to dimethoate and monocrotophos.

The scope for avoiding situations, as just described, may lie in the avoidance of producing resistance and de-stabilisation of populations by increasing their fertility. My suggestions are therefore, that we:

1. Avoid developing toxicologically highly specific insecticides – they are liable to be good selectants for resistance;

2. Develop instead, "population modulators" rather than "killers" – they can be expected to select less efficiently for resistance;

3. Recognise typical "accelerator-molecules" and avoid them; and

4. Monitor resistance not only in primary but also in secondary pests. Once resistant they are prime candidates for the role of major pests.

REFERENCES

Dittrich, V., Streibert, P., and Bathe, P.A. (1974). An old case
    reopened: mite stimulation by insecticide residues.
    Environmental Entomology 3, 534–540.
Ghobrial, A. (1972). Laboratory and Field Toxicological Studies on
    Spider Mites infesting Cotton Plants in A.R.E. PhD thesis,
    Faculty of Science, University of Cairo, Egypt.
Joyce, R.J. V. (1955). Cotton spraying in the Sudan Gezira. II.
    Entomological problems arising from spraying. FAO Plant
    Protection Bulletin 3, 266–273.
Luckey, T.D. (1968). Insecticide hormoligosis. Journal of Economic
    Entomology 61, 7–12.
Putman, W.L. (1963). Lack of effect of DDT on fecundity and dispersion
    of the European Red Mite, Panonychus ulmi (Koch) (Acarina:
    Tetranychidae) in peach orchards. Canadian Journal of
    Zoology 41, 603–610.
Van der Laan, P.A. (1961). Stimulatory effect of DDT treatment on
    cotton and whiteflies (Bemisia tabaci Genn.: Aleyrodidae) in
    the Sudan Gezira. Entomologia Experimentalis et Applicata
    4, 47–53.

14. STRATEGIES FOR PREVENTION OF HERBICIDE RESISTANCE IN WEEDS

J. Gressel
Department of Plant Genetics, The Weizmann Institute of
Science, Rehovot IL-76100, Israel

INTRODUCTION

More than half the pesticides currently used in agriculture
are herbicides. Herbicides far surpass mechanical cultivation as the
most cost effective method of preventing weed competition in crops.
Soon after the introduction of phenoxy-type herbicides, warnings were
issued about the possibilities of impending evolution of resistant weeds
(Abel, 1954; Harper, 1956). Resistance never appeared to the phenoxy
herbicides and the warning voices were ignored. Indeed, after the first
triazine-resistant weeds appeared, they were discounted as "not being a
serious problem, nor does it appear to be a major threat in the future"
(Parochetti, 1979). Since then use of s-triazine herbicides has had to
be discontinued (except in mixtures) in Hungary, an important maize-
growing country, after over 75% of agricultural land became infested
with newly-evolved triazine-resistant Amaranthus. Triazine resistance
has become a serious problem in areas of Switzerland, and there is an
almost exponential increase yearly in infestation world-wide. Because
of the lack of novelty, new cases and even new species becoming
resistant are not now being reported. In 1981 there were 29 species
with well documented resistance to s-triazine herbicides (Le Baron and
Gressel, 1982), in 1983 there were 37 species (Le Baron, pers. comm.)
and my count from the literature now reaches 42 species. Herbicide
resistance is recognised as being a "growing problem" (Murphy, 1983),
and the number of reviews is keeping pace with the problem. In
addition, we now have reports of clear resistance to paraquat,
diclofop-methyl, trifluralin and other dinitro-anilines in agricultural
situations (cf. Le Baron and Gressel, 1982; Gressel, 1983). The ease
with which researchers are selecting for resistant genes in the
laboratory (cf. Gressel, 1984b) suggests that resistance to other
herbicides will soon appear in the field.

Simple population genetics explain why there has been no resistance to
phenoxy herbicides, and the resistance to triazines has become so widespread (Gressel and Segel, 1978, 1982). The same models, along with
close attention to crop and weed ecologies, agronomy and common-sense
can help us develop strategies to prevent resistance to effective,
inexpensive herbicides such as the s-triazines. The situation with the
s-triazines is at present the major problem, and will thus be used here
as a model. What we learn must be extrapolated to other herbicides,
especially as they become more cost-effective.

Costs

The decrease in the price of atrazine when the patents
expired made it much more competitive with 2,4-D and other herbicides in
maize, and thus farmers used more atrazine per year and decreased the
use of rotation. Indeed, it is now estimated to be 20% cheaper to treat
maize with atrazine than 2,4-D (Ammon and Irla, 1984). More areas went
into monoculture maize as more farmers left full-time farming, or looked
for simpler cropping systems. Resistance has appeared in these
agricultural areas alone. Levels of triazine carry-over have become so
high in the cooler climates that only maize can be grown the year after
an atrazine treatment in maize.

Farmers and their advisers should consider another perspective: what if
all major maize weeds should become resistant to triazines? Ammon and
Irla (1984) calculate that while it now costs SF (Swiss Francs) 26/ha to
treat sensitive weeds with atrazine, alternative treatments totalling
SF 273/ha would be needed when the major weed species become resistant.

Triazine resistance usually appears suddenly without forewarning. It
seems inevitably to occur where-ever triazines are used over long
periods. If the farmer does not watch his fields for resistance and
fails to use expensive mid-season treatments when it appears, the crop
can be lost. We must try to delay that appearance and spread it over as
long a period as possible, so as not to lose this excellent group of
herbicides from our arsenal.

THE APPEARANCE OF RESISTANCE: SIMPLE POPULATION GENETICS

Before we can develop strategies to control the enemy, we
must understand how the enemy first appeared. The occurrence of
resistance is a function of characteristics of the different weeds and
herbicides, which can be mathematically modelled.

For resistance to build up, resistant mutants must be present initially
at some low frequency in the population. We now know that there is a
plastid gene that can readily mutate to give triazine resistance through
lack of triazine binding (cf. Gressel, 1984a). There need not be genes
that degrade herbicides in all populations. When resistant biotypes are
grown in competition with susceptible (wild-type) biotypes of the same
species without herbicides their seed yield is about half that of the
wild type (cf. Gressel, 1984b). This lack of "fitness" is meaningless
when herbicides persist through the season, as with atrazine. Fitness
will considerably decrease the rate of enrichment for resistance when
non-persistent herbicides such as 2,4-D are used, as much of the season
will be available for the remaining susceptible individuals to exert
their superiority. This competition is especially fierce at the time of
seedling establishment (Aikman and Watkinson, 1980).

Persistence of herbicides interrelates not only with fitness, but also
with a special characteristic that separates weeds from crops as well as
from other pests - the 'spaced out' germination of weed seeds. Because
of a multitude of dormancy properties, weed seeds germinate throughout

the season.   Susceptible seeds can germinate after a rapidly degraded
herbicide has disappeared, and then can produce more seeds before the
season is over, considerably lowering the effective selection pressure.
Selection pressure is a result of "effective kill", which is
differentiated from the actual "knock down" kill that the farmer
measures after herbicide treatment.   Effective kill is a measure of the
number of surviving seeds, or propagules at the end of the season.

Weed seeds can also germinate over a period of years.   Every time we
enrich for resistant individuals by the use of herbicides, the
'resistant' seeds are diluted by a seed bank of susceptible seeds from
previous years which also exert a 'buffering effect' and delay the
appearance of resistance.   Indeed, the first weed reported to evolve
triazine resistance, Senecio vulgaris, does not have an appreciable seed
bank.   The effects of the selection pressure, the herbicide persistence,
and the seed bank on the rates of enrichment for resistance must be
quantified to allow us to see how modifying each parameter will affect
the rate at which resistance will appear.   Such modelling has been done
for the evolution of insecticide resistance by Georghiou and Taylor
(1977), for fungicide resistance by Delp (1981), and for resistance of
cancer cells to anti-tumour agents by Goldie and Coldman (1979).   These
non-plant systems have no equivalent to the large soil-seed reservoir.
We have integrated the factors governing the rates of evolution of
herbicide-resistant weeds, including the effects of the seed bank
(Gressel and Segel, 1978, 1982).   A series of mathematical considerations has culminated in the simple expression:

$$N_n = N_o \ (1 + f \ \alpha/\bar{n})^n$$

where n is the number of years of treatment; N_n is the proportion of
resistants of a given species in the nth year of continued treatment of
a given herbicide; and N_o is the initial frequency in the field prior to
herbicide treatment. N_o itself is a function of the frequency of
natural mutation to the resistant biotype, and the fitness of such a
biotype.   The factor in parentheses governs the rate of increase of
resistance.   It contains the overall fitness (f) of the resistant
compared to the susceptible biotype, which in the known cases of
herbicide resistance is between 0.3 and 0.5.   The selection pressure (α)
is defined as the proportion of the remaining resistants divided by the
proportion of remaining susceptibles.   For example, if no resistants are
killed and 95% of the susceptibles are killed, α = 1/0.05 = 20.
Selection pressure and fitness are divided by $\bar{n}$, the average life-span
of the species in the soil seed-bank.   In weeds that germinate
immediately such as Senecio, $\bar{n}$ equals one year.   With most weed species,
$\bar{n}$ is between two and five years, and an increase in $\bar{n}$ depressesthe rate
at which resistance will increase.

The interrelations are clearer when we use the equation to generate
hypothetical curves from different selection pressures and from an
average seed-bank life-span, with different fitnesses (Fig. 1).   We have
started in year zero with a frequency of 10^{-10} as might occur if
resistance was inherited by a single recessive gene in a diploid weed.

It is possible to move the frequency scale in Figure 1 to fit any other
initial field frequency. From the slopes, it is clear that the
proportion of herbicide-resistant individuals would increase year by
year. The slopes indicate that it will take many years to reach a
frequency of resistant weeds that will be noticeable (i.e. more than the
1-10% that usually remain after a herbicide treatment). Thus, we will
not realise that we are enriching for herbicide resistance until it is
upon us.

Figure 1. Effects of various combinations of selection pressure
(α) (measured as effective kill, EK) and of soil seed-bank
longevity ($\bar{n}$) on the rates of enrichment of herbicide-resistant
individuals over many seasons of repeated treatment. The values
are plotted for fitnesses which would be allowed to develop
after the herbicide becomes degraded. With the persistent
triazine-type herbicide the fitness (f) would be near to unity,
as the fitness differential has no time to become apparent.
With the phenoxy type herbicide, fitness differentials (f =
0.4-0.6) will have time to be influential. Resistance (R) would
become apparent in the field only when more than 30% of the
plants are resistant. The scale on the right indicates the
increase in resistance from any unknown initial frequency of
resistant weeds in the population, whereas the scale on the left
starts from a theoretically expected frequency of a recessive
monogene. (Plotted from equations in Gressel and Segel (1978).)

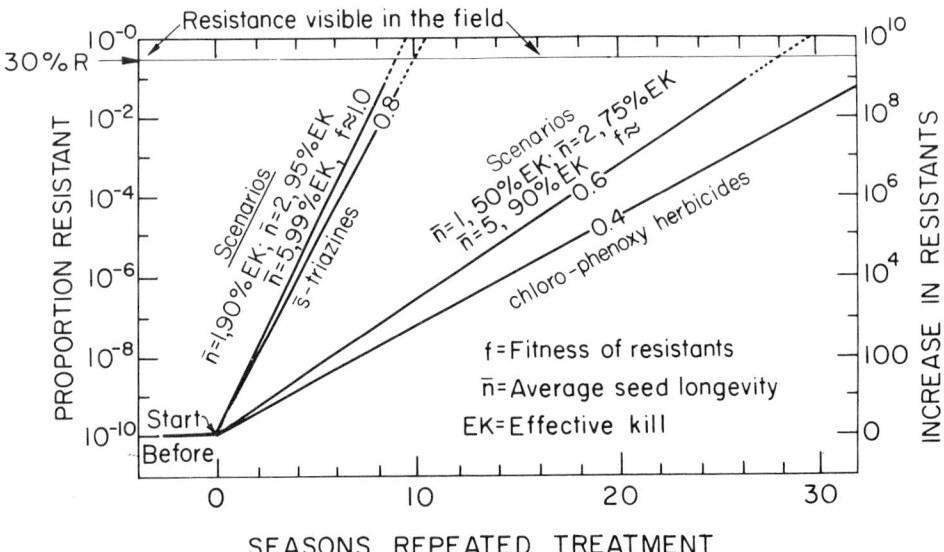

It is unclear how different the effective selection pressures are for
the phenoxy and the triazine groups. One very large scale mid-season
survey of North Dakota wheat fields showed that the best control of
Amaranthus retroflexus, Chenopodium album and Brassica campestris with a
phenoxy herbicide gave 0.2 plants of each weed per m^2 (Dexter et al.,

1981). Considering the plasticity of these species, there would be sufficient seeds to produce a good stand of weeds enriched for sensitivity the following year. I have not found similar data for triazine-treated fields.

Weed and crop ecologies

It has been easy for the farmer growing maize to get excellent and cheap control of most weeds by giving a single massive treatment with atrazine. This required little knowledge of weeds but clearly contributed to the appearance of resistance. The farmer who understands the weeds in his fields as well as the crops can design optimal treatment regimes, usually with considerable savings in herbicide use, a factor which will also delay the appearance of resistance. The reader is referred to reviews on weed ecology (Zimdahl, 1980; Mortimer, 1983) and to an excellent text by Radosevich and Holt (1984).

A concept broadly developed by Zimdahl (1980) and based on studies throughout the world, shows that weeds only compete with crops during certain critical periods. From the time of crop germination until such a period starts, weeds will not reduce yield, and after this period newly-germinated weeds will also not affect yield. Soybean is one of several examples given by Zimdahl (1980); despite some seeming contra-dictions, arising from the collation of data from different climatic areas, it appears that for the first 3 weeks, most weeds do not compete with the soybean crop, and that those appearing after 8-9 weeks are without effect on yield. Thus, the ideal approach would be a post-emergence treatment given 3 weeks after germination and remaining effective for 5-6 weeks. From the point of view of yield, we need only cover this window; such limitations of treatment would lower the selection pressure for resistance on weeds.

Another concept that the farmer must relearn is that there are thres-holds of weed infestations in many crop situations below which yield is not reduced. As pointed out above a minor infestation can greatly reduce the selection pressure of a given herbicide. The farmer also has other considerations beyond yield, and must consider each particular case. Minor contamination by certain weed seeds in crop seed can greatly lower value (e.g. poisonous Brassica campestris in rape seed). There can also be effects on the following crop in rotation if a minor infestation is allowed to develop in this year's crop. Other consider-ations include picker and combine jamming by weeds during harvest, and the status of certain weeds as hosts of crop pests or diseases. The weed-free threshold can vary within a crop depending on why it is grown. The allowable threshold of weed infestation is far lower for seed corn than for maize grown for silage. The farmer needs to know which weeds will have which effects on his crop, and when he may require spot treat-ments. Some weeds may not be tolerated at all after the initial period and may have no allowable threshold. But generally, when all factors are integrated, viable alternatives to the single highly persistent pre-emergence herbicide can be found.

Rights of way – the right way to infest farmers' fields
with resistant weeds
        Most of the triazine-resistant weed populations have evolved
in maize fields and orchards, where most of the triazines are used.
However, simazine was also the herbicide of choice for weed control on
railway beds and roadside verges because of its low cost, very high
persistence and broad spectrum activity. The total areas treated are
small but the selection pressure is very high due to the sterilant
levels of the herbicide being used. Because of this, many previously
innocuous weeds have evolved resistant biotypes, even in a small country
such as Israel (Table 1). Some of these resistant species are
especially worrysome, e.g. Amaranthus blitoides which is an alternate
host to Cuscuta spp. (dodder), a parasite of many vegetable crops.
Dodder is very hard to control, and the transfer of seed to fields
adjoining the railroads is inevitable. The occurrence of triazine-
resistant Kochia scoparia along thousands of miles of railroad tracks in
the USA is probably not due to concurrent evolution, but to trains
moving the seed.

Alternatives are being sought because of the increase in resistance to
triazines along the rights of way. One such alternative suggested is
diuron, mixed with atrazine. Diuron-resistant mutants have been
isolated in algae, where the inheritance has been shown to be maternal,
as in triazine resistance, i.e. the gene is located on the plastid DNA
(cf. Gressel, 1984a). The data from algae show that triazine and diuron
plastid resistance are inherited on different alleles of the same gene;
this conclusion was supported by DNA sequence work (Erickson et al.,
1984 and pers. comm.). Some mutant alleles have been found with a
modicum of both triazine and diuron resistance. The frequency of plants
with mutated recessive plastid genes such as those with triazine or
diuron resistance is exceedingly low within populations. It has been
postulated recently that the frequency of resistant individuals was
increased by a recessive nuclear gene which increased plastome mutations
(Arntzen and Duesing, 1984). Plants bearing this gene were found in
populations of triazine-resistant Amaranthus sp. and Solanum nigrum, and
are probably prevalent within triazine-resistant populations. If one
treats triazine-resistant populations with diuron, the plastome mutator
gene in the population should enhance the rapidity of evolution of
diuron resistance. If it took ten years to get triazine-resistant
populations along roadsides, it may take only a few years to evolve
diuron resistance in the triazine-resistant populations containing high
frequencies of a plastome mutator gene.

Thus, it seems inadvisable to use two herbicides in a mixture which seem
to have the same interrelating modes of action on photosystem II. It
would be far more logical to choose two herbicides with totally different modes of action to delay the appearance of resistance.

EVALUATING STRATEGIES
        Because of the potential problems of resistance, it is well
worth while to evaluate the implications of various agronomic
procedures.

Table 1. Weeds that have evolved triazine resistance along roadsides and railways.

Species	Country	Reference
Kochia scoparia	Mid-west,Northwest USA	Bandeen et al. 1982
Bromus tectorum	" " " " "	" " "
Panicum capillare	Michigan, USA	" " "
Poa annua	France	Ducruet and Gasquez, 1978
Brachypodium distachyon	Israel	Gressel et al. 1983
Phalaris paradoxa	Israel	Yaacoby et al. 1984
Lolium rigidum	Israel	" " "
Alopecurus myosuroides	Israel	" " "
A. ultriculatus	Israel	Nir, 1982
Lophochloa phleoides	Israel	Yaacoby et al. 1984
Polypogon monspeliensis	Israel	Nir, 1982
Amaranthus blitoides	Israel	Kleifild and Gressel, unpub.
Erigeron canadensis	Switzerland	Ammon and Beuret, 1983
Senecio vulgaris	"	" " "

Herbicide rotations
    The mathematical models leading to Figure 1 can be used to predict what happens when a mono-herbicide culture is not used (Gressel and Segel, 1978, 1982). Simply stated, it is the number of weed generations which are treated that affects enrichment. If it took 20 years in mono-herbicide culture to obtain resistance, it would take 40 or 60 years in 1 in 2, or 1 in 3 rotations, respectively. Indeed, all s-triazine resistance confirmed to date has come from mono-herbicide culture. If the theories are true, we shall soon start seeing triazine resistance in parts of the USA corn-belt where maize with atrazine is grown in a 1 in 2 year rotation.

Herbicide mixtures
    The use of herbicide mixtures is becoming more prevalent. These mixtures may be divided into two (often overlapping) types: those where a full rate of each herbicide is used and the weed spectrum killed

by each is mutually exclusive, and those where both control the same
weed spectrum. The first case is easier to analyse within the model;
the model merely has to be applied separately to each weed species. The
use of mixtures may increase the proportion of years over which a given
herbicide is used. This should cause a greater rate of enrichment of
resistance in the weeds that each herbicide controls. Thus, care must
be taken in considering the long term necessity to use such mixtures.

The second situation with overlapping spectra is harder to analyse
through the models. Some complex mathematical models have already been
suggested to help ascertain if components of such mixtures have only
additive effects or have synergistic effects (Morse, 1978; Streibig,
1981).

Figure 2 can be used to help predict effects of mixtures when the
effects are additive. If there is a true synergy, there exists the
possibility that only plants genetically resistant to both herbicides
will survive, which would further delay resistance. If two additive
herbicides are used at their normal rates, the frequency of individuals
resistant to both should be the compounded frequency of each used
separately. If the frequency of resistant individuals to one herbicide
was 10^{-10} and to the other 10^{-6}, then when used together the frequency
of double resistance should become 10^{-16} (if each herbicide has a
different mode of action). This should considerably lengthen the time
to reach resistance; but it may not. If each herbicide has a low
effective kill and together they strongly increase the rate of kill,
they may well cancel out the effect of lower frequency, as can be seen
in Figure 2. If the two herbicides in this example when used separately
are on the curve $\alpha f = 10$, it would take 5.5 years of use of the first
herbicide, and 9 years of use of the other to develop 30% of resistance.
If together they exert a pressure of $\alpha f = 100$, the 30% resistance would
be apparent in 7.5 years because of the strongly increased selection
pressures, and not in $5.5 + 9 = 14.5$ years.

Often, when two components in a mixture are used to kill the same
spectrum of weeds, each is used at a lower dosage. The mixture then
gives the same kill as when each herbicide is used separately. The
lower dosages should effect a much lower selection pressure for each
herbicide, considerably delaying the appearance of resistance. This
possibly could compensate for the effect of using the herbicide more
often in a rotation.

Herbicide extenders
Recently, compounds have been developed to enhance the
action of herbicides. They all increase selection pressure of the
herbicide whether they are used, (a) to decrease the rate of soil
degradation of the herbicides, as with compounds being developed for use
with EPTC, or (b) to inhibit glutathione-s-transferase (GST) activity,
as with tridiphane and atrazine. The latter case has many implications
vis-a-vis triazine resistance. Rates of atrazine as low as 30 g ha^{-1}
are sufficient to kill the sensitive biotypes of the dicotyledonous

plants that have become resistant. Rates as high as 300 g ha^{-1} are
needed to kill the Poaceous grasses, which have higher levels of
triazine-degrading GST. The use of an extender to decrease this
degradation by GST will strongly enhance selection pressure on these
grasses and more resistance to them will appear. The effects would then
approach those of the sterilant levels of triazines used on roadsides,
mentioned earlier. Conversely, lowering the levels of triazines used
with extender should give about the same selection pressure now being
exerted, with existing rates of build-up of resistance.

Figure 2.   The effects of varying selection pressure and fitness
on the years to apparent resistance with $\bar{n}=1$ (no seed bank).
This method of plotting is especially useful in evaluating
levels of herbicides used in herbicide mixtures with the same
spectrum of weed control.   Initial frequencies for various types
of presumed mutation frequencies are given, although there is
some debate as to whether frequencies of diploid recessive are
actually as low as has been calculated (Williams, 1976).
(Plotted from equations in Gressel and Segel, 1978, reprinted
with permission from Gressel and Segel, 1982.)

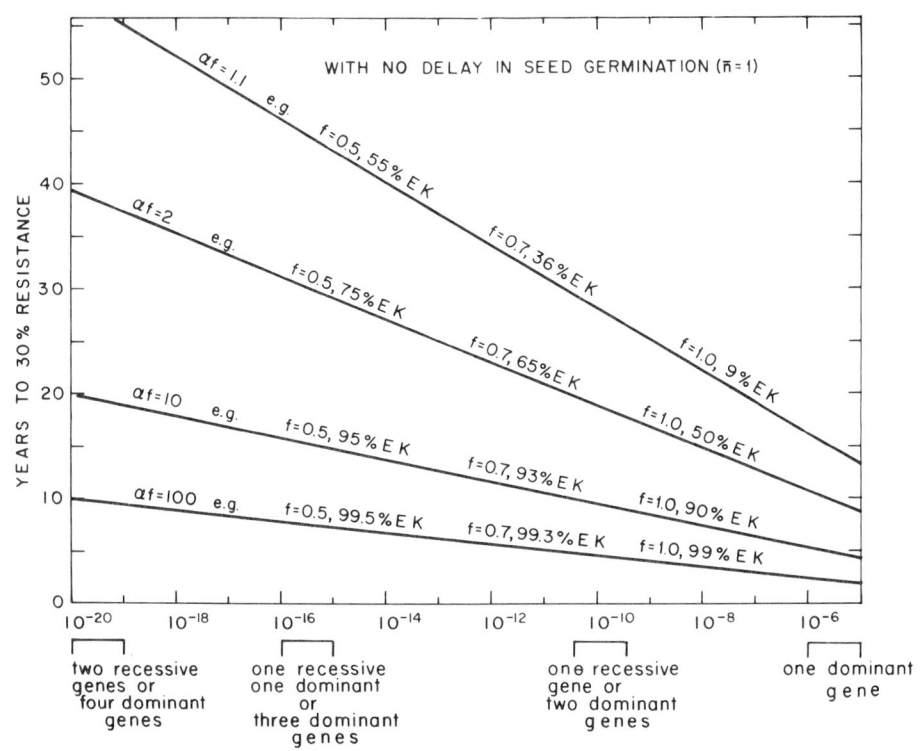

Herbicide protectants

There is an increasing interest in compounds that protect crops from the action of herbicides that would otherwise be toxic to them. These compounds have been termed "safeners", "protectants" or, with semantic accuracy, "antidotes" (cf. Hatzios, 1983). They are used to treat crops before, or along with, the herbicide application. Two such compounds, a diallyldichloroacetamide, used in combination with EPTC, and cyometrinile, applied to sorghum seed to protect it from chloroacetamide herbicide injury, are already marketed, and others are at various stages of development. These compounds will broaden the utilisation of some herbicides with expected implications; there will be enrichment for resistance to those particular herbicides if they are used more frequently or in more crops. If the herbicides are highly persistent or are made more persistent by using extenders and are already used with other crops in the rotation, their long-term useful-ness may be jeopardised. If the herbicides concerned have a low effect-ive selection pressure, as when protectants are used against paraquat and diquat injury (Lewinsohn and Gressel, 1984), they will pose a far less serious problem, especially if they are to replace highly persistent herbicides.

'Engineering' herbicide-resistant crops

Efforts are being made to confer resistance to various herbicides by various novel cell culture and genetic engineering techniques (cf. Gressel et al., 1983; Gressel, 1984b). This will allow the use of a new selective herbicide for that crop. In all cases, enrichment for resistance to that herbicide may well occur. If the herbicide is not already in the rotation used in those fields, the effect may be inconsequential for a very long time. The suggestion has been made to engineer triazine-resistant soybeans (cf. Marx, 1983). This would lead to continuous use of atrazine in maize/soybean rotations. The USA corn belt is almost totally free of triazine resistance, because herbicide mixtures and rotations were used in maize. Clearly, the planting of triazine-resistant soybeans would change that picture.

Germination stimulators

Synchronous germination of weed seeds by chemical stimulators has been proposed as a sort of "shaving cream" strategy: "get them up so we can mow them down". If the mowing is done mechanic-ally or by frost, and a fixed proportion of seed of all ages germinates, this strategy should have little effect on the rate of appearance of herbicide-resistant weeds. If resistants and susceptibles are equally stimulated, there would be no change in fitness and thus no effect on the population ratios. If fitness is decreased because resistants suffer relatively more from frost, f is smaller and resistance would be delayed. If the stimulant strongly induces the germination of older weed seed in the soil, the proportion of herbicide-resistant seed germinating could be even greater than that introduced giving a situation akin to $\bar{n} = 1$, which is certainly not desirable.

It could be counter-productive if the stimulated weeds are killed with a herbicide as part of the normal herbicide rotation. The germination induced will strongly increase the effective selection pressure of rapidly degraded herbicides, as more seed will germinate when the herbicide is still active. This will advance the time at which resistance appears. In the terms of the model, a totally effective "shaving cream" strategy used with herbicides means that the average span of existence in the seed bank will become one year, with all the implications of achieving n = 1 seen in Figure 1.

The use of persistent herbicides and slow-release formulations

The "positive" importance of high selection pressure has been heavily emphasised by some weed control specialists. It has been stated that: "herbicide applications aimed at control but not elimin-ation are especially dangerous" (Harper, 1957). There has never been total elimination of a pest with pesticides, except in a miniscule number of isolated or localised infestations. It must thus be decided which level of selection pressure (i.e. control) is desirable. Practitioners and theoreticians of pesticide and antibiotics' use in agriculture and medicine have concluded that to prevent resistance and help the patient or crop, it is best to control the pest just to the level where the desired individuals can naturally compete.

There can be two kinds of persistence; that of atrazine which retains long soil activity, and that conferred by farmers through frequency of use. The latter case is the cause of the appearance of resistance to paraquat, a herbicide with no residual activity. The farmers sprayed at monthly intervals until resistance was achieved (cf. Le Baron and Gressel, 1982). Close to total kill has given us totally diminished returns in too many cases to date. The same caution must apply to efforts to achieve slow-release formulations of less persistent herbicides: they will tend to hasten the onset of resistance. From the crop/weed ecological considerations outlined earlier, we have seen that it is possible to achieve sufficient weed control without persistence.

CONCLUDING REMARKS

It is clear that any treatment that will decrease the extent and duration of selection pressure will be a useful strategy for delay-ing the appearance of resistant biotypes. The farmer does not need fantastic control, just adequate control. Such strategies will keep the most cost-effective herbicides such as the triazines in our arsenal longer. The same can be said for speciality herbicides such as gly-phosate. When glyphosate-resistant crops are available, and the price of glyphosate drops, much more will be used. If Man has selected for resistance, Nature soon will. One must stall Nature as much as possible.

ACKNOWLEDGEMENTS

Great indebtedness is due to Professor L.G. Holm who first

whetted my interest in this subject and to the fruitful collaborations
with Dr H.M. Le Baron which greatly enhanced my knowledge about resistance, and with Professor L.A. Segel who helped develop the mathematical
aspects of the subject. The author is the Gilbert de Botton Professor
of Plant Sciences.

REFERENCES
Abel, A.L. (1954). The rotation of weedkillers. Proceedings of the 1st
    British Weed Council Conference, 249–255.
Aikman, D.P. and Watkinson, A.R. (1980). A model for growth and self-
    thinning in even-aged monocultures of plants. Annals of
    Botany 45, 419–427.
Ammon, H.U. and Beuret, E. (1984). Verbreitung triazin-resistenter
    unkrauter in der schweiz und besherige bekampfungser-
    fahrungen. Zeitschrift für Pflanzenkrankheiten Pflanzen-
    pathologie und Pflanzenschutz 10, 183–191.
Ammon, H.U. and Irla, E. (1984). Bekamfung resistenter unkrauter in
    maiserfahrungen mit mechanischen und chemischen verfahren.
    Die Grune (Schweiz. Lansw. Z.) 112(12), 18–26.
Arntzen, C.R. and Duesing, J.H. (1983). Chloroplast-encoded herbicide
    resistance. In Advances in Gene Technology. Molecular
    Genetics of Plants and Animals, eds K. Downey, R.W. Voellmy,
    F. Ahmand and L. Schultz, 273–293. New York: Academic
    Press.
Bandeen, J.D., Stephenson, G.R. and Cowett. E.R. (1982). Discovery and
    distribution of herbicide resistant weeds in North America.
    In Herbicide Resistance in Plants, eds H.M. Le Baron and
    J. Gressel, 9–27. New York: Wiley-Interscience.
Delp, C.J. (1981). Resistance to plant disease control agents – how to
    cope with it. Proceedings of the 9th International Congress
    on Plant Protection, ed. T. Konnendahl, 1, 253–161.
    Minneapolis: Burgess.
Dexter, A.G., Nalewaja, J.D., Rasmusson, D.D. and Buchli, J. (1982).
    Survey of wild oats and other weeds in North Dakota: 1978
    and 1979. North Dakota State Extension Service Research
    Report No. 79 Fargo (North Dakota).
Ducruet, J.M. and Gasquez, J. (1978). Observation of whole leaf
    fluorescence and demonstration of chloroplastic resistance
    to atrazine in Chenopodium album L. and Poa annua L.
    Chemosphere 8, 691–696.
Erickson, J.M., Rahire, M., Bennoun, P., Delepelaire, P., Diner, B. and
    Rochaix, J.-D. (1984). Herbicide resistance in
    Chlamydomonas reinhardtii results from a mutation in the
    chloroplast gene for the 32kD protein of photosystem II.
    Proceedings of the National Academy of Sciences, USA 81,
    3617–3621.
Georghiou, G.P. and Taylor, C.E. (1977). Operational influences in the
    evolution of insecticide resistance. Journal of Economic
    Entomology 70, 653–658.
Goldie, J.H. and Coldman, A.J. (1979). A mathematical model for
    relating drug sensitivity of tumors to their spontaneous
    mutation rate. Cancer Treatment Reports 63, 1727–1733.

Gressel, J. (1983). Spread and action of herbicide tolerances and uses
         in crop breeding.. In Proceedings of the 10th International
         Congress of Plant Protection, 608-615. Croydon: The
         British Crop Protection Council Publications.
Gressel, J. (1984a). Herbicide tolerance and resistance: Alteration of
         site of activity. In Weed Physiology, ed. S.O. Duke, Vol.
         2, 159-189. Boca Raton, Florida: CRC Press.
Gressel, J. (1984b). Biotechnologically conferring herbicide resistance
         in crops: The present realities. In Molecular Form and
         Function, ed. L. van Vloten-Doting. New York: Plenum Press.
Gressel, J., Ezra, G. and Jain, S.M. (1983). Genetic and chemical
         manipulation of crops to confer tolerance to chemicals. In
         Chemical Manipulation of Crop Growth and Development, ed.
         J.S. McLaren, 79-91. London: Butterworths.
Gressel, J., Regev, Y., Malkin, S. and Kleifeld, Y. (1983).
         Characterization of an s-triazine resistant biotype of
         Brachypodium distachyon. Weed Science 31, 450-456.
Gressel, J. and Segel, L.A. (1978). The paucity of plants evolving
         genetic resistance to herbicides; possible reasons and
         implications. Journal of Theoretical Biology 75, 349-371.
Gressel, J. and Segel, L.A. (1982). Interrelating factors controlling
         the rate of appearance of resistance. The outlook for the
         future. In Herbicide Resistance in Plants, eds H.M. Le
         Baron and J. Gressel, 325-347. New York: Wiley-Interscience.
Harper, J.L. (1956). The evolution of weeds in relation to resistance
         to herbicides. Proceedings of the 3rd British Weed Control
         Conference 1, 179-188.
Harper, J.L. (1957). Ecological aspects of weed control. Outlook on
         Agriculture 1, 197-205.
Hatzios, K.K. (1983). Herbicide antidotes: development, chemistry and
         mode of action. Advances in Agronomy, 36, 265-316.
Lewinsohn, E. and Gressel, J. (1984). Benzyl viologen mediated counter-
         action of diquat and paraquat phytotoxicities. Plant
         Physiology 76, 125.
Le Baron, H.M. and Gressel, J. (eds) (1982). Herbicide Resistance in
         Plants. New York: Wiley-Interscience.
Marx, J.L. (1983). Plants' resistance to herbicide pinpointed. Science
         220, 41-42.
Morse, P.M. (1978). Some comments on the assessment of joint action in
         herbicide mixture. Weed Science, 26, 58-71.
Mortimer, A.M. (1983). On weed demography. In Recent Advances in Weed
         Research, ed. W.W. Fletcher, 3-40. Slough, UK: Commonwealth
         Agricultural Bureau.
Murphy, K.J. (1983). Herbicide resistance in weeds: a growing problem.
         Biologist, 30, 211-219.
Nir, A. (1982). Is only Brachypodium distachyon resistant to triazines?
         Alei Asev (Newsletter Israel Weed Sci. Soc.) (in Hebrew) 8
         Dec., 2-4.
Parochetti, J.V. (1979). Herbicide resistance found in some weeds.
         Crop Soils Magazine (July), 9-10.
Radosevich, S.R. and Holt, J.S. (1984). Weed Ecology. New York: Wiley-
         Interscience.

Streibig, J.C. (1981). A method for determining the biological effect
        of herbicide mixtures. Weed Science 29, 469-473.
Williams, K.L. (1976). Mutation frequency at a recessive locus in
        haploid and diploid strains of a slime mould. Nature 260,
        785-787.
Yaacoby, T. Schonfield, M. and Rubin, B. (1984). Atrazine resistance
        developed in several grass weeds following repeated
        applications of roadsides in Israel. Weed Science Society
        of America Annual Meeting, Abstract 265.
Zimdahl, R.L. (1980). Weed-Crop Competition: A Review, pp. 195.
        International Plant Protection Center, Oregon State
        University, Corvallis, Oregon.

15. PESTICIDE RESISTANCE: STRATEGIES AND CO-OPERATION IN THE AGROCHEMICAL INDUSTRY

C.N.E. Ruscoe
ICI Plant Protection Division, Jealotts' Hill Research
Station, Bracknell, Berkshire, RG12 6EY, England

INTRODUCTION
The extent of the phenomenon of resistance by pests
(specifically insects, mites and disease pathogens) to chemical crop
protection agents, is frequently reviewed and highlighted. It is
therefore enough to say here that the problems are sufficiently common
and important to provide a major threat to the quantity and quality of
world agricultural production. The situation is undoubtedly such that
major and concerted efforts need to be made, by all parties, to contain
resistance problems within manageable bounds.

In this context, "all parties" means the end user, adviser, research
scientist and of course the pesticide industry, incorporating as it does
the different components of research and development, and central and
local marketing and sales, of each company.

This paper deals with the role of the agrochemical industry in combating
resistance, and the problems associated with this role, given the often
different objectives not only of different companies, but also of the
different components of any one company.

COMMERCIAL IMPLICATIONS OF PESTICIDE RESISTANCE
Benefits
A static, unevolving market is unattractive to any industry;
once market sectors have been satisfied with a product from one's own
company, or worse from a competitor, it becomes increasingly difficult
to obtain new market shares without major technical improvements or
price reductions.

The agrochemical market, based as it is on maintaining disequilibria in
dynamic biological systems, is never likely to fit this model, since any
particular chemical force applied will sooner or later result in an
effective biological counter-reaction. Pesticide resistance is one such
reaction. Historically, this has provided a rejuvenation of business in
a number of agrochemical markets - particularly insecticides and
acaricides - since the products which have suffered have generally been
commodities of low profitability, and their successors newer and of
higher business value.

In the early years of the agrochemical industry - the 1940s to the 1960s - resistance could therefore be regarded as of overall benefit to the industry - if not to the grower!

Drawbacks
During the 1970s, however, it became increasingly clear that major change was occurring within the industry. A number of forces, particularly the costs of energy, chemical feedstock based on oil, and labour, and the increasing demands and time-scales of registration processes, were all combining to escalate the costs of pesticide development. In the mid-1970s, the average cost of developing a new product to the market was quoted as $ (US) 10 million (Braunholtz, 1977), and this cost has now doubled. This was in addition to the already major and increasing costs of capital investment for pesticide production, and marketing costs.

These factors, coupled with an increasing incidence of occurrences of laboratory or field resistance (acquired resistance in 305 species of insects and mites were documented in 1975, compared with 185 in 1965, and around 50 in 1955, Georghiou and Taylor, 1977), in increasingly abbreviated time-scales, were significant influences on the industry view. The overall drawbacks of resistance were now outweighing the usually more transient benefits.

ANTI-RESISTANCE ACTIVITIES IN THE AGROCHEMICAL INDUSTRY
Pre-1970s
Up to the early 1970s, reactions to resistance had, in the main, been essentially tactical; on the appearance of resistance, the immediate reaction of growers, abetted by the industry, was to switch to alternative products, which were normally available, albeit at higher cost! Extreme examples of the "pest treadmill" this created are to be found in the use and disuse patterns of insecticides in intensive, continuous, treatment situations, such as against diamond-back moth (Plutella xylostella) in brassicae in Malaysia (Fig. 1), and houseflies (Musca domestica) on Danish animal farms (Fig. 2).

There was relatively little attempt, other than in a few enlightened academic circles, to understand and so manage the problem by more circumspect use of the compounds under threat, or of the compounds being considered as replacements.

1970s
The 1970s, as well as seeing marked changes in the costs of R and D, registration, production and marketing of agrochemicals, saw a very marked increase in cases of resistance. In particular, there was a rapid development of problems associated with relatively short periods of use of a number of compounds both in broad-acre and high-value crops.

Prominent amongst these were a number of modern fungicides, such as benomyl on vegetables and fruit, kasugamycin on rice, and pyrimidines on cereals and vegetables, in addition to the slower, often less

Figure 1. Use and disuse of insecticides against diamond-back moth (_Plutella_ _xylostella_) on brassicae in Malaysia (Cameron Highlands).

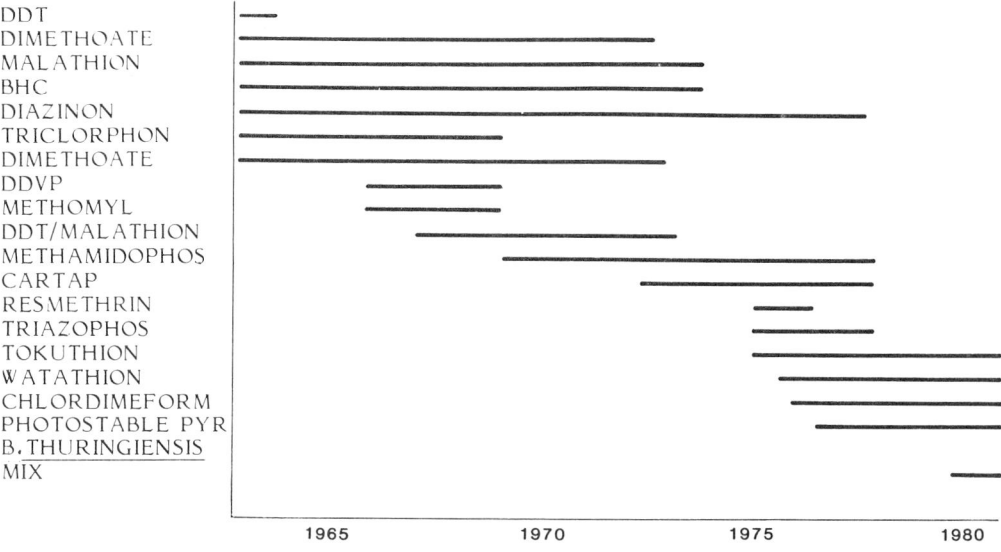

Figure 2. Use and disuse of insecticides as a result of housefly (_Musca_ _domestica_) resistance in Danish animal farms.

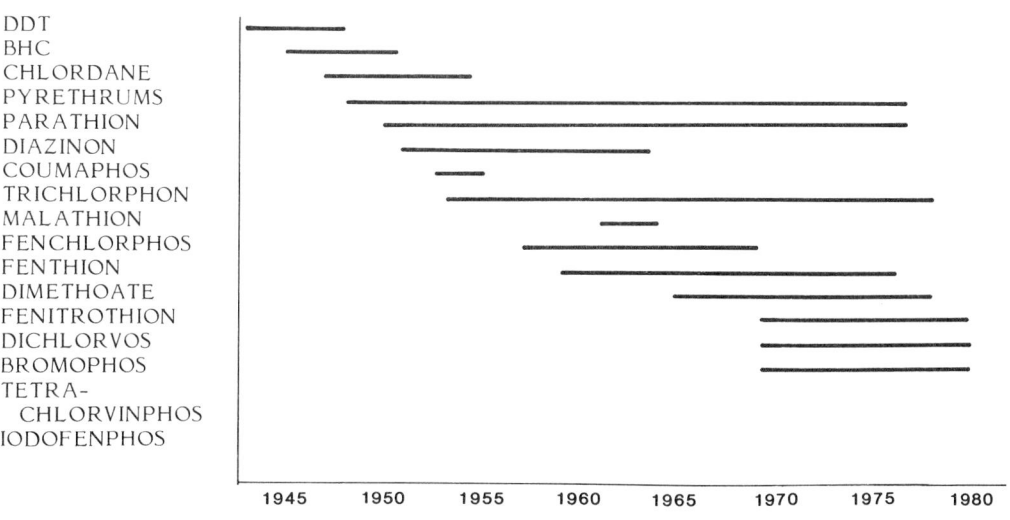

perceptible, but still significant decline of other products, such as the earlier organophosphorus insecticides on a variety of pests, but most significantly on cotton Lepidoptera.

The first resistance problems with benomyl came in the second season of widespread commercial use, and those with dimethirimol in the first season, whilst use of ethirimol on cereals provoked sensitivity decreases in the third season of use. Against the time-scale required for a new pesticide to pay back its R and D and early marketing costs - 10 years from discovery in the 1970s, (Braunholtz, 1977) and longer now - such events were highly significant.

Together with the increasing incidence of acaricide and insecticide resistance, they stimulated a different response from the industry in the 1970s, which in any event had become more mature, responsible and scientifically based than in its early years.

Significant company research and development resources were allocated to the fungicide resistance problems, and joint projects with appropriate government advisory bodies and academic institutions were evolved and sponsored. This work was aimed at determining the nature and magnitude of the problems and so at establishing practical use strategies to contain them.

In aiming to minimise further build-up of resistance to fungicides, while still retaining maximum benefit, the resultant strategies inevitably involved reduced usage of the compounds at risk, and implementation of more complex treatment schedules. Examples were the withdrawal by ICI of its recommendation for the use of ethirimol on winter barley, to provide a longer break in the annual cycle of usage (Shephard et al., 1975), and the recommendation of Du Pont that benomyl treatment should be combined with the use of non-specific protectant fungicides (Delp, 1979). Even with the support of advisory services, there was, not unexpectedly, unwillingness at the grower level to implement certain elements of these strategies, and so extensive educational publicity efforts were made to promote them.

Clearly, these activities exemplified a considerable change in the company's marketing efforts with the compounds at risk, involving publicising the problems and reduction of use strategies, in an effort to prolong product life at the expense of short term sales.

CO-OPERATIVE ANTI-RESISTANCE STRATEGIES

It is therefore possible for a single company to promote use of a product in a way likely to extend its useful life. However, considerable problems arise when it becomes necessary for a number of companies, each with different but related products, and each with its own technical and commercial risk:benefit assessments, to make com- patible, supportive recommendations for anti-resistance strategies. This is particularly true when these recommendations have to be promul- gated in an environment in which the companies are otherwise in direct sales competition.

The situation is made even more difficult when national legislation prevents any collaboration between companies which results in a restriction of freedom of choice of registered product by the user, as is the case in the USA.

Yet, despite these difficulties, the late 1970s saw the start of real attempts by the industry to face up to this situation in a responsible and concerted manner, not only to counteract resistance problems as they arose, but also to anticipate and prevent future problems.

The first initiative is this area came on the insecticide front with the setting up of the Pyrethroid Efficacy Group.

The Pyrethroid Efficacy Group (PEG)

In November 1979, ICI made an approach to the nine companies which were actively involved in the development and marketing of the new photostable pyrethroid insecticides, and hence particularly influential in their usage.

These companies were:

FMC
ICI Americas
Imperial Chemical Industries
Mitchell Cotts
Roussel Uclaf
Shell Development
Shell Research
Sumitomo
Wellcome Foundation
(American Cyanamid was included in 1984)

ICI suggested that representatives of these companies should meet to consider the possibility of setting up technical liaison on the subject of pyrethroid resistance, based on the following premises:

1. The new pyrethroids were proving to be particularly valuable to the grower in view of their effectiveness, to registration authorities because of their safety and low environmental impact, and of course to the industry itself. It was therefore in the common interest to cause their effective life to be as prolonged as possible. As with other insecticides, development of resistance was an obvious threat to that life.

2. Existing knowledge of insecticide (particularly pyrethrin and pyrethroid) cross-resistance patterns suggested that all the compounds were likely to be affected to broadly the same extent by resistance, and certainly by rumours of it.

3. A joint industry approach to aspects of the subject was likely to prove better informed, and more effective, than the divergent recommendations and unconcerted action that might otherwise arise. The most important aspects which might benefit from pooling of information and a concerted approach were:

a. Monitoring of resistance development - selection of the right pests, compounds and countries, the harmonisation of techniques, and the interpretation and publication of results.

b. Joint industry approaches to research workers, advisers and registration authorities, to assist and influence work being done, and pronouncements made, on the subject of resistance.

c. Recommendations of strategies for use which might be considered to be of benefit in delaying onset of resistance.

d. Countering unsubstantiated rumours of resistance, of "reports" based on inappropriate work, which might otherwise damage registration or product image or life.

At a meeting of representatives of the companies in December 1979, it was agreed that there was a strong basis for such a joint industry approach, while recognising that such a forum would need to face up to some severe problems if it was to be effective. The most significant of these problems were identified as follows:

1. Freedom to share information might be limited if a company's technical knowledge of a particular aspect of resistance development was seen by that company to give it a commercial advantage over the others.

2. Agreed technical responses to actual or potential resistance problems might not prove commercially acceptable to one or all of the participants, especially given their different business positions and objectives.

3. Strategies for use of the pyrethroids, which might be recommended to delay resistance development, might not be in the shorter-term interest of distributors and especially growers, and therefore might prove difficult to implement, even if agreed by the companies concerned.

Further progress

Since November 1979, PEG has continued to meet regularly. Experience has shown that the problems identified are surmountable, and the following is a summary of the aspects of the Group and its operation which have been most important in this:

1. Benefits of shared knowledge. It is evident to the
Group that knowledge about resistance is most effective if
employed with reference to the pyrethroids as a whole. Data
on development of resistance to just one of the compounds
for example, are better displayed, so that it is investigated and then acted on in a manner which will cause
minimal damage to the market for that compound and so to the
pyrethroids as a whole.

2. Mismatch of technical and commercial requirements. The
acceptance in commercial practice of technical policies
generated within the Group is made easier if the representatives in the Group are technical, but are sufficiently
senior, and therefore have considerable credibility with
their colleagues. The members of the Group may need the ear
and understanding of their Company Boards in cases where
their recommendations - which are likely to be related to
the longer-term benefit of the business as a whole - are at
variance with shorter term or geographically limited company
market objectives. Such credibility is highly important,
since recommendations for insecticide use with the objective
of delaying resistance will almost certainly be based on
incomplete data and a number of expert assumptions, given
the limited knowledge of this complex subject.

3. Implementation of recommendations. It is recognised
that even if the difficulties of acceptance of strategies at
the company commercial centres are overcome, problems exist
in getting these strategies implemented by distributors,
advisers and growers. The key here lies in the need for
advisers and, in extreme cases, Government registration
authorities, to assist and even legislate to implement
agreed resistance-delaying strategies. Thus, if the Group
believed that a particular rotation pattern, or restriction
of use of the compounds to certain crops and parts of the
year, is likely to delay resistance, it would invariably
need to work with the appropriate Government agencies to
agree and spearhead the implementation of those
recommendations.

It is critical that such authorities be fully involved at an
early stage for another important reason. Anti-resistance
strategies, which are likely to involve some direction and
restriction of use, can be interpreted as contrary to the
principles of free competition, even if all the companies
were to agree to the restriction. It is therefore important
that such strategies receive the backing of Government
advisers - and of course this is imperative where control of
use by restriction of registration is sought.

Even given support by the company commercial centres, it is
recognised that gaining acceptance and implementation by

their regional marketing units is often a difficult step,
charged as these units are with local sales and profit
targets. It was therefore seen to be critical to conduct
internal as well as external promotion of anti-resistance
strategies following agreement between companies at their
technical and commercial centres.

In 1984, PEG became part of a broader inter-company organisation, the
Insecticide Resistance Action Committee (IRAC). This has objectives
similar to those of FRAC (see below), and has set up working groups to
address problems of insecticide resistance in major crops such as
cotton, rice, fruit, field crops and vegetables, and in public health.

The Fungicide Resistance Action Committee (FRAC)
Following discussions initiated in August 1980 within the
agrochemical industry, FRAC was formed in late 1981, with the general
objective of prolonging the effectiveness of those fungicides likely to
encounter resistance problems (Delp, 1984). To do this, FRAC
established Working Groups for each of the acylalanine, benzimidazole,
dicarboximide and sterol inhibitor types (Table 1). Each group thus had
a sphere of operation, guidelines and working methods - and indeed
problems - very similar to those detailed for PEG.

Table 1. FRAC Working Groups and member companies.

Acylalanines	Chevron Ciba-Geigy Farmoplant Schering
Benzimidazoles	BASF DuPont Hoechst Merck Penwalt
Dicarboximides	BASF Bayer Hoechst Rhone-Poulenc Speiss and Sohn
Sterol inhibitors	Geographical sub-committees

EFFECTIVENESS OF CO-OPERATIVE INDUSTRY ANTI-RESISTANCE
ORGANISATIONS

The formation of groups such as FRAC, PEG and IRAC is a
critical step towards the development of anti-resistance strategies.
However, given the considerable problems – commercial and sometimes
legislatory – besetting the implementation of strategies formulated and
promoted by the Groups, it is important to examine their record to see
if they are or are likely to be effective instruments for promoting the
use of anti-resistance strategies.

FRAC

Although in some cases initiated before the formalisation of
FRAC, the following co-operative initiatives have been taken and
expanded:

1. Acylalanines. Joint resistance monitoring programmes,
and the marketing of the compounds only as preformulated
mixtures with residual preventative fungicides, having a
different mode of action.

2. Benzimidazoles. Joint monitoring programmes for eyespot
resistance.

3. Dicarboximides. Industry support for use of a maximum
of two sprays/season on grape-vines in Germany and France.

4. Sterol inhibitors. Monitoring for the incidence of
resistance problems, and increasing development of mixtures
with alternative mode of action compounds.

PEG

1. Egypt/Spodoptera. United company support for Egyptian
authority initiatives to prevent the use of pyrethroids on
non-cotton crops, thus stopping the use of pyrethroids on
Spodoptera throughout the year. This was a proactive
attempt to prevent the development of resistance, by
restriction of use, prior to its appearance.

2. UK/houseflies. Inter-company agreement, together with
MAFF, to advise against the use of pyrethroids as residual
treatments in intensive animal rearing houses since in these
situations there were strong signs of development of
housefly resistance to this method of use.

3. Australia/Heliothis armigera. This case deserves some
detailed mention, as an example of good industry, Government
and grower co-operation in a difficult situation. At the
end of the 1982/3 season, control failures with synthetic
pyrethroids against Heliothis armigera (the cotton bollworm)
were experienced in a small number of isolated situations in
Australia. Local monitoring revealed significant resistance
levels in these locations, and signs of development of

resistance in others. Accordingly, meetings of the
companies involved were held centrally and locally, and an
agreed strategy for implementation in the 1983/84 season
worked out. The main recommendations were:

    a. Total ban of pyrethroid use in the main area
experiencing resistance problems (Emerald, Queensland).

    b. Restriction of pyrethroid use on cotton in other
areas to the middle part of the season (10 January - 20
February), a 42 day period corresponding to one average
Heliothis life-cycle, allowing a maximum of 3 pyrethroid
sprays out of a possible total of 10-12.

    c. During this period, rotation or mixture of
pyrethroid sprays with other compounds.

These recommendations were backed with extensive publicity
and label warnings. Grower compliance was as a result very
high, and no commercial failings were directly attributed to
pyrethroid use in the 1983/84 season.

It is clear from these cases that both PEG and FRAC have become
influential, in a co-operative and effective manner, on the use of their
pesticides in anti-resistance strategies.

CONCLUSIONS: THE FUTURE
    The setting up of groups such as PEG, FRAC and IRAC has
meant that a mechanism is now in place for inter-industry discussion,
ratification and action on resistance problems as, or even before, they
emerge. The existence of such a mechanism means that any problems can
be dealt with relatively rapidly, by the groups or appropriate local
sub-groups, along the following lines:

    1. Highlighting of the problem (often local).

    2. Review of facts (local and international), using the
assistance of advisory services, academics and others.

    3. Generation of recommendations acceptable to all
companies at a technical level.

    4. "Selling" of this recommendation to company commercial
centres.

    5. "Selling" by company commercial centres to local
commercial units.

    6. Co-operative local action - restrictive labelling,
publicity, legislation.

7. Grower implementation.

Problems of course still beset the successful processing of any one
problem through such a system. Facts are sure to be scarce, and
informed technical opinion will often be the basis for making robust -
and often commercially hurtful in the short-term - resistance avoidance
recommendations. Acceptance by company commercial centres will
therefore often be difficult to obtain, and there will probably always
be need for early confirmed resistance reports to trigger the
implementation of as strategy. However, the early generation of a
strategy by anticipatory co-operation between the industry groups, and
others involved, will mean it can still be impemented in a timely and
effective manner. Such strategies are being developed, for example, by
the PEG USA sub-group with key academics and advisers, in order to be
prepared for any onset of pyrethroid resistance by <u>Heliothis</u> in cotton
in the USA.

It is in this area that the co-operation and help of academics and
advisory services is vital; in the past there has been too much
emphasis, especially in academia, on identifying isolated, possibly
atypical, instances of resistance, and the creation of resistant pests
and pathogens by laboratory pressuring. There has been insufficient
emphasis on the much more difficult, but essential, activity of overall
resistance level monitoring, and research into practical field tecniques
for resistance containment.

The assistance of advisory services, and if appropriate governmental
backing, will always be vital to the success of strategies based on use
restrictions, and here the support of organisations such as Groupement
International des Associations Nationales de Fabricants de Produits
Agrochimiques (GIFAP) and National Agricultural Chemicals Association
(NACA) to groups such as IRAC and FRAC is important. In the USA in
particular, the attitudes of the Environmental Protection Agency to
restriction of products use to extend their commercial life in the face
of resistance needs to be changed.

Even given wholehearted industry and governmental support and publicity,
implementation of any strategy will ultimately be in the hands of the
grower. It is clear however, that with appropriate co-operative action
and explanation, a majority of growers in risk situations can be
influenced to adopt sensible strategies, and this can generally be
adequate to contain a developing resistance situation. The excellent
co-operation of the Japanese industry, prefectural advisory stations and
growers to implement anti-resistance strategies, with rice fungicides
and fruit acaricides, is an example.

Overall, therefore, I am confident that an effective start has been made
by the industry to develop technical co-operation and strategies to
overcome or prevent pesticide resistance, and that this must provide a
major contribution to the more rational use of pesticides.

REFERENCES

Braunholtz, J.T. (1977). Pesticide development and the chemical manufacturer. Proceedings of the XVth International Congress of Entomology, Washington DC 1976, 747-755.

Delp, C.J. (1979). Resistance to plant disease control agents - How to cope with it. Proceedings of the IXth International Congress of Plant Protection, 253-261.

Delp, C.J. (1984). Industry's response to fungicide resistance. Crop Protection 3, 3-8.

Georghiou, G.P. and Taylor, C.E. (1977). Pesticide resistance as an evolutionary phenomenon. Proceedings of the XVth International Congress of Entomology, Washington DC 1976, 759-785.

Keiding, J. (1974). The development of resistance to pyrethroids in field populations of Danish houseflies. Proceedings of the 3rd International Congress on Pesticide Chemicals, Helsinki.

Shephard, M.C., Bent, K.J., Woolner, M. and Cole, A.M. (1985). Sensitivity to ethirimol of powdery mildew from UK barley crops. Proceedings of the 1984 British Crop Protection Conference - Pests and Diseases, 59-65. British Crop Protection Council Publications.

Sudderukkin, K.I. and Kok, P.F. (1978). Insecticide resistance in Plutella xylostella collected from the Cameron Highlands of Malaysia. FAO Plant Protection Bulletin, 26, 53-57.

16. DECISION THEORY AND THE ECONOMICS OF CROP PROTECTION
MEASURES

J.P.G. Webster
School of Rural Economics, Wye College, University of
London, Wye, Ashford, Kent, TN25 5AH, England

INTRODUCTION
        Rationality in decision-making is a topic which has
intrigued many groups of thinkers over the ages. Individual decisions
are at the heart of the study of economics and so, numerous economists
have given thought to the whys and wherefores of decisions. Crop
protection decisions are largely taken by farmers and growers in their
interests of efficient production but other factors intrude. The aim of
this paper is to investigate the characteristics of crop protection
decisions and to see what contributions theory might make towards their
solution. The first section sets the scene by looking at the reasons
behind the concern of crop protection scientists with farmer decision-
making. The second section categorises the characteristics of these
decisions, whilst the body of the paper deals with the ways in which
these characteristics may be more or less tackled by adopting some
decision-theoretic stance or other. It will be seen that the multifarious nature of the consequences of pesticide decisions leads us into
many areas of theory, which have often yet to see applications in
agriculture.

INTEREST IN DECISION-MAKING BY RESEARCHERS IN PEST
MANAGEMENT
        As applied scientists, researchers in pest management have
naturally been interested in seeing the application of their ideas at
the farm level. Indeed as resources for research become scarce, the
pressure to demonstrate impact at the farm level increases. When future
lines of research are being planned, interest turns to the adoption by
farmers of the output of the research. In particular, it appears that
three issues make themselves apparent to the concerned scientist.

Firstly, he is aware that the use of his research results by farmers
implies not simply automatic application of whatever technique is being
proposed, but the processing of the knowledge generated, and application
in individual situtions by individual farmers. Farmers have to trans-
late impacts demonstrated on the research station or experimental farm
into impacts within the perhaps very different environment of their own
farms (Davidson and Martin, 1965). In addition, our researcher is aware
that farmers act in their own perceived interests, which may or may not
coincide with those farmer's interests as perceived by outsiders such as
himself.

Secondly, he may be concerned by apparent evidence of under-use or over-use of pesticide materials in particular circumstances. Some surveys have been carried out which suggest consistent biases amongst farmers towards or away from the use of pesticides (Tait 1978). Surveys of trends in cereal diseases and in fungicide application also suggest some inconsistencies (Cook, 1983), but various technical factors make it difficult to draw hard conclusions from aggregate data (King, 1977). From the farmer's point of view, these 'mistakes' may or may not be economically damaging.

The third concern of the scientist is his increasing awareness of a community interest in many of the consequences of pesticide decisions. He is aware of the interests of the industry rather than those of individual farmers when it comes to certain recommendations, for example, those concerning materials towards which pest resistance can arise. But he is increasingly aware also of pressures from outside the industry with regard to the impact of pesticide materials upon the environment.

The question thus arises as to how to organise research and its extension such that crop protection decisions (CPD) are in some sense rational. If all the consequences of CPD were known and were easily valued, then such decisions would be relatively trivial. Difficulty arises because of the various characteristics of these consequences.

SOME CHARACTERISTICS OF THE CONSEQUENCES OF CROP PROTECTION DECISIONS
Varying degrees of complexity and sizes of costs and benefits
Whilst some decisions involve only a binary choice between 'treat' and 'do not treat', most are considerably more complex. They may involve the choice amongst many slightly different materials, and amongst combinations of treatments. Crop protection decisions also may be highly time-specific in order to have maximum effect upon a developing problem within the crop. At the same time, the size of the costs and benefits can also vary considerably. If potentially high levels of loss can be controlled using relatively cheap methods, then treatment may become routine and other environmental consequences may result. But if costs of treatment are high, then farmers may be expected to invest more in the achievement of better decisions.

Impact over time
Many CPD have impacts over time which should be accounted for when choice is made. These impacts may affect future decisions in the medium or long term. They may be upon future levels of pest population either directly or indirectly via damage to beneficial predators. Other longer-term affects are discussed below.

Existence of uncertainty
Uncertainty surrounds all CPD to a greater or lesser extent. The unpredictability of biological systems, either because of gaps in

scientific knowledge or because of the major impact of changing weather, conspires to make the outcome to some extent unknown at the time the decision is made. In addition, as pointed out earlier, the fact that the grower must infer the consequences from a research environment which may be more or less unlike his own, means uncertainty is usually involved. Instead of having to choose between single-valued estimates of costs and benefits, he has to choose between probability distributions of these estimates.

Multiple attributes to the consequences of CPD
Whilst the primary aim of crop protection is to further efficient crop production, there may be many dimensions (usually referred to as 'attributes' by decision theorists) to the results. For instance, as well as impacts upon net revenues, there may be working capital requirements. There may be impacts upon other non-crop species of plants or animals. There may be impacts by residues upon other aspects of the local ecology or on more-or-less distant ecologies.

There may be impacts upon the health of spray operatives who apply the materials, upon neighbours down-wind of operations, and perhaps upon the final consumer of the product. There may finally be an impact upon the future viability of the treatment since the pest may become resistant. Thus, in the absence of regulations which may circumscribe the farmers choice, he has to balance all these factors in some rational way.

Consequences beyond the farm gate
It is also clear that many of these consequences of CPD have impacts on people other than the decision-maker. There is no reason why these factors should necessarily be accounted for by individual action. Unless all farmers co-operate to reduce pesticide usage when pest resistance is a risk, the actions of a single farmer will have no benefits in this direction. Only group action by all decision-makers will have the desired action of maximising the benfits to the agricultural sector.

Furthermore, society as a whole has interests in some of the impacts of pesticides. If the benefits to society, of which farmers of course are a part, are to be maximised, then some method must be found by which these values are incorporated into the decision process.

This section can be summarised by noting that pesticide decisions involve individuals making choices which have many varied consequences. The remainder of the paper attempts to investigate what part the theory of decision-making might play in ensuring that good choices are made.

ALTERNATIVE THEORIES OF DECISION-MAKING
Methods of studying decisions have included surveys of actual decisions, laboratory studies where the experimental environment may be better controlled, and theoretical studies. A distinction may be made between positive studies and normative studies. Positive or descriptive studies aim to describe the decision process. They

investigate what factors are important and try to draw conclusions about what are the major influences. They may be quantified (e.g. Tait, 1978) or they may be qualitative (e.g. Norton, 1982). Normative studies aim to show how decisions should be taken in the light of given objectives. They usually involve some assumptions about the objectives.

A model of decision-making is often involved, and the aim is to draw conclusions about what ought to be done. Obviously a normative model must have positive underpinnings in the sense that the realism or otherwise of its assumptions may be debated, but the emphasis is on drawing up recommendations for action.

The general stance is that many decisions have a common structure which can be abstracted from a particular application. A decision is conceived as consisting of choice between a number of alternative actions each of which has a range of outcomes which may occur with varying degrees of likelihood. The question addressed is whether we can provide recommendations such that the decision-maker would choose them had he all the information at our disposal concerning the various dimensions and values of consequences.

Simple profit maximisation
Economic theory has provided rules for calculating optimal dosages of farm inputs (Dillon, 1977). But the assumptions of smooth differentiable relationships between application of pesticide and benefits from control do not generally apply. Norgaard (1977) provides a good critique of some of the problems. Nonetheless the marginalist idea of comparing the benefits of additional control measures with their cost has found relevance in the concept of the economic threshold (Headley, 1972). The notion has been discussed, criticised and refined (e.g. Hall and Norgaard, 1973). Its attraction is its simplicity, wherein also lies its problem. Assumptions have to be made about changes in the expected pest population over time, the pest damage function, and the pesticide kill function. Problems in the setting of economic thresholds were outlined by Way and Cammell (1973).

As knowledge has improved about the interactions between crop, pest and pesticide, more complex models have been constructed. Shoemaker (1979) reports an application using multidimensional thresholds dependent upon crop factors and weather factors as well as pest density. Talpaz et al. (1978) report a simulation model which delivers optimal spraying strategies based on profit maximising objectives. The emphasis in both cases was upon the use of better biological data.

Nevertheless the approach is usually deterministic in the sense that important parameters (such as the pest migration rates) are assumed to be known. If the analyst assumes perfect knowledge and can value all the inputs and outputs in monetary terms, then the problem of specifying a 'best' solution is much simplified. Things are not often so easy.

Decision-making under uncertainty
As indicated earlier, some degree of uncertainty is inherent

in most crop protection decisions. The position can be illustrated
using a pay-off table as shown in Table 1. In this Table, the grower is
faced with the choice between two actions, namely to treat or not. The
outcomes of these actions are determined by the severity of the disease,
which is unknown when the decision has to be taken, and they are
specified in terms of gross margin for the crop in question.

Table 1.   A decision under uncertainty.

Actions	Outcome in £ gross margin	
	Disease severe	Disease not severe
Treat	6,000	8,000
Do not treat	4,000	10,000

In passing, we can note that if we knew (or assumed we knew) that the
disease was always severe, the choice then becomes trivial. But the
fact that uncertainty surrounds the disease means that whichever action
we choose, there will always be those years when, with hindsight, we
chose 'wrongly'. This is a fact of life which has to be lived with, and
means that the best we can offer the farmer is a 'best-bet' recommendation if we wish to recommend to him one of the alternatives. Much of
decision theory concerns itself with these types of situation, and it is
to the development of the theory that we now turn.

Early approaches. Agriculturalists in particular have been well aware
of the impact of uncertainty for many years. Ways of attempting to
minimise this impact were suggested (Heady, 1953,) which mainly
consisted of devising additional actions (e.g. insurance, forward
contracting) or compound actions (e.g. diversification) in which the
variability of the outcomes could be reduced. Many of these devices are
not relevant to the crop protection decision.

Some inspiration was taken from developments in the theory of games (von
Neumann and Morgenstern, 1947), which deals with the strategies to be
adopted by two (or more) competing players. In the agricultural case,
'Nature' was regarded as the adversary and the aim was to find rules by
which the grower should play 'the game against Nature'.

One typical rule was the maximin rule which recommended that the player
should choose that action which maximises the minimum possible income.
This was a conservative rule which protected people from extreme losses.
In Table 1, the maximin choice would always be to treat since the gross
margin would never fall below £6,000. There were other similar rules
based on different assumptions about the decision-makers.

Whilst in some circumstances these rules could be seen to be relevant (e.g. a heavily mortgaged farmer wishing to avoid the possibility of losses), it is easy to see why they break down. In the maximin case, choice is entirely dependent upon the outcomes in the poorest circumstances. The rule ignores any information about the likelihood of severe disease. As the disease becomes less and less likely we may expect the farmer to shift his choice to 'do not treat'. Moreover, even if the farmer found himself in a situation where the £10,000 outcome were multiplied by, say, ten times, the maximin rule would still recommend 'treat'!

Thus, whilst occasionally relevant in some restricted circumstances, these rules are not widely used since they ignore important information about the likelihood of the different outcomes. Dillon (1962) provides a review of applications of game theory in agriculture.

Expected utility. To date, probably the most generally accepted theory of risky decision-making is that of expected utility, or Bernoulli's principle, which postulates that when people take risky decisions they take account of: (1) their beliefs about the relative likelihoods of the uncertain events; and (2) their preferences for the outcome. It stemmed from Bernoulli's observation in the 18th Century that gamblers would pay only small amounts of money to participate in a gamble whose statistically expected outcome was infinite.

Then von Neuman and Morgenstern (1947) showed that maximisation of expected utility was rational in the sense that it could be derived from several axioms of behaviour which appeared to be reasonable. For further details see Anderson et al. (1977). The point is that if these axioms are accepted then the maximisation of expected utility becomes the rational objective.

The implications for our spraying decision in Table 1 are two-fold. First, we now admit that people use subjective probabilities in their assessment of the alternatives. As the probabilities tend to favour severe disease, so the farmer is more likely to choose to spray and vice versa. Second, we also admit the possibility that farmers preference for the amounts of the outcomes may not be strictly linear with money. This is illustrated by taking particular probabilities of disease severe or not severe, say 0.5 and 0.5, and calculating the actuarial or expected value of each of two alternatives.

The expected values in this example are found to be equal but many decision-makers, when faced with such a choice would have no hesitation in choosing the 'treat' alternative. Thus, many farmers have non-linear preferences for money or non-linear 'utility' functions. Decreasing marginal utility for money implies risk aversion in the sense that the decision-maker prefers a smaller more certain set of outcomes as compared with a higher more dispersed set of outcomes.

There has been considerable discussion as to the validity of the expected utility model. Writers have attacked it because it does not describe the way decisions are taken. Simon (1955) suggested that it is too complex since psychological experiments have shown that people tend to simplify complex decisions; they have limited memory for all the factors and they tend to process information sequentially rather than 'holistically', or in a lump, as the theory would suggest. Likewise Shoemaker (1982) reported that there appear to be 'context' effects in the sense that people alter their choices, depending upon the context in which the decision is made.

Whilst these are cogent criticisms of expected utility as a descriptive theory, they are less serious from a normative viewpoint because the stance is taken that people would maximise expected utility if they could. More serious from this viewpoint, are attacks on the axiomatic basis of the theory which have been made by Allais (1953) and others. Suffice it to say that whilst a number of difficulties have been exposed (Shoemaker, 1982), no generally accepted improved theory of risky decision-making has emerged.

There has developed a considerable literature in agricultural economics which deals with expected utility theory. Surveys have been carried out which have attempted to quantify farmers' preference functions (Young, 1979; Binswanger, 1980). Generally the conclusion is that most farmers are risk averse for gains and losses over at least some of the range of outcomes which face them. Some farmers appear to be indifferent to risks. Thus, farmers do appear to differ in respect of their attitudes to risk. The question arises as to whether we must take this into account when pesticide recommendations are made.

The concept of subjective probability has also received attention. Studies have shown that people typically under-estimate objective evidence suggesting high probabilities, and over-estimate evidence which points to low probabilities. People may be over-confident in that they under-estimate variability. Judgements seem to be over-influenced by recent events (Slovic, 1966; Lee, 1971).

There is a calculus of learning provided by Bayes Theorem which shows how subjective probabilities should be updated on the occurrence of new information. In some situations, people seem to revise their judgements conservatively (Francisco and Anderson, 1972), whilst in others people over-react on the basis of the new information (Shoemaker, 1982). But from a normative point of view, recommendations should be generated on whatever evidence provides the soundest basis for action. Carlson (1972) outlined a method which combined the judgements of the farmer with those of specialists. Webster and Cook (1979) used a Delphi method to elicit subjective distributions from plant pathologists. There seems to be little doubt that in the (frequent) absence of controlled trials, replicated over a number of years, such judgemental methods can provide a way of formalising the experiences of specialists on whose shoulders lies the task of interpreting the patchy evidence.

Some studies have examined the impact of various degrees of risk aversion on utility maximising choices. Two studies (Webster, 1977;

Thornton, 1983) showed that for particular fungicide options, the range
of attitudes found in the respective samples of farmers was not great
enough to make much difference to the recommended strategy. A third
study showed how varying degrees of risk aversion affected the optimal
situation in a maize/lucerne system with rootworm control by pesticides
(Lazarus and Swanson, 1983).

The question as to whether risk attitudes differ sufficiently or whether
variability is great enough to produce differences in utility-maximising
choices is important from the extension point of view. If no differences are seen, then the provision of recommendations is much simplified
since these can be based upon generalisations (for instance, the
assumption of risk-indifference, if appropriate) about farmers' preferences. Otherwise, theory would suggest that the farmers need to know
the complete probability distributions (or some relevant simplification)
of the alternatives before they can decide on the basis of their own
preferences.

One criticism of the foregoing studies might be that they all used
probabilities based on their published information or specialists'
knowledge, rather than on farmers' beliefs after they had been advised
by the specialists. If farmers modify the probabilities to allow for
the transition to their own situation then the variability may be more
or less under-estimated by the experimental data. Such a case would
argue for more complete specification of the outcomes of the alternatives, allowing the farmer to insert his own judgements.

In summary, expected utility theory, despite its heavy data requirements, is probably the most complete guide to normative decision-making
under uncertainty that we have. There are severe methodological
problems in making inferences from observed decisions (e.g. those of
farmers) since choice is dependent upon the perceived outcomes, their
subjective probabilities, and the farmer's utility function embodying
his attitude to risk. Experimental control is very difficult in these
circumstances. Even so, the evidence would suggest that where
variability is important, farmers should be provided with a summary of
the risks involved in each option. Only then can they make rational
choices on the basis of their own preferences.

Decision-making with multiple attributes
Following the developments in risky decision-making, the
analysis has been extended to situations in which multiple types of
outcomes are involved. For example, in addition to making changes in
gross margin, pest management decisions may have consequences for the
farmers' overdraft limit. The question is whether the decision can or
should be simplified by adopting some formal procedure which takes all
these factors into account in some way acceptable either to the farming
community or to society at large.

Most of the methods developed by decision theorists look at the way
people attach weights to the various levels of the attributes. A simple
system of weighting different attributes is shown for example, by

carcass classification schemes in which different characteristics of
the carcass (e.g. back fat thickness, total length) are scored and some
weighting procedure is applied to provide a single index of carcass
value.

Important concepts here include linearity of preference (or otherwise)
for the individual attributes, and the stability of preferences for one
attribute given different levels of other attributes. Thus, trade-offs
between the various attributes are handled. Utility functions under
risk for the attributes may be specified and the relationships between
these functions must be quantified.

Keeney and Raiffa (1976) provide a lengthy discussion of these issues
which have yet to be applied widely in agriculture. Herath et al.
(1980) report an application relating to choice of rice variety when two
attributes were at issue. Many of the non-agricultural applications
approach the problem from the point of view of the analyst who is
attempting to generate an agreed set of rules for aggregating the values
of the attributes to a single index which provides a basis for choice.

An alternative approach is where the decision-maker used some method of
sequential elimination by attribute. This approach is simpler in that
the decision-maker is now not prepared to allow trade-offs between
attributes. His priorities are expressed by an ordering of the
attributes in terms of hierarchy. There may be constraint levels on the
attributes (e.g. spray only if net volume revenue is greater than
$10 ha^{-1}) or in terms of categories (e.g. spray only if no residues will
be left after four weeks).

These methods are often known as lexicographic methods. Gladwin (1976)
applied such concepts to farmers' choices of fertiliser rates and plant
populations. Again, the role of the analyst might be to investigate
whether, in a particular pesticide decision, there was enough agreement
or technical constraint upon the decision such that a set of rules could
be generated. One obvious application of lexicographic rules is seen in
the flow-charts provided by the Agricultural Development and Advisory
Service for managed disease control in wheat and barley. However, these
only deal with crop characteristics and make various assumptions about
farmers' objectives as discussed earlier.

Although there have been so few applications yet in agriculture, we can
draw a few tentative conclusions about the impact of the multiple
attributes of pesticide decisions. First, they exist. Second, it seems
unlikely that we could get an accepted set of weights even for those
attributes which are internal to the farmer himself. Thus it behoves us
to list the alternative actions with their consequences so far as they
are known and relevant, so that farmers can insert their own trade-offs.

Handling externalities
But a more serious issue in relation to many of the
consequences of pest management decisions is that they can extend across
farm boundaries to affect other parts of the community. The problem is

how to circumscribe the farmer's decision and situation in such a way
that society's needs are met in a rational way. This leads us into the
arena of the study of public decision-making, which is outside the remit
of this paper. Nevertheless, a few observations may be made.

Firstly, one can conceive a hierarchy of levels of constraints. At the
community level, there are trade-offs between the benefits to the
agricultural sector and the benefits and costs to the rest of the
community. At the sector level, there are trade-offs between benefits
to some groups of farmers and costs to other groups of farmers (Siebert,
1980). At the farm level, there are the trade-offs within the enterprise itself, and which have been dealt with in preceding paragraphs
concerned with individual farmers' decisions.

Secondly, we can observe that the mechanisms by which those community
level and sector level trade-offs are made is often by way of enquiries,
commissions or working parties. The calculus of cost-benefit analysis
attempts to address some of these issues but, given the conflicting
objectives and values, often in a necessarily imprecise fashion.

Thirdly, it is the role of scientists to provide evidence of the size
and importance of the multifarious consequences for the environment and
for human health of using pesticides. A characteristic of public
decision-making in the United Kingdom is the adversarial system whereby
evidence provided by one group is pitted against evidence provided by
another. Where judgements are involved about possible or probable
events, many different interpretations may be made (Kunreuther et al.,
1984). Even though the process is sometimes alien to the natural
scientist, decisions about these issues must be taken. The scientist
has to be involved.

        CONCLUSIONS
        Earlier sections of the paper have shown that pest manage-
ment decisions can be uncertain and can have a variety of attributes to
their consequences, some of which may appear 'beyond the farm gate'.

Particular decisions may not involve all of these characteristics.
Nevertheless, apparently simple recommendations imply predictions of the
losses foregone by treatment and the costs of treatment. Because not
all the costs of treatment will be apparent to the adviser, e.g. the
opportunity costs of labour and working capital, the expected monetary
consequences of both treatment and no treatment should be listed. This
will enable the farmer to decide whether he thinks in his circumstances
treatment is worthwhile. Without this qualification, the information
content of the advice is very low.

A number of theories of normative decision-making have been described
which attempt to incorporate other aspects of decision. Theory would
suggest that farmers' behaviour will be important in determining what
choices they take. Enough is known about farmers' attitudes to risk to
conclude that attitudes do vary, and that they can be important in

determining choice. Thus, advice should make the farmer aware of both the range of outcomes and the likelihood for the best 'treat' and 'no treat' options. Merely quantifying the gains or losses for 'favourable', 'average' and 'unfavourable' weather conditions will often be a help, since the farmer can then use his judgement as to the weather.

It is not yet possible to generalise about farmers' attitudes to the wider consequences of pesticide use. In so far as such consequences exist, they should be made known in as complete a manner as possible. In the absence of this information, farmers are in no position to take what the scientist might regard as a rational decision.

Finally, it has to be recognised that people other than farmers have an interest in pest management decisions. There is a hierarchy of interested groups, including the totality of farmers, the agricultural industry as a whole, and the community at large. The mechanism by which these groups impose their interests is by setting a framework of agreements, regulations and laws around the farmers' decisions. It is the responsibility of the scientist to ensure that these regulatory instruments, which imply trade-offs amongst the various groups, are arrived at with full knowledge of the costs of all the impacts.

REFERENCES

Allais, M. (1953). Le comportement de l'homme rationnel devant le risque. Econometrica 21, 503–546.

Anderson, J.R., Dillon, J.L., and Hardaker, J.B. (1977). Decision Analysis in Agricultural Management. Ames: Iowa State University Press.

Bennett, D. (1984). Criteria for farm decision aids. In Computers in Agriculture, 181–197. University of Western Australia.

Binswanger, H.P. (1980). Attitudes toward risk: experimental measurement in Rural India. American Journal of Agricultural Economics 62, 395–407.

Carlson, G.A. (1970). A decision theoretic approach to crop disease prediction and control. American Journal of Agricultural Economics 52, 216–223.

Cook, R.J. (1983). Prediction of crop protection needs. In Factors Affecting the Accumulation of Exploitable Reserves in the Cereal Plant. London: HMSO.

Davidson, B.R. and Martin, B.R. (1965). The relationship between yields on farms and in experiments. Australian Journal of Agricultural Economics 9, 129–140.

Dillon, J.L. (1962). Applications of game theory in agricultural economics: Review and requiem. Australian Journal of Agricultural Economics 6, 20–35.

Dillon, J.L. (1977). The Analysis of Response in Crop and Livestock Production, 2nd edn. Oxford: Pergamon.

Francisco, E.M. and Anderson, J.R. (1972). Chance and choice west of the Darling. Australian Journal of Agricultural Economics 16, 82–92.

Gladwin, C.H. (1976). A view of the plan Puebla; the application of
        hierarchical decision models. American Journal of
        Agricultural Economics 58, 881-887.
Hall, D.C. and Norgaard, R.B. (1973). On the timing and application of
        pesticides. American Journal of Agricultural Economics 55,
        198-201.
Headley, J.C. (1972). Defining the economic threshold. In Pest Control
        Strategies for the Future, 100-108. Washington: National
        Academy of Sciences.
Heady, E.O. (1953). Economics of Agricultural Production and Resource
        Use. Englewood Cliffs, New Jersey: Prentice Hall.
Herath, H.M.G., Hardaker, J.B. and Anderson, J.R. (1982). Choice of
        varieties by Sri Lankan rice farmers; comparing alternative
        decision models. American Journal of Agricultural Economics
        64, 87-93.
Keeney, R.L. and Raiffa, H. (1976). Decisions with Multiple Objectives.
        New York: Wiley and Sons.
King, J.E. (1977). Surveys of diseases of winter wheat in England and
        Wales, 1970-75. Plant Pathology 26, 8-20.
Kunreuther, H., Linneroth, J. and Vaupel, J.W. (1984). A decision
        process perspective on risk and policy analysis. Management
        Science 30, 475-485.
Lazarus, W.F. and Swanson, E.R. (1983). Insecticide use and crop
        rotation under risk. American Journal of Agricultural
        Economics 65, 738-747.
Lee, W. (1971). Decision Theory and Human Behaviour. New York: Wiley
        and Sons.
Norgaard, R.B. (1976). The economics of improving pesticide use.
        Annual Review of Entomology 21, 45-60.
Norton, G.A. (1982). A decision analysis approach to integrated pest
        control. Crop Protection 1, 147-164.
Shoemaker, C. (1979). The optimal management of an alfalfa ecosystem.
        In Pest Management, eds G.A. Norton and C.S. Holling,
        301-315. Oxford: Pergamon Press.
Shoemaker, P.J.H. (1982). The expected utility hypothesis; its
        variants, purposes, evidence and limitations. Journal of
        Economic Literature 20, 529-563.
Siebert, J.W. (1980). Beekeeping, pollination and externalities in
        California agricultre. American Journal of Agricultural
        Economics 62, 165-171.
Simon, H.A. (1955). A behavioural model of rational choice. Quarterly
        Journal of Economics 69, 174-183.
Slovic, P. (1966). Value as a determiner of subjective probability.
        Transactions on Human Factors in Electronics, HFE-7, 22-28.
        Institute of Electrical and Electronic Engineers.
Tait, E.J. (1978). Factors affecting the usage of insecticides and
        fungicides on fruit and vegetable crops in Great Britain.
        II. Farmer specific factors. Journal of Environmental
        Management 6, 143-151.
Talpaz, H., Currey, G.L. and Sharpe, P.J. (1978). Optimal pesticide
        application for controlling boll weevil in cotton. American
        Journal of Agricultural Economics 60, 469-475.

Thornton, P.K. (1983). Information System Design for the Rationalis-
        ation of Fungicide Use: The Control of Puccinia hordei.
        Ph.D. thesis. Lincoln College, New Zealand.
von Neuman, J. and Morgenstern, O. (1947). Theory of Games and Economic
        Behaviour, 2nd edn. Princeton University Press.
Way, M.J. and Cammell, M.E. (1973). The problems of pest and disease
        forecasting, possibilities and limitations as exemplified by
        work on the bean aphid. Proceedings of the 7th British Crop
        Protection Council Conference 3, 933-940. British Crop
        Protection Council Publications.
Webster, J.P.G. (1977). The analysis of risky farm management
        decisions; advising farmers about the use of pesticides.
        Journal of Agricultural Economics 28, 243-259.
Webster, J.P.G. and Cook R.J. (1976). Judgemental probabilities for the
        assessment of yield response to fungicide applications
        against Septoria in winter wheat. Annals of Applied Biology
        92, 39-48.
Young, D.L. (1979). Risk preferences of agricultural producers; their
        use in extension and research. American Journal of
        Agricultural Economics 61, 1063-1070.

17. RATIONALITY IN PESTICIDE USE AND THE ROLE OF FORECASTING

E. J. Tait
Systems Group, Faculty of Technology, The Open University,
Walton Hall, Milton Keynes, MK7 6AA, England

THE CONCEPT OF RATIONALITY IN PESTICIDE USE

Rational pesticide use is one of a range of pest management strategies which can be loosely categorised as indicated in Figure 1 (Tait, 1985). Routine pest management implies the use of pesticides on a prophylactic basis, as an insurance against pest attack, and biological pest control avoids the use of pesticides altogether. There is generally little disagreement about the nature of these two types of strategy. It is in the middle ground, in the distinction between rational and integrated pest management, that terminological confusion arises, with many scientists and practitioners assuming that the two are synonymous. There is, however, some value in making a distinction which places rational pest management in the reductionist paradigm and integrated pest management in the holistic paradigm, denying the possibility of a smooth transition from one to the other (Tait, 1981).

Figure 1. Categorisation of Pest Management Systems

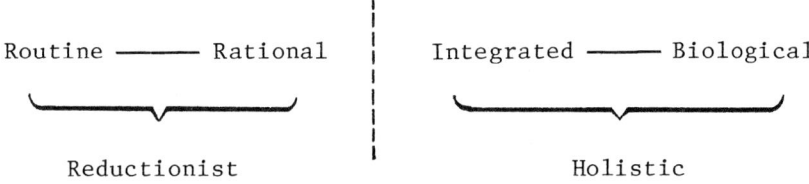

Routine ——— Rational　｜　Integrated ——— Biological

Reductionist　　　　　　　Holistic

Broadly defined, rationality implies any behaviour designed to maximise the achievement of a particular set of goals or objectives. Thus, rationality can be described as scientific, technical, economic, social, legal, political or ecological, depending on the goals which are appropriate to each disciplinary boundary. The common theme underlying all types of rationality is the rejection of aesthetics and value judgements. In relation to pest control, scientific, technical and economic rationality are usually the assumed basis of decision-making, and each has acquired some special meanings in addition to being simply goal-directed.

Scientific rationality is closely allied to the idea of the scientific method itself (Perkins, 1982). Scientifically rational decisions can only be made on the basis of knowledge which is established as factual. The scientific method of attaining this knowledge is to study the part of the system which is of interest in isolation from disturbing influences, as in the laboratory or in a controlled field trial. This has the advantage of enabling one to make very precise statements about the part of the system under study, but only at the expense of ignoring interactions with the rest of the system.

Technical rationality arises from the application in practice of scientific rationality. It has been described by Schon (1983) as the dominant epistemology of the professions.

The economist has imbued the idea of rationality with some demanding, and somewhat controversial assumptions about human behaviour, in order to analyse such behaviour on a more scientific basis. Individuals are assumed to be motivated only by self-interest, maximising their utility, subject to income and price constraints (Hill et al., 1978; Hollis, 1979). A more extreme view assumes that the grower's primary objective is to maximise profit (Varley, 1979).

Rationality in pesticide use embraces this claim to scientific, tech- nical and economic rigour, which is why it has been placed on the reductionist side of the divide in Figure 1. Integrated pest manage- ment, on the other hand, is holistic because it sacrifices some of this rigour for a greater emphasis on the inter-connectedness of many variables to create a whole which is greater than the sum of its individual parts.

There are two major reasons why pest management decisions can be irrational. There may be insufficient information on which to make a properly rational assessment of options, or alternatively, decision makers may allow their values to influence their decisions. Forecasts can contribute to the scientific, technical and economic rationality of pest management by reducing the uncertainty surrounding some aspects of a decision, but like all scientific endeavour, they are subject to distortion by the values and attitudes of those who create and use them.

FORECASTS AND THE REDUCTION OF UNCERTAINTY
        A forecast consists of information about various states of the real world which, by reducing some of the uncertainty surrounding a decision, contributes to its rationality. The types of information which are most commonly relevant to pest management decisions are weather data, pest infestation levels and economic data related to crop yields and prices.

Norton and Mumford (1983) have discussed four categories of data relevant to pest management:

        1. Fundamental information, consisting of knowledge of the basic relationship among biological and technical variables, principles of pest management and decision rules such as economic thresholds;

2. __Historical information__ such as past events and trends in levels of pest attack and damage;

3. __Real-time information__ on the current state of such variables as weather, crop microclimate, pest attack or crop prices, obtained by monitoring at the farm or regional level; and

4. __Forecast information__ on the future state of important variables.

Figure 2 shows how these categories of information interact in the development of more rational systems of pest management.

Figure 2. Use of information in pest management decision-making.

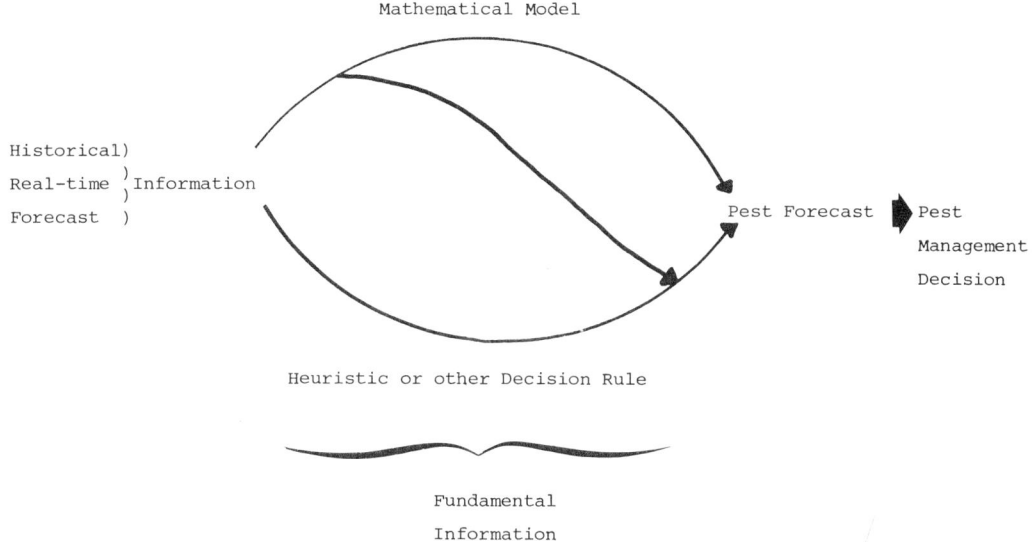

Decision rules or mathematical models, both of which are in the category of fundamental information, are used to interpret either historical, real-time or forecast information to arrive at a pest forecast and hence a decision. Heuristic decision rules are based on mental subjective probability models of future outcomes (Tversky and Kahneman, 1974), or a type of mental "simulation" where the decision maker imagines various possible outcomes and the ease with which they could occur (Kahneman and Tversky, 1982). Such heuristics can be a useful basis for decision-making but they are often subject to severe and systematic errors. Mathematical models, although not altogether free from subjective bias, are a much more rational basis for a forecast. Sometimes a model is used to specify a more precise decision rule such as an economic threshold, based on insect pest or disease incidence, or an action threshold, based on weather variables, and the model itself has no further involvement in the decision-making process. Alternatively, the model can be used directly to recommend the optimal course of action for each decision as it arises.

Decisions made by the farmer
Where decisions are made by farmers themselves, a simple readily-available forecast, such as a weather forecast, can with the aid of a heuristic decision rule, sometimes lead directly to a crude farm-level pest forecast and hence to a pest management decision. A forecast of heavy rain can lead to the decision to postpone or cancel the application of a pesticide. A forecast period of hot, dry weather could lead to the decision to apply pesticide, in the belief that the development of some pest problems on the crop could be accelerated. The heuristic "simulation" involved in the decision could include factors like the knowledge that the farmer is about to go on holiday and does not trust his spray operator to make a rational decision in his absence.

Farmers often develop their own heuristic decision rules which may or may not be based on scientific fact, but which they feel are contributing to the rationality of their decisions. For example, farmers in Lincolnshire felt that it was unnecessary to treat vegetable brassicas for cabbage root fly before "cow parsley" (presumably Anthriscus sylvestris) was in bloom on the dyke banks. Less rationally, timing of the first caterpillar sprays was sometimes based on the appearance of cabbage white butterflies, neglecting altogether the diamond-back moth; also timing of the first aphid sprays was based on the appearance of aphids (unspecified species) on the farmer's car windscreen (Tait, 1983)!

Many farm level pest management decisions are "rational" only to the very limited extent possible with such a restricted information input. Incorporating real-time information on pest levels and/or microclimate, perhaps along with a weather forecast, can greatly increase the rationality of the decision.

Advances in micrometeorology and microcomputer technology are increasing the quantity and quality of real-time information which can be collected on farms, leading to more rational predictions of, for example, the development of fungal diseases (Thompson, 1982). Teaching packages are also being produced to improve the skills of monitoring and assessing pest and disease levels in those involved in the day to day management of pests, including farmers (Anon., 1984a). The farm-specific information which these methods will make available is an essential complement to real-time data collected on a regional basis.

Advanced real-time monitoring of microclimate and crop pest and disease levels can only be justified if the decision rules employed match them in sophistication. Many economic thresholds were suggested initially as heuristic decision rules, on the basis of the experience of scientists and crop protection advisers, and refined as a result of their implementation in practice. Others were based initially on experiments involving mathematical models (Way and Cammell, 1979). However,

according to Tatchell (1982), the quality of economic thresholds as
decision rules still lags considerably behind the quality of the real-
time information with which they are compared.

In the above examples, the pest management decision is made at the farm
level, based on forecast and/or real-time information without the intervention of an outside agent. For many monitoring and forecasting
systems, the complexity of the decision rule, model or information
inputs, or their interactions, are such that technical experts must
intervene in the process. On the basis of their calculations, a
recommendation is made to farmers, who admittedly still have the freedom
to reject the recommendation, but who are no longer deciding for themselves on the recommended course of action. Pest management decision-
making in these circumstances is "supervised".

Decisions made by experts: supervised pest control
Given the technical and economic expertise of those involved
in supervised pest control, and the greater sophistication of their
resources, one would expect the reduction in uncertainty, and hence the
increase in rationality, to be greater than for decisions made at the
farm level. It is, however, arguable whether this is always the case.
The gains may be offset by the greater generality of the data, which
cannot take account of the variation among individual farms or fields,
and by the time taken to disseminate information.

The role of information in generating forecasts for supervised pest
control systems will be illustrated here with reference to two very
different systems, the Rothamsted Insect Survey (Tatchell, 1982) and the
EPIPRE system (Rijsdijk, 1982).

For the Rothamsted Insect Survey, the raw data come from a nationwide
network of suction traps, continuously monitoring the aerial insect
fauna, i.e. real-time information. Computers are used to aid in interpreting and collating the data, but no complex mathematical modelling is
involved in its analysis.

To generate a forecast these data must be interpreted in relation to
particular agricultural problems, and this requires fundamental knowledge of insect biology and ecology and of the relationships between
aerial sampling numbers and numbers on the crops, in addition to further
real-time information on local crop conditions. Accordingly, the data
collected centrally at Rothamsted are disseminated directly or indirectly to a wide variety of advisers who then pass their recommendations on to farmers.

As the Insect Survey has been in operation for over ten years,
historical information, collected in earlier years can be helpful in
interpreting the real-time data. If sample counts are expressed as
"higher than" or "lower than" last year or than an average figure,
decision makers can have a clearer idea of the likely levels of infestations on the current year's crops. Similarly, historical information
from previous weeks can be used in a crudely predictive manner, for

example to forecast the northwards progress of pest infestations in spring and early summer.

The Rothamsted Insect Survey thus has a strong emphasis on real-time information, which is collected at a central point and disseminated to a wide variety of farming advisers. The information is then processed by these expert advisers, on the basis of relatively simple decision rules, into a pest forecast which can improve the rationality of farmers' pest management decisions.

The EPIPRE system on the other hand, involves farmers themselves in collecting very detailed real-time and historical information about individual fields, including crop variety, soil type, seed rate, previous crop, field observations of insect pests and diseases and the use of fertiliser, growth regulators and pesticides. The information is stored in a central data bank and processed by a mathematical model, giving a pest and disease forecast and hence a recommended course of action to each individual farmer. The cycle is repeated at intervals throughout the growing season with recommendations issued as either:

> 1. No spraying, with a request to the farmer for new field observations by a defined date;
>
> 2. Spraying for control of specified pests or diseases with a request for new field observations by a defined date; or
>
> 3. Towards the end of the season, a further message indicates that no further spraying should be needed.

Being collected by individual farmers, who are not experts in the identification and monitoring of pests and diseases, the scientific validity of the real-time information for the EPIPRE system is less than that collected for the Rothamsted Insect Survey. However, in practice serious mistakes have rarely occurred, the most common difficulty being that farmers tended to overlook very low disease incidences. This was compensated for by the frequency of monitoring, so that farmers usually identified a problem well before the pest or disease reached the threshold value. Experience in the commercial introduction of the EPIPRE system in the UK is discussed by Wilson in Chaper 24.

The mathematical models in the EPIPRE system were originally complex dynamic models of population growth developed to aid scientific understanding. These have been modified and simplified to be useable for the large numbers of farmers involved in EPIPRE. The major simplification is the assumption that the relationship between severity of infestation and resulting crop loss is linear in the region between maximum and minimum economic thresholds. Experience indicates that this is a technically rational assumption.

The major improvement in the rationality of decision-making with the EPIPRE system arises from the use of the mathematical model which can, at least in The Netherlands, allow considerable reductions in pesticide use with no corresponding decline in yields. There may also be a long-

term and less tangible increase in the overall rationality of pest
management decisions by farmers participating in the system – as they
are required to monitor fields closely and regularly, their fundamental
scientific knowledge of crop protection will improve and will probably
be carried over to other crops not covered by EPIPRE.

SUBJECTIVE INFLUENCES ON RATIONAL DECISION-MAKING
The quality and quantity of real-time, historical and forecast information available to those involved in pest management are
increasing all the time. There has only been space here to give a few
brief illustrative examples of the many ways in which this is helping to
improve the rationality of pest management decisions. However, like all
decision making, there remains a subjective value-laden aspect to the
process. It would be dangerous to assume that this can, or even should,
be removed entirely by improvements in scientific, technical and
economic rationality. It is important to be aware of and to understand
how the various subjective influences can bias decisions or recommendations and affect their implementation.

Mathematical models and decision rules
The susceptibility of heuristic and other decision rules to
subjective bias has been studied for a wide range of decisions (Kahneman
et al., 1982). From the small amount of research on pest management
decision-making, there is nothing to suggest that farmers, scientists or
farming advisers behave any more rationally, as defined above, than the
rest of society. The effects of individual and group biases are
discussed below. When these biases are incorporated in decision rules
or mathematical models, with their aura of scientific rationality, they
can be very influential, and at the same time difficult to detect.

Models and decision rules inevitably require simplifying assumptions,
each of which imparts a particular bias to the output. Generally the
assumptions are scientifically rational and their effect on management
outcomes is carefully checked (Shoemaker, 1983). However, they can also
be influenced by the values of the scientists making them. A series of
apparently valid assumptions, each biased slightly, "just to be on the
safe side", can lead to a significant cumulative bias. This will only
be detected if the scientists who develop a model or decision rule take
the trouble to explain their assumptions and to evaluate their effect on
the use of the rule or model in practice.

For example, the economic threshold used in monitoring Aphis fabae on
field beans (Cammell and Way, 1977) is admittedly conservative (i.e. it
is set at a lower level of pest infestation than should rationally be
the case) because farmers are assumed not to be prepared to tolerate
crop losses, even when these are less than the cost of insecticide
treatment, and advisers are concerned to protect their credibility with
clients. In this case, with an economic threshold of 5% of stems
infested, the maximum crop loss experienced by farmers was 0.7% of the
total yield, which was equivalent to only 10% of the cost of treatment
in most years. Under these circumstances, it is probable that many
farmers needlessly treated their crops with insecticide (Tait, 1977a).

In common with most other pest management models, that used
for the EPIPRE system requires farmers to estimate the yield potential
of the field in question. This is used as a basis for calculating the
economic value of a pesticide treatment - if the yield potential is
high, a pesticide treatment will be recommended at a lower level of pest
incidence than otherwise. No research has been done on farmers'
abilities to estimate the yield potential of their crops, or on how they
interpret the meaning of the term. However, it could be postulated that
they are more likely to over-estimate than to under-estimate this parameter. The effect would once again be to encourage pesticide treatment
where this is not a rational course of action.

Farming advisers
Many schemes for rationalising pesticide use rely on
recommendations being transmitted to farmers by advisers, and, to an
increasing extent, non-commercial advisers do not have the resources to
compete with those who are selling pesticides. In one survey of fruit
and vegetable growers in East Anglia, 62% favoured a commercial source
of advice, generally the pesticide salesman (Tait, 1978). A later
survey of cereal and sugar beet growers in Norfolk and Suffolk in 1981
indicated an even greater reliance on commercial advisers, with a
majority having their fields monitored at weekly intervals in spring and
early summer by a pesticide salesman. Under these circumstances, the
monitoring and forecasting information generated in the non-commercial
sector was reaching the farmer almost exclusively through the commercial
network (Tait, E. J., unpublished data). Lawson (1982) found a similar
pattern of information flow, particularly for verbal information, among
oilseed rape growers.

Contrary to expectations, there was no significant difference in
pesticide usage levels between farmers who relied on commercial and on
non-commercial advisers (Tait,1978). Lawson (1982) has interpreted this
as evidence that commercial sources of advice were not acting to
increase pesticide usage by farmers. An alternative interpretation
(Tait, 1977a) is that non-commercial advisers have not been given the
policy guidelines or the financial resources to enable them to influence
farmers more effectively and encourage more rational use of pesticides.
There is also the problem of risk aversion in the advisers, who must
retain their credibility with their clients. Even with non-commercial
advisers, there is a reluctance to be proved wrong, and the recommendation not to apply a pesticide carries greater risks of this happening.
Further research would be needed to clarify these points, but it is
against current economic and social theory to claim that an organisation
with clear financial interests in selling pesticides will, given the
opportunity, do nothing to further these interests.

Assuming that the bias towards commercial sources of advice is general,
that most of the forecasting and other information generated in the
non-commercial sector is being filtered through this medium, and that
the resulting effect is to increase the use of pesticides, the
implications for rational pesticide use could be serious. Much of the
scientific and advisory effort of the non-commercial sector could be
dissipated.

Farmers

From the literature on subjective influences on farmers'
pest management decision-making, there are several findings which are
relevant to the use of forecasting and monitoring information.

In a survey of brassica growers and their attitudes (Tait, 1983), the
presumed risk aversion of many farmers was demonstrated. Most of the
opinions which were significantly correlated with pesticide usage
related to financial risks. It is interesting to speculate on what the
end result may be if a risk-averse decision rule or model is further
biased by the risk aversion of the farming community. Any potential
benefits in terms of rationality may be cancelled out.

Farmers expressed negative attitudes to pesticides on the environmental
and personal health risk dimensions but these were not often correlated
with pesticide usage (Tait, 1982). Beliefs, rather than values, were
correlated with pesticide usage and to change peoples beliefs is a
simpler and less traumatic process than to change their value systems,
so this may indicate that a more rational approach to pest management
could be readily encouraged by changes in the information available to
farmers.

In the above survey, there was a cluster of values which significantly
correlated "good conscientious farming" with high pesticide usage. If
this situation applies to other crops than brassicas, it may constitute
a more serious blockage to the rationalisation of pesticide use.

The encouragement and increasing use of tank mixes also militates
against scientific and technical rationality. On potatoes, for example,
the incorporation of an insecticide in fungicide sprays for blight is
unlikely to be rational since the same weather conditions are not
favourable to both pests (Tait, 1977b). One of the least rational uses
of tank mixes found has been on vegetable brassicas where DDT was added
to sprays of demeton-S-methyl for aphids "because it is a good spreader"
(Tait, 1983).

The adoption of standard operation procedures (Cyert and March, 1963) is
a device for cutting down on the time and effort involved in decision-
making which could be even less rational than insurance treatment of
crops. It leads some farmers to treat all crops in a given category
equally. It has been suggested as an explanation for the greater
variance in pesticide use in early, as opposed to late, potato crops
(Tait, 1977b) and brassica crops (Tait, 1983). Another standard operating procedure in use on brassica crops was related to the three-week
harvest interval of demeton-S-methyl. The farmer's spray programme was
planned so that the last application of pesticide would take place three
weeks before the expected harvesting date for the crop, and sprays would
be applied at two week intervals up to that point (Tait, 1983).

The reason given for this last strategy was the avoidance of "fire-
engine jobs". In the minds of the Lincolnshire farmers surveyed, a
fire-engine job was the application of pesticide in response to a pest

infestation on the crop. This was seen as a "bad thing", an admission of failure on their part. This is another attitude which, if found to be widespread in the farming community, could militate against the rationalisation of pesticide use.

The concept of the dominance of a pest in crop protection decision-making has relevance for forecasting and rational pesticide use, particularly where tank mixes are prevalent. On potatoes, blight was the dominant pest, with the application of pesticide for aphids being subservient to this (Tait, 1977a). On vegetable brassicas, aphids were dominant over caterpillars (Tait, 1983). When a pest forecast is transmitted to farmers, the use made of the information may depend on whether the pest is dominant or not. If it is not dominant, the information is more likely to be ignored. If the pest is dominant, the information may be acted upon, but the increase in rationality may be counteracted by the tendency to include in the tank mix pesticides for other non-dominant pests.

Overall effect of subjective influences
There have been sustained pressures from the academic and non-commercial advisory communities for improvements in the rationality of pest management decision-making, but considerable disappointment in the uptake at the farm level. At the present time, as described above, subjective influences act mainly to increase pesticide use, their source depending on the system concerned. For example, the important sources of subjective bias are likely to be the model itself and farmers for EPIPRE, and farming advisers for the Rothamsted Insect Survey.

When planning any public investment in more rational pest managment systems, it is important to be aware of the possible influences on decision-making, and to incorporate in the planned system tactics for ensuring its implementation. This means conducting the relevant market research for each system. The financial outlay will be amply rewarded, as the agrochemical industry has known for some time.

Awareness of possible changes in the environment of a system can be invaluable in anticipating changes in attitude to it. An example of this is the conflict currently simmering between the agricultural community and the rest of society. Evidence from a survey in Bedfordshire suggests that the issues associated with this conflict in the minds of farmers are straw burning, hedge and woodland removal, wetland drainage and public access to farmed land. Farmers do not spontaneously associate pesticides with this conflict and even when they are prompted, the commonest responses are that present levels of pesticide use are indispensable, and that the people who regulate these matters know what they are doing. There is, however, a significant undercurrent of concern in the farming community for the acute and chronic side effects of pesticide use (S. Carr, unpublished data). The misuse of pesticides is also much more common than officially admitted (Tait, 1983), and this is increasingly being brought to the attention of the public and the government. The British Agrochemicals Association, in commenting on the new legislation to regulate the supply and use of

pesticides in the UK, emphasised the need to control the user more
effectively (Anon., 1984b).

If the issue of pesticide use is added to other pressures on the farming
community, the climate may rapidly become more encouraging of rational
pest control. Changes in the EEC price support system, which have been
widely advocated but so far not implemented, could also radically alter
farmers' attitudes to high input systems of crop protection.

DECISION SUPPORT FOR RATIONAL PEST MANAGEMENT
For rational pest management, some form of decision support
for the farming community is required. At the moment this is generally
provided in the form of supervision of the decision by an expert. Under
most circumstances, the aim should be to transfer responsibility for
informed decision-making to the farmers themselves, by providing them
with a complete decision support system (DSS). This could ensure more
rapid turn-round of information, using more detailed farm-level data,
avoiding some sources of subjective bias and making others more
explicit.

If a problem can be completely specified, the human decision-maker can
be replaced by an algorithm, and decision support is not an issue.
Likewise, if the problem is totally unstructured, decision support is
not relevant (Ginzberg and Stohr, 1982). As the above analysis has
shown, rational pest management falls between these two extremes,
providing ideal circumstances for the introduction of a DSS. The
medical profession has been developing them for some time, and it is
surprising that there has not been a similar interest in the agricultural community, given that the ethical problems are not so great.

A DSS should be designed with a particular context of use in mind.
However, a hypothetical system for pest management would include the
following components (Fig. 3):

1. A training and educational component to teach farmers
how to recognise pest problems and assess their severity;

2. A diagnostic information system, or "expert system" to
back up the training component;

3. A data base containing historical, real time and forecast information, qualitative and quantitative facts, some
long-term and some updated regularly, collected by the
farmer and supplied by the advisory network;

4. An inference analyser consisting of models to provide
quantitative routines suggesting solutions to problems,
diagnosed with the help of (1) and (2) and in the light of
additional information in (3); and

5. A knowledge base which provides for the declarative
representation of mainly qualitative knowledge.

Figure 3.   Decision support system for pest control.

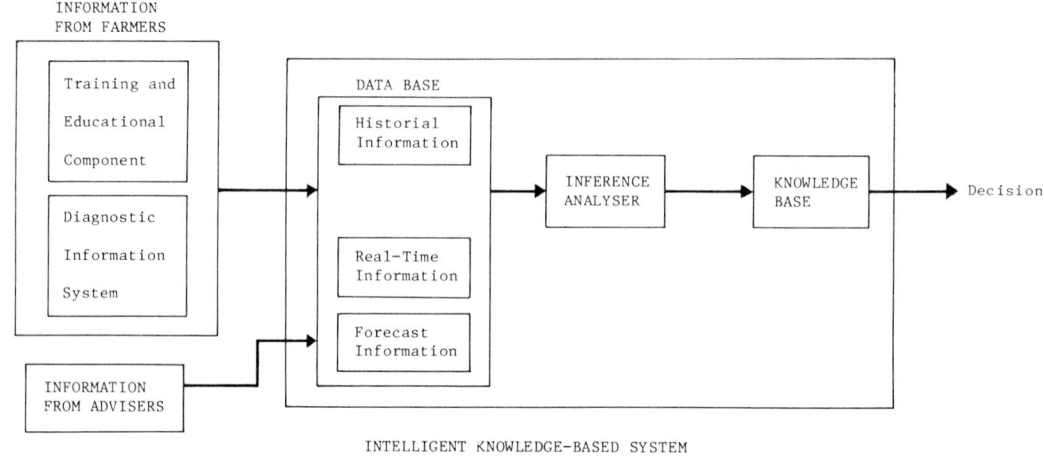

The data base, inference analyser and knowledge base together constitute an **intelligent knowledge based system** (IKBS) which is open-ended (Lee, 1983), explains to the user the assumptions behind suggested solutions and allows him to reject the solution and ask for another. Sophistic-ated systems will also allow the user to change the rules on which the knowledge base operates.

Such systems are already technically possible and are being developed for much more complex decision-making situations than pest control. The improvements in scientific, technical and economic rationality to which they could lead may also become increasingly rational in a social and political sense as norms and policies change. Their development for pest management could help to achieve the desirable aim of transferring responsibility for decision-making to the farmer. Rather than have a farming adviser impart his own risk aversion and the farmer's presumed risk aversion to a recommendation, the farmer would have the inform-ation, the knowledge and the support to make his own assessment of the need for a pesticide application, and to decide in some circumstance to take the risk of not applying a pesticide.

REFERENCES
Anon. (1984a). Pest and Disease Management in Oilseed Rape. Milton Keynes: Centre for Continuing Education, Open University. Open University Press.
Anon. (1984b). Chemistry and Industry $\underline{7}$, 594.
Cammell, M.E. and Way, M.J. (1977). Economics of forecasting for chemical control of the black bean aphid, Aphis fabae, on the field bean, Vicia faba. Annals of Applied Biology $\underline{85}$, 333-343.
Cyert, R.M. and March, J.G. (1963). A Behavioural Theory of the Firm, 33. New Jersey: Prentice Hall Inc.

Ginzberg, M.J. and Stohr, E.A. (1982). In Decision Support Systems, eds M.J. Ginzberg, W. Reitman and E.A. Stohr, 9-31. Amsterdam: North Holland Publishing Company.

Hill, P.H. et al. (1978). Making Decisions: a Multi-disciplinary Approach, pp. 111. Reading, Massachusetts: Addison Wesley Publishing Company.

Hollis, M. (1979). In Rational Action: Studies in Philosophy and Social Science, ed. R. Harrison, 1-15. Cambridge: Cambridge University Press.

Kahneman, D. and Tversky, A. (1982). In Judgment Under Uncertainty: Heuristics and Biases, eds D. Kahneman et al., 201-208. Cambridge: Cambridge University Press.

Lawson, T.J. (1982). Information flow and crop protection decision-making. In Decision Making in the Practice of Crop Protection, Monograph No. 25, ed. R.B. Austin, 21-32. British Crop Protection Council Publications.

Lee, R.M. (1983). In Processes and Tools for Decision Support, ed. H.G. Sol, 25-36. Amsterdam: North Holland Publishing Company.

Norton, G.A. and Mumford, J.D. (1983). Decision making in pest control. In Advances in Applied Biology, ed. T. H. Coaker, Vol. VIII, 87-119. London: Academic Press.

Perkins, J.H. (1982). Insects, Experts and the Insecticide Crisis: the Quest for New Pest Management Strategies. New York: Plenum Press.

Rijsdijk, F. (1982). The EPIPRE system. In Decision Making in the Practice of Crop Protection, Monograph No. 25, ed. R.B. Austin, 65-76. British Crop Protection Council Publications.

Schon, D. (1983). The Reflective Practitioner: How Professionals Think in Action, 21. London: Temple Smith.

Shoemaker, C.A. (1983). Integrating economic analysis with population modelling in pest management problems. Proceedings of the 10th International Congress of Plant Protection: Plant Protection for Human Welfare 1, 154-159. British Crop Protection Council Publications.

Tait, E.J. (1977a). The use of forecasting as a method of rationalising pesticide applications. Proceedings of the 1977 British Crop Protection Conference - Pests and Diseases, 235-240. British Crop Protection Council Publications.

Tait, E.J. (1977b). A method of comparing pesticide usage patterns between farmers. Annals of Applied Biology 86, 229-240.

Tait, E.J. (1978). Factors affecting the usage of insecticides and fungicides on fruit and vegetable crops in Great Britain. II. Farmer specific factors. Journal of Environmental Management 6, 143-151.

Tait, E.J. (1981). The flow of pesticides: industrial and farming perspectives. In Progress in Resource Management and Environmental Planning, eds T. O'Riordan and R.K. Turner, Vol. 3, 219-250. Chichester: John Wiley and Sons.

Tait, E. J. (1982). Farmers' attitudes and crop protection decision-
     making. In Decision Making in the Practice of Crop
     Protection, Monograph No. 25, ed. R.B. Austin, 43–52.
     British Crop Protection Council Publications.

Tait, E.J. (1983). Pest control in brassica crops. In Advances in
     Applied Biology, ed. T.H. Coaker, Vol. VIII, 121–188.
     London: Academic Press.

Tait, E.J. (1985). In Insect Pest Management, ed. T.H. Coaker, London:
     Academic Press.

Tatchell, G.M. (1982). Aphid monitoring and forecasting as an aid to
     decision-making. In Decision Making in the Practice of Crop
     Protection, Monograph No. 25, ed. R.B. Austin, 99–112.
     British Crop Protection Council Publications.

Thompson, N. (1982). Meteorology as an aid to crop protection. In
     Decision Making in the Practice of Crop Protection,
     Monograph No. 25, ed. R.B. Austin, 55–61. British Crop
     Protection Council Publications.

Tversky, A. and Kahneman, D. (1974). Judgement under uncertainty:
     heuristics and biases. Science 185, 1124–1131.

Varley, J.E. (1979). High yield systems – the need for programmed pest
     control. Proceedings 1979 British Crop Protection
     Conference – Pests and Diseases, 673–681. British Crop
     Protection Council Publications.

Way, M.J. and Cammell, M.E. (1979). Optimising cereal yields – the role
     of pest control. Proceedings 1979 British Crop Protection
     Conference – Pests and Diseases, 663–672. British Crop
     Protection Council Publications.

18. ADVANCES IN DISEASE FORECASTING

W.E. Fry
Department of Plant Pathology, Cornell University,
Ithaca, New York, 14853, USA

INTRODUCTION

Plant pathologists have been developing disease forecasts for many years, e.g. for prediction of potato late blight (Van Everdingen, 1926). Major motivations for forecasts include better disease control through more timely application, reduced use of pesticide with equivalent control, and the avoidance of disease. During the last half century we have made some progress in achieving these goals, but the motivations persist.

Disease forecasting means different things to different people. Some distinguish between forecasts and predictions. Zadoks (1984) used the term "disease warning" to avoid specific connotations associated with the words "forecasting" and "prediction". Some agriculturalists view forecasts as techniques to determine whether a spray is required tomorrow or not. Jones (1983), who works with apple scab, described disease predicting as "... the science of monitoring the physical conditions of the environment and declaring after "disease weather", but before symptoms are visible, that infection has occurred." Clearly, many techniques regarded as forecasts by other plant pathologists do not fit this definition. The prediction technique for Stewart's wilt of maize indicates before planting, whether Stewart's wilt is likely to be a problem during the forthcoming season (Stevens, 1934). Other techniques identify the time during the season when pesticide applications should be initiated. Many such techniques have been devised for potato late blight (Beaumont, 1947; Grainger, 1953; Hyre, 1954; Schrödter and Ullrich, 1966; Krause et al., 1975; Cupsa et al., 1983; Forsund, 1983). In this paper, I define forecasts broadly as techniques which enable predictions of disease occurrence. I do not distinguish between forecasts, predictions or warnings.

Reliable forecasts generally depend on accurate knowledge of disease epidemiology. Forecasts are useful for diseases which occur with variable intensity or at variable times. If variation in disease occurrence or severity results from variation in the amount of initial inoculum, then the forecast should incorporate factors which influence the amount of initial inoculum. Typically, analysis of weather prior to the growing season has been used. If the efficacy of the initial inoculum is largely responsible for variation in disease occurrence or severity, then factors which influence initial inoculum efficacy must be considered. If variation in disease occurrence is due to rates of

pathogen growth, then factors influencing growth rates must be considered. Variation in weather has been the primary determinant in most
disease forecasts.

Forecasts often do not involve measurements of disease, because of the
variable relationship between visible disease and total disease.
Assessments of visible disease do not include pre-symptomatic
infections. However, some reliable forecasts do include assessments of
the pathogen population.

There have been significant advances in disease forecasting in recent
years. Some advances are extensions of previous activity and emphasise
the influence of weather on diseases not previously forecast; some
incorporate factors other than weather into new or existing forecasts,
some involve new technologies, such as computers, in the development and
evaluation of forecasts, and some require increased understanding of
factors affecting implementation of disease forecasts.

NEW DISEASE FORECASTS
        By the early 1980s, forecasts had been developed for many
different diseases. For some "popular" diseases, such as potato late
blight, several forecasts had been developed (Table 1). Unfortunately,
one cannot determine from the research literature whether or not a
forecast has been adopted by growers. However, some forecasts are well-
known, and in the United States for example, those for apple scab and
for Stewart's wilt of maize, have been widely adopted.

Newer forecasts (Table 2) differ from each other in the time period
given for effective grower response (the "response time"), in the land
area for which the forecast is appropriate, and in the types of measurements required. The forecast for bean root rot is typical of that
possible for many diseases caused by soil-borne pathogens (Kobringer and
Hagedorn, 1983). The potential for root rot in an individual field is
assessed prior to the season by growing beans in the glasshouse in soil
from that field. The glasshouse conditions favour disease development.
Root rot indices in the glasshouse tests are related to root rot
potential in the field. If the disease potential is high, growers are
cautioned to plant resistant cultivars or alternative crops. The
response time can be weeks or months, the forecast is specific to a
given field, and measurements involve soil samples and glasshouse
plantings.

In contrast, a forecast for wheat stripe rust, Puccinia striiformis, in
Northwestern United States applies to a much larger region and depends
only on the weather (Coakley et al., 1983). Appropriate control
measures for seasons in which stripe rust is likely to be important are
not yet well-publicised, but will probably require a response time of
days or weeks.

Table 1. Forecasts developed before 1983.

Crop	Disease	Reference
Apple	Scab	Mills, 1944
Barley	Powdery mildew	Channon, 1981
Celery	Early blight	Berger, 1969a,b
Cherry	Leafspot	Eisensmith and Jones, 1981
Grape	Downy mildew	Tarr, 1972
Lima bean	Downy mildew	Hyre, 1957
Maize	Leaf blight	Berger, 1970
"	Stewart's wilt	Stevens, 1934
Onion	Leaf blight	Shoemaker and Lorbeer, 1977
Pea	Root rot	Reiling et al. 1960
Pear	Fireblight	Thomson et al. 1982
Peanut	Leafspot	Jensen and Boyle, 1965, 1966; Parvin et al. 1974
Potato	Early blight	Harrison et al. 1965
"	Late blight	Beaumont, 1947; Grainger, 1953; Hyre, 1954; Wallin, 1962; Schrödter and Ullrich, 1966; Krause et al. 1975; Forsund, 1983
Rice	Wet stem rot	Webster, 1974
Sugar beet	Sclerotium rot	Backman et al. 1981;
"	Yellows virus	Hull, 1976; Watson et al. 1975
Tea	Blister blight	DeWeille, 1960
Tobacco	Blue mould	Miller, 1958
Tomato	Early blight	Madden et al. 1978; Pennypacker et al. 1983
Wheat	Leaf rust	Burleigh et al. 1972
"	Stem rust	Eversmeyer et al. 1973
"	Stripe rust	Rabbinge and Rijsdijk (pers. commun.)

Table 2.   Recent disease forecasts.

Crop	Disease	Reference
Apple	Powdery mildew	Butt et al. 1983
"	Scab	Jeger et al. 1983
Bean	Root rot	Kobringer and Hagedorn, 1983
Cantaloupe	Leaf blight	Thomas, 1983
Cereals	Barley yellow dwarf virus	Kendall and Smith, 1983; Plumb, 1983
Potato	Early blight	Pscheidt and Stevenson, 1982
"	Late blight	Cupsa et al. 1983; Fry et al. 1983
Soybean	Leaf spots	Backman et al. 1984
"	Mosaic virus	Reusink and Irwin, 1983
Turfgrass	Anthracnose	Danneberger et al. 1984
"	Pythium blight	Nutter et al. 1983
Wheat	Septoria	Royle et al. 1983
"	Stripe rust	Coakley et al. 1983

Recent forecasts proposed for Pythium blight of turfgrass, soybean anth-racnose, turfgrass anthracnose, cantaloupe leaf blight, apple powdery mildew, potato early blight and Septoria of wheat (Table 2), all have a short response time. They identify the need for applications of chemicals from analysis of weather and inoculum factors, and vary in terms of the land area for which they are appropriate.

INCORPORATION OF NON-WEATHER FACTORS
     In attempts to increase the accuracy of disease forecasts, researchers have investigated the impact of cultivar resistance, patho-gen population size and composition, and/or pesticide characteristics. In their regional forecast for wheat stripe rust, Coakley et al. (1983) devised different forecasts (regression equations) for three different wheat cultivars differing in resistance. Forecasts for barley yellow dwarf virus depend on assessments of the pathogen population and its virulence in plants or in aphid vectors (Kendall and Smith, 1983; Plumb, 1983).

Our research group at Cornell University has developed a potato late blight forecast which includes host resistance and fungicide character-istics as well as weather, to identify fungicide application frequency. Construction of the forecast was done from analysis of computer simulation models, hence it is termed the simulation forecast. The influence of host resistance on disease development was quantified in field experiments. Fungitoxicity of various protectant fungicides was assessed in laboratory experiments, while tenacity and redistribution characteristics of various fungicides were determined in glasshouse and field experiments (Bruhn and Fry, 1982a,b; Spadafora et al., 1984). The forecast is such that resistant cultivars will be sprayed less frequently than suceptible ones (Fry et al., 1983). Blight-flavourable weather is considered more frequently between sprays on resistant cultivars than on susceptible ones, and fungicide residues between sprays are assumed to decline further on resistant than on susceptible cultivars. A fungicide spray can be recommended on the basis of the decline of fungicide residues, or by accumulation of blight-favourable weather. In our experience, spray recommendations have been based on both criteria in approximately equal numbers. In field experiments during dry seasons, fungicide decline has more frequently determined the need for a spray, and during wet seasons blight-favourable weather has been the more frequent criterion (unpublished field results).

In addition to weather-dependent spray frequencies, we have also identified cultivar-specific frequencies which, on average, will provide optimal disease suppression. Susceptible cultivars should be sprayed every 6-7 days, moderately susceptible and moderately resistant cultivars every 8-9 and 10-12 days, respectively.

These fixed intervals were determined through the analysis of computer simulations. Costs of fungicide application and costs due to disease-induced yield suppression were calculated and plotted as a function of the interval between applications (Fig. 1). The point of minimum costs identified the optimum spray frequency. In limited field experiments, the cultivar-specific fixed-intervals were as effective as a weather-dependent forecast (Spadafora et al., 1984).

We are also attempting to incorporate host resistance into predictions of the timing of the first fungicide application of the season. We expect that applications can be started later on resistant cultivars than on susceptible ones. In our first three field experiments, late blight in potato foliage appeared in susceptible cultivars at about the time predicted by BLITECAST (Krause et al., 1975, see below). However, blight appeared later in the foliage of moderately resistant cultivars. It is as yet too early for us to identify with confidence a technique to predict the first occurrence of late blight in foliage of moderately resistant cultivars.

APPLICATIONS OF COMPUTERS
The rapidly increasing availability of computers has had, and will continue to have, a large impact on the development, implementation, and evaluation of disease forecasts. One of the first applications in the USA was in the development of BLITECAST, a computerised forecast for potato late blight (Krause et al., 1975; MacKenzie, 1981). During the early stages, BLITECAST was available on a central computer at the Pennsylvania State University. Growers or agents telephoned

Figure 1.   Effect of fungicide application
interval on simulated potato late blight
severity (area under the disease progress
curve) and production costs for chlorothalonil applied to a susceptible potato
cultivar.   (Reprinted from Spadafora et al.,
(1984) with permission from Phytopathology.)

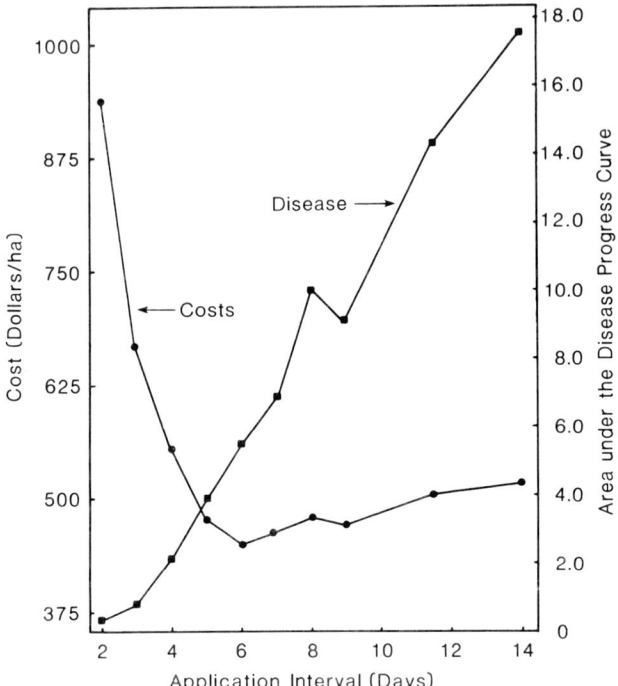

weather data to an operator at the University, who put the data into the
computer and subsequently transmitted a recommendation to the grower or
agent during the same telephone call. Use of a central computer is
important in other disease management systems (Teng and Rouse, 1984),
for example, the EPIPRE system (see Wilson, Chapter 24) uses a central
computer.   In EPIPRE, much data concerning crop development are observed, and transmitted to the computer by the grower.   These data serve as
the basis for recommendations concerning crop (including disease)
management.   In the USA, several integrated pest management (IPM)
programmes use central computers in similar, though less comprehensive,
systems (Teng and Rouse, 1984).

Instead of relying on centralised computers, some researchers have
developed micrometeorological data loggers on microcomputers as single-
purpose disease forecasters.   Jones et al. (1984) favoured a small
dedicated forecaster for apple scab.   Growers use this forecaster
directly and do not depend on operator availability at the University.

Jones and Fisher (1984) indicated that during the early development of
their integrated orchard pest management programme, the University's
computer was not available until about 9.00 h, several hours after spray
recommendations were needed!

This interest in small dedicated forecasters provides an opportunity for
the private sector. One company is attempting to commercialise the
apple scab forecaster, and University research and extension personnel
have co-operated with the company on technical details (Jones et al.,
1984). An advantage of this interaction is that individuals in the
private sector are urging adoption of a forecasting technology which has
been developed and recommended by University scientists.

Computer simulators are likely to become more useful in the development
and analysis of forecasts. The potato late blight forecast which we are
testing was developed from analysis of simulation experiments.
Agricultural economists and plant pathologists at Cornell employed
several computer simulation models to evaluate part of BLITECAST (Fohner
et al., 1984). We simulated epidemics under the common grower practice
of weekly spray applications and under applications timed according to
BLITECAST, and then compared the two strategies in terms of disease
control and numbers of protectant fungicide applications. In the
simulations, spray programmes were initiated on the same date in each
strategy, and the protectant fungicides used had no curative action.
The comparison was made for ten years of real weather data recorded at
Geneva, New York, and for ten years for each of two microclimates
artificially less favourable than that recorded. The first artificial
microclimate (designated moderately favourable) was created by subtracting one hour from each period of high relative humidity ($\geq 90\%$) in
the real data, and the second (designated unfavourable) was created by
subtracting two hours from each period of high relative humidity in the
real data.

Contrary to our expectations, and to the experience of others with
systems in which different fungicides are used, BLITECAST-scheduled
applications of fungicide in the simulation experiments on average did
not suppress late blight more effectively than did the weekly applications (Table 3). For the less favourable microclimates, BLITECAST
scheduled fewer applications than did the common grower practice of
weekly applications. However, these BLITECAST schedules were usually
associated with higher levels of disease. In the favourable microclimate (real weather), BLITECAST scheduled slightly more applications
than the weekly schedule, but these were not associated with decreased
disease.

These unexpected results made us re-evaluate our previous field experiments. Almost all field experiments were conducted under favourable
micro-environments, and they corroborate the trends identified in the
simulation experiment. There was no clear advantage in using BLITECAST
to schedule application frequency. However, BLITECAST seems to predict
accurately that time during the season when fungicide programmes should
be initiated on susceptible cultivars. In response to these results, we

Table 3. Number of applications and disease (%) in diverse microclimates observed in simulation analysis of potato late blight
suppression achieved with protectant fungicide applied weekly or
according to BLITECAST*.

	Favourable		Moderately favourable		Unfavourable	
	No. of applications	Disease (%)	No. of applications	Disease (%)	No. of applications	Disease (%)
Weekly applications	10	15.2 ± 6.7	10	10.0 ± 6.1	10	3.9 ± 3.4
BLITECAST	10.6	16.3 ± 6.0	8.7	16.5 ± 9.1	6.6	9.5 ± 4.6

* Data obtained from Fohner et al. (1984). Disease resulted from moderate exposure to inoculum.

BLITECAST is described in Krause et al. (1975).

modified our recommendations for New York State potato growers and we no
longer urge the use of a weather-dependent forecast for scheduling
application frequency of protectant fungicides.

We are in the process of analysing our simulation forecast in a similar
manner, and are comparing weekly applications with those scheduled by
the simulation forecast system.

Realistic mathematical models as computer simulators should help in the
development of other disease forecasts. An approach that has always
intrigued me is to develop, from analyses using simulated weather, predictions of probable disease levels for a soil-borne pathogen. Inputs
would be initial size (and possibly distribution) of the pathogen population, soil type, weather and host resistance. A simulator could be run
iteratively over different seasons to develop expected disease levels as
a function of pathogen population size and weather. The availability of
a weather simulator which provides stochastic weather typical of a
region (Bruhn et al., 1980) should enable many simulation experiments to
be done without repeating the specific weather of a particular season.
General predictions could then be made from knowledge of the initial
pathogen population and probability of general weather.

IMPLEMENTATION
       In some cases, growers do not adopt a forecasting system as
widely as expected by the scientist who originated it. One factor
contributing to such disappointing uptake may be insufficient follow-up.
Biological scientists may lose interest in a forecast system after its
essential elements have been developed, but additional effort is inevitably required to promote its adoption. The additional effort may
take the form of developing instrumentation, grower education, and/or
preparation of the forecast in a form readily useable by a farmer (Jones

and Fisher, 1984). It appears that EPIPRE has benefitted from considerable implementation effort. The instrumentation involves a central
computer, but does not involve esoteric equipment at the farm, and the
farmer does the monitoring. Farmers are aware of the economic benefits
from participation, and the system provides timely recommendations for a
specific farm.

Instrumentation has been a major feature of some other systems. Use of
computer technology may have enhanced adoption of EPIPRE, and may have
contributed to the initial excitement concerning BLITECAST. Use of
independent, self-contained forecasting micro-computers may enhance
forecast adoption. Commercialisation of the apple scab forecaster
should provide an interesting test of this type of machine. Adoption of
similar forecasters for use with BLITECAST in the USA was not great,
probably due to the limited benefit. The next step in evolution of
instrumentation may be to develop systems compatible with the farmer's
personal computer.

The costs of forecast implementation may also limit adoption by growers.
Costs are those of managerial effort, labour, equipment or pesticide.
Incompatibility with other control tactics may also be considered a type
of cost. If the implementation costs are high, then the benefits of a
forecast must be correspondingly high.

Sometimes forecast costs are determined only after an implementation
effort. In some potato-growing areas, early blight (caused by
Alternaria solani) can be as much a yield limitation as late blight. A
forecast for either disease alone may be difficult for growers to adopt
because the diseases are influenced differently by several factors.
Forecasts for late blight are inappropriate for early blight suppression
(Nutter and MacHardy, 1981). Prior to evaluation of late blight forecasts, however, the importance of early blight was less well-known,
because the protectant fungicides applied regularly for late blight also
suppressed early blight. In response to the inadequacy of current recommendations there are now efforts to develop forecasts which deal with
both diseases.

Consideration of more than one disease increases the complexity of forecasting. If fungicide type and host resistance are included, an overall
best strategy may require the use of optimisation techniques (Shoemaker,
1980). Some research is attempting to identify optimal disease management strategies. In a first step, our research group hopes to identify
an optimal use of protectant fungicide on potatoes when both early
blight and late blight are considered. We plan to include host resistance to each pathogen, fungicide weathering characteristics and
fungitoxicity in a simulation analysis of both diseases. We are
currently developing the simulation model of early blight. Optimal
strategies will be identified either with a trial and error approach or
by optimisation methods. To apply optimisation techniques, it may be
necessary to develop a simpler version of the complex simulation models.
Accurate determination of an optimal strategy for both diseases should
aid grower adoption of the forecast system.

CONCLUSIONS

The potential for disease forecasts to aid the rational use of pesticides has not yet been realised, but advances in several systems move us closer to that goal. New forecasts continue to be developed, and the inventory of diseases for which forecasting systems are being developed increases. Use of computers will continue to aid forecast development and implementation. The next steps appear to include identification of optimal management strategies for crops affected by more than one significant disease or pest.

ACKNOWLEDGEMENTS

Several persons at Cornell are responsible for the progress in forecasting in potatoes. They include: A.E. Apple, J.A. Bruhn, G.R. Fohner, V.J. Spadafora, and G.B. White. I also thank M.G. Milgroon and J.R. Pelletier for helpful comments on the manuscript.

REFERENCES

Backman, P.A., Crawford, M.A. and Hammond, J.M. (1984). Comparison of meteorological and standardized timings of fungicide applications for soybean disease control. Plant Disease 68, 44–46.

Backman, P.A., Rodriquez-Kabana, R., Caulin, M.C., Beltramini, E. and Ziliani, N. (1981). Using the soil-tray technique to predict the incidence of Sclerotium rot in sugar beets. Plant Disease 65, 419–421.

Beaumont, A. (1947). The dependence on the weather of the dates of outbreak of potato blight epidemics. Transactions of the British Mycological Society 31, 45–53.

Berger, R.D. (1969a). A celery early blight spray program based on disease forecasting. Proceedings of the Florida State Horticultural Society 82, 107–111.

Berger, R.D. (1969b). Forecasting Cercospora blight of celery in Florida. Phytopathology 59, 1018 (Abstr.).

Berger, R.D. (1970). Forecasting Helminthosporium turcicum attacks in Florida sweet corn. Phytopathology 60, 1284 (Abstr.).

Bruhn, J.A. and Fry W.E. (1981). Analysis of potato late blight epidemiology by simulation modelling. Phytopathology 71, 612–616.

Bruhn, J.A. and Fry W.E. (1982a). A statistical model of fungicide deposition on potato foliage. Phytopathology 72, 1201–1305.

Bruhn, J.A. and Fry W.E. (1982b). A mathematical model of the spatial and temporal dynamics of chlorothalonil residues on potato foliage. Phytopathology 72, 1306–1312.

Bruhn, J.A., Fry W.E. and Fick, G.W. (1980). Simulation of daily weather data using theoretical probability distributions. Journal of Applied Meteorology 19, 1029–1036.

Burleigh, J.R., Eversmeyer, M.G. and Roelfs, A.P. (1972). Development of linear equations for predicting wheat leaf rust. Phytopathology 62, 947–953.

Butt, D.J., Jeger, M.J. and Swait, A.A.J. (1983). Biometeorological forecasting scheme for apple mildew. Proceedings of the 10th International Congress of Plant Protection, Brighton, 184. British Crop Protection Council Publications.

Channon, A.G. (1981). Forecasting barley mildew development in west Scotland. Annals of Applied Biology 97, 43-53.

Coakley, S.M., Line, R.F. and Boyd, W.S. (1983). Regional models for predicting stripe rust on winter wheat in the Pacific Northwest. Phytopathology 73, 1382-1385.

Cupsa, I., Ignat, V. and Pamfil, G. (1983). Possibilities of forecasting primary infection of potato late blight. Proceedings of the 10th International Congress of Plant Protection, Brighton, 164. British Crop Protection Council Publications.

Danneberger, T.K., Vargas, J.M. and Jones, A.L. (1984). A model for weather-based forecasting of anthracnose on annual bluegrass. Phytopathology 74, 448-451.

DeWeille, G.A. (1960). Blister blight (Exobasidium vexans) in tea and its relationship with environmental conditions. Netherlands Journal of Agricultural Science 8, 183-210.

Eisensmith, S.P. and Jones, A.L. (1981). Infection model for timing fungicide applications to control cherry leaf spot. Plant Disease 65, 955-958.

Eversmeyer, M.G., Burleigh, J.R. and Roelfs, A.P. (1973). Equations for predicting wheat stem rust development. Phytopathology 63, 348-351.

Fohner, G.R., Fry, W.E. and White G.B. (1984). Computer simulation raises question about timing protectant fungicide application frequency according to a potato late blight forecast. Phytopathology 74, 1145-1147.

Forsund, E.F. (1983). Late blight forecasting in Norway 1957-1980. EPPO Bulletin 13, 255-258.

Fry, W.E., Apple, A.E. and Bruhn, J.A. (1983). Evaluation of potato late blight forecasts modified to incorporate host resistance and fungicide weathering. Phytopathology 73, 1054-1059.

Grainger, J. (1953). Potato blight forecasting and its mechanisation. Nature 171, 1012-1014.

Harrison, M.D., Livingston, C.H. and Oshima, N. (1965). Control of potato early blight in Colorado. I. Fungicidal spray schedules in relation to the epidemiology of the disease. American Potato Journal 42, 319-327.

Hull, R. (1976). Research on the sugar beet crop. Annals of Applied Biology 82, 1-10.

Hurst, G.W. (1965). Forecasting the severity of sugar beet yellows. Plant Pathology 14, 47-53.

Hyre, R.A. (1954). Progress in forecasting late blight of potato and tomato. Plant Disease Reporter 38, 245-253.

Hyre, R.A. (1957). Forecasting downy mildew of lima bean. Plant Disease Reporter 41, 7-9.

Jeger, M.J., Butt, D.J. and Swait, A.A.J. (1983). Monitoring host, weathering and inoculum in the control of primary apple scab. Proceedings of the 10th International Congress of

Plant Protection, Brighton, 185. British Crop Protection
Council Publications.

Jensen, R.E. and Boyle, L.W. (1965). The effect of temperature,
relative humidity and precipitation on peanut leafspot.
Plant Disease Reports 49, 975-978.

Jensen, R.E. and Boyle, L.W. (1966). A technique for forecasting
leafspot on peanuts. Plant Disease Reporter 50, 810-814.

Jones, A.L. (1983). Disease predictions: current stastus and future
direction. In Challenging Problems in Plant Health,
eds T. Kommedahl and P.H. Williams, 362-367. St. Paul,
Minnesota: American Phytopathological Society.

Jones, A.L. and Fisher, P.D. (1984). Implementation of predictive
disease control. Plant Disease 68, 87.

Jones, A.L., Fisher, P.D., Seem, R.C., Kroon, J.C., VanDeMotter, P.J.
(1984). Development and commercialization of an in-field
microcomputer delivery system for weather-driven predictive
models. Plant Disease 68, 458-463.

Kendall, D.A., and Smith, B.D. (1983). Damage thresholds in the
forecasting of barley yellow dwarf virus. Proceedings of
the 10th International Congress of Plant Protection,
Brighton, 170. British Crop Protection Council Publications.

Kobriger, K.M. and Hagedorn, D.J. (1983). Determination of bean root
rot potential in vegetable production fields of Wisconsin
central sands. Plant Disease 67, 177-178.

Krause, R.A., Massie, L.B. and Hyre, R.A. (1975). Blitecast: a
computerized forecast of potato late blight. Plant Disease
Reports 59, 95-98.

MacKenzie, D.R. (1981). Scheduling fungicide applications for potato
late blight with BLITECAST. Plant Disease 65, 394-399.

Madden, L., Pennypacker, S.P. and MacNab, A.A. (1978). FAST, a forecast
system for Alternaria solani on tomato. Phytopathology 68,
1354-1358.

Miller, P.R. (1958). Plant disease forecasting. In Plant Pathology:
Problems and Progress 1908-1958, eds C.S. Holten, G.W.
Fischer, R.W. Fulton, H. Hart and S.E.A. McCallan, 557-567.
Madison, Wisconsin: University of Wisconsin Press.

Mills, W.D. (1944). Efficient use of sulfur dusts and sprays during
rain to control apple scab. Cornell Experimental Bulletin,
630. Cornell University, Ithaca, New York, USA.

Nutter, F.W., Cole, H. and Schein, R.D. (1983). Disease forecasting
system for warm weather Pythium blight of turfgrass. Plant
Disease 67, 1126-1128.

Nutter, F.W. and MacHardy, W.E. (1981). Timing of additional sprays for
control of potato early blight (Alternaria solani) when
following a late blight forecasting spray schedule in New
Hampshire. Protection Ecology 3, 47-54.

Parvin, D.W., Jr., Smith, D.H. and Crosby, F.L. (1974). Development and
evaluation of a computerized forecasting method for
Cercospora leafspot of peanuts. Phytopathology 64, 385-388.

Pennypacker, S.P., Madden, L.V. and MacNab, A.A. (1983). Validation of
an early blight forecasting system for tomatoes. Plant
Disease 67, 287-289.

Plumb, R.T. (1983). The infectivity index and barley yellow dwarf
        virus. Proceedings of the 10th International Congress of
        Plant Protection, Brighton, 171. British Crop Protection
        Council Publications.
Pscheidt, J.W. and Stevenson, W.R. (1982). Forecasting potato early
        blight in relation to timing fungicide sprays in Wisconsin.
        Phytopathology 72, 1193 (Abstr.)
Reiling, T.P., King, T.H. and Fields, R.W. (1960). Soil indexing for
        pea root rot and the effect of root rot on yield.
        Phytopathology 50, 287-290.
Royle, D.J., Cook, R.J., and Obst, A. (1983). A European approach to
        forecasting Septoria in winter wheat. Proceedings of the
        10th International Congress of Plant Protection, Brighton,
        173. British Crop Protection Council Publications.
Ruesnik, W.G. and Irwin, M.E. (1983). A model for the impact of soybean
        mosaic virus in soybean. Proceedings of the 10th
        International Congress of Plant Protection, Brighton, 163.
        British Crop Protection Council Publications.
Schrödter, H. and Ullrich, J. (1966). Further research on the
        biometeorology and epidemiology of Phytophthora infestans
        (Mont.) de By. A new proposal for the solution of the
        problem of epidemiological prognosis. Phytopathologische
        Zeitschrift 56, 265-278.
Shoemaker, C.A. (1980). The role of systems analysis in integrated pest
        management. In New Technology of Pest Control, Environ-
        mental Science and Technology ed. C.B. Huffaker, 25-49.
        New York: John Wiley and Sons Inc.
Shoemaker, P.B. and Lorbeer, J.W. (1977). Timing initial fungicide
        application to control Botrytis leaf blight epidemics on
        onions. Phytopathology 67, 409 (Abstr.).
Spadafora, V.J., Bruhn, J.A. and Fry, W.E. (1984). Influence of
        selected protectant fungicides and host resistance on simple
        and complex potato late blight forecasts. Phytopathology
        73, 519-523.
Stevens, N.E. (1934). Stewart's disease in relation to winter
        temperatures. Plant Disease Reporter 12, 141-149.
Tarr, S.A.J. (1972). Principles of Plant Pathology, pp. 423. New York:
        Winchester Press.
Teng, P.S. and Rouse, D.I. (1984). Understanding computers:
        applications in plant pathology. Plant Disease 68, 539-543.
Thomas, C.E. (1983). Fungicide applications based on duration of leaf
        wetness periods to control Alternaria leaf blight of
        cantaloupe in South Texas. Plant Disease 67, 145-147.
Thomson, S.V., Schroth, M.N., Moller, W.J. and Reil, W.O. (1982). A
        forecasting model for fireblight of pear. Plant Disease 66,
        576-579.
Van Everdingen, E. (1926). Het verband tusschen de weergesteldheid en
        de aardappel zeikte (Phytophthora infestans). Tijdschrift
        over Plantenziekten 32, 129.
Wallin, J.R. (1962). Summary of recent progress in predicting late
        blight epidemics in the United States and Canada. American
        Potato Journal 39, 306-312.

Watson, M.A., Heathcote, G.D., Lauckner, F.B. and Sowray, P.A. (1975).
        The use of weather data and counts of aphids in the field to
        predict the incidence of yellowing viruses of sugar beet
        crops in England in relation to the use of insecticides.
        Annals of Applied Biology 81, 181-198.
Webster, R.K. (1974). Relationships between inoculum level, disease
        severity and yield reduction in stem rot of rice.
        Proceedings of the American Phytopathological Society 1,
        106-107.
Zadoks, J.C. (1984). A quarter century of disease warning, 1958-1983.
        Plant Disease 68, 352-355.

19. POTENTIAL OPPORTUNITIES FOR RATIONAL PEST CONTROL IN DEVELOPING COUNTRIES

C.A.J. Putter
Plant Protection Service, FAO, 00100 Rome, Italy

INTRODUCTION

Developing countries are usually discussed in terms of the extent and nature of the problems that impede the process of rural improvement. These problems are important and deserve the attention that they receive. However, there are many positive aspects of agricultural production patterns in developing countries that should also be considered. Germane to the theme of this book are the potential opportunities for rational pest control inherent in traditional farming practices.

In an extensive review of pest management practices of small farmers, Matteson et al. (1984) have pointed out that few technical packages aimed at improving the well-being of small farmers have been successful. A frequent criticism of technology packages intended for developing countries in general, is that they display a lack of understanding and appreciation of the ecological and socio-economic milieu in which farmers in these countries operate; that they exclude the small farmer as both collaborator and beneficiary, and that inept and inappropriate technology is promoted as a result (de Janvry, 1981). On the other hand, there is well-documented though sparse literature that describes the wealth of ecological knowledge that underlies traditional, subsistence farming methods (Anderson, 1952; Barrau, 1955, 1958, 1961; Conklin, 1957; Spencer, 1966; Panoff, 1969; Allen, 1976; see also the review by Matteson et al., 1984). In addition to this ecological soundness, traditional agriculture is also closely intertwined with the cultural values and social norms of these societies. Thus, crop protection demands consideration of social and cultural values integrated with an ecological perspective superimposed on the usual technical and scientific skills associated with the science (Putter, 1976, 1978, 1980a,b).

This paper reviews potential opportunities for rational pest control strategies that recognise these needs. The situation is first described in general and then followed by a detailed discussion of a particular pest management procedure. Next, a pest control programme in a developed country is discussed and compared with an example from a developing country, with an emphasis on the value of farmer participation.

THE POTENTIAL FOR PESTICIDES IN DEVELOPING COUNTRIES

Discussions of agriculture in developing countries frequently focus only on agricultural production patterns that may be described as developed agriculture practiced in a developing country. Usually, the emphasis is on cash crops such as coffee, cacao, rubber and oil palm, or on food crops such as maize, sorghum and cassava that are produced intensively and frequently on an extensive scale. The sophistication of such production systems, as practiced in plantation crops for example, is comparable to the level of management in orchard crops in developed countries.

The significance of this observation is that the technology and basic scientific knowledge used to manage farming systems in developed countries may often be transferred to developing countries after appropriate adaptation. Thus, rural development based on plantation cash cropping can proceed from the assumption that the value changes which Enke (1964) defined as pre-requisites for rural development have either already taken place or could be brought about relatively easily.

Once a farmer or estate manager is locked into plantation cash cropping, he is committed to yield maximisation and cost-benefit optimisation; usually he will be literate and will have had at least some experience in manipulating technological products such as tractors and spraying equipment. He will be served by a reasonable communications infrastructure and transport system which will enable him to participate in a cash economy via his purchases of raw materials such as fertilisers, fuel, machinery and pesticides. His apparent favoured socio-economic position may leave much to be desired, either in real terms or by "Western" standards, but in comparison with his subsistence farming counterparts, the plantation crop farmer can readily be shown to be prepared for "take-off" economic forces to catapult him into the mainstream of rural development programmes.

In such situations, pest management becomes a financial aspect of farm management. It is possible to think in terms of "economic thresholds" and to trade off cost of control against potential financial returns. Crop protection can thus be viewed as an aspect of rural development economics as much as it is a technical research activity. The role of sociology may be limited to considerations at the level of the State, and to those aspects of sociology that enter extension programmes aimed at convincing farmers to adopt those optimum, economic pest control procedures developed in the crop protection research programmes. Thus, significant rural development may, under these circumstances, develop from a concentration on improving the managerial skills of the farmers and depends on the ability of the extension service to achieve this.

There are notable exceptions to this general description. In several countries, coffee crops are tended and managed only when the price warrants inputs of time and effort. When prices decline, plantation management, as reflected in routine pruning for example, is neglected. During these periods, the coffee crop is left to its own devices and may then be regarded as a crop that is "gathered" rather than farmed.

Plantation cropping and cash farming as described here, have an important role to play in the future of the developing countries. Indeed, some may argue that it is an essential and perhaps the only means of achieving meaningful rural improvement because it represents the traditional evolutionary route of rural societies from subsistence farming towards industrialisation. Whether there are other routes or whether this particular route is the most desirable falls outside this discussion. Important here is the consideration that pesticide use in developing countries is confined to what may be referred to as "industrial" agriculture where cash-market economics prevail. The purchase of pesticides requires not only money but foreign exchange and whether the buyer is an individual farmer, a plantation manager or a government co-operative, pesticide purchase and application are only possible in a system where monetary principles are applied. As far as rational pesticide use is concerned, it has already been mentioned that concepts such as "economic thresholds" employed in integrated pest management (IPM), only have meaning in farming systems where production efficiency is measured in monetary terms.

This raises the question as to what proportion of the agricultural activities in developing countries is currently compatible with crop protection strategies based on pesticide use as whole, let alone rational use of these chemicals. This issue may be discussed with reference to the classification presented in Table 1.

Compiled from several sources of demographic data, Table 1 is intended to place the current market for pesticides and "developed" crop protection technology in perspective. The data on which this analytical classification is based are real, as are the different indicator categories. On the other hand, the agricultural production systems shown are convenient rather than axiomatic. The boundaries between these activities are not usually clear-cut. These categories should, therefore, be viewed as points in a continuum rather than as discrete entities. The argument presented here will not be invalidated by pointing out that farmers in developing countries may be involved simultaneously in farming activities in more than one of these categories. Table 1 is aimed at alerting those involved in crop protection, so that on a global basis they may be confining their efforts to a limited portion of agricultural activities in developing countries. It emphasises that the major activity of developing countries involves peasant farmers, whose main concern is to ensure food security for their families.

Some cropping patterns and systems which may not conform to this classification or are too complex are not placed in Table 1. An example from coffee production has already been quoted. Rice cropping in Asia would be another example. Here, one can find some families that use pesticides even though they grow the rice solely for subsistence purposes, while others are locked into intensive, profit-maximising farming that rivals wheat farming in developed countries in terms of sophistication. With such a range of over-lapping social and economic forces, rational pest management demands a dynamic and adaptable

approach.   In the FAO Intercountry Programme on Integrated Pest Control
(IPC) in rice in Asia, under the leadership of J. Lowe and P. Kenmore,
IPC principles have been modified so that they can be subsumed by the
value systems of the farmers.   The success of this programme may, <u>inter
alia</u>, be attributed to the presentation of IPC concepts and knowledge in
such ways that they can be assimilated by and blended with existing
knowledge systems, even through access to these knowledge systems may be
gained initially via interaction with social, cultural or economic
values.   This is very different from programmes that expect the farmer
to adapt to the new knowledge; and that subsume him and replace his
traditions by a set of foreign production criteria.

Table 1.   A typology of agriculture in developing countries.

Input indicator	Farming category		
	Subsistence farming	Commercial food production	Plantation agriculture
Population distribution (%)	70–95	5–30	0–5
Nutritional status	Adequate	Fragile	Good
Economic system	Bartered exchange	National markets	International commodity markets
Marketing pattern	Dictated by supply and demand	Controlled by market forces and dependent on political decisions	
Available technology	Indigenous	Indigenous and imported	Sophisticated and imported
Cropping pattern	Diverse polyculture	Mixed cropping	Approaching monoculture
Ecological understanding	Extensive	Being undermined by growing reliance on technology-dependent solutions	
Ecological stability	Stable	Robust	Fragile

Population pressures coupled to the sophistication demanded
of primary food production and delivery systems to feed urbanised and
industrialised societies will inevitably accelerate the movement of
procedures along a technology gradient from left to right in Table 1.
Subsistence farming will have to evolve into cash cropping in a market
economy. This will have many far reaching consequences most of which
fall outside the scope of this presentation. However, there is one
aspect that has important implications and potential benefits for the
evolution of rational pest management strategies in developing
countries. It may best be explained first by making a general comparison of pest management in developed and developing countries and
then by discussing a specific example that illustrates the difference.

A COMPARISON OF PEST MANAGEMENT TRENDS IN DEVELOPED AND DEVELOPING COUNTRIES

In the developed countries, the problem in recent years has
been to shift the emphasis in pesticide use away from regular calendar-
based applications to need-based applications; away from routine
pesticide use which evolved from the axiom that prevention is better
than cure. Now the aim is to promote programmes wherein ecological
awareness, knowledge of the environmental impact of indiscriminate
pesticide use and careful monitoring of pest densities play an important
role. One of the vehicles used to develop a more rational approach has
been Integrated Pest Management (IPM). Nevertheless, pesticides are
still the major defence tool in our crop protection armoury and the task
of integrating their use with other control measures remains a daunting
challenge.

In terms of the changes required by an individual farmer, the swing away
from sole reliance on routine pesticide use may be viewed as an attempt
to create an ecological awareness and conscience that is compatible with
the economic demands of industrial agricultural production. This and
other related, more abstract aspects of IPM are what make it a science
with socio-economic and cultural components as much as it is an agricultural research and technical activity.

In developing countries, the trend is in the opposite direction: away
from pest management based on ecological principles towards reliance on
routine pesticide application. Here, the majority of the population is
involved with subsistence farming and is renowned for its ecological
awareness. Even non-farmers, senior administrators and politicians have
a recent rural background and most, if not all, will have experience of
the labour involved in food production. These officials thus have a
tremendous advantage over their counterparts in developed countries
where a person may become the Minister of Agriculture without knowing
anything about farming, rural societies or nature. He may be a third
generation urban dweller who only experiences nature in his leisure
hours when it is not raining. This difference between agricultural
policy makers in developing and developed countries presents an
important opportunity for the implementation of rational pesticide use
strategies in developing countries. However, the emphasis here will be

on the potential value of exploiting the ecological common-sense that
underlies traditional farming methods.

OPPORTUNITIES FOR RATIONAL PEST MANAGEMENT PRESENTED BY TRADITIONAL FARMING METHODS

Integration as the cornerstone of IPM is a common, assumed
virtue of subsistence farming procedures. Thus, the objective of rural
developers should be to prevent the adoption of calendar-based pesticide
applications and to conserve and dove-tail with, the ecological knowledge of subsistence farmers. Compared with the other two categories of
agricultural production patterns listed in Table 1, it will be considerably easier to ensure rational pest management based on ecological principles among farmers who have a recent history of subsistence farming.
It is not the intention here to present a cavalier dismissal of the real
problems faced by plantation cash cropping in developing countries.
Many problems exist and will be with us for a long time. However, they
are due to distributional problems associated with resources, knowledge
and managerial skills rather than the lack of information on which to
base appropriate rural improvement strategies. For example, the
plantation- and cash-cropping production systems may require only
relatively minor modification of the IPM "doctrine". These changes
would be sociological rather than technical and would involve extension
programmes aimed at delivering IPM ideas and technical inputs. This
process of technology transfer will probably not require major, technical shifts in the IPM concept.

The sophistication of subsistence farming practices is exemplified by
the crop protection procedures developed by certain Tolai banana farmers
on the Gazelle Peninsula of East New Britain in Papua New Guinea (PNG).

East New Britain is regarded as one of the major centres of origin and
diversification of bananas (Argent, 1976). Consequently, the evolution
of the indigenous agricultural system and social customs is inseparably
intertwined with banana production. This is further reflected by the
position of bananas as the second most important food crop in PNG
(Walters, 1963; Kimber, 1972;) and their significance as a primary and
preferred food crop among the Tolai people on the Gazelle Peninsula.

Tolai banana crops are attacked by several pests. Those pertinent to
this presentation are the banana scab moth, Nacoleia octesema; the rhino
beetle, Oryctes rhinoceros; and the Sigatoka leaf disease complex caused
by Mycosphaerella spp.

Scab moth damage is caused by the feeding of N. octesema larvae on
banana fruit. In mild infestations the damage to banana skins is super-
ficial and largely cosmetic. However, when the larvae invade and feed
on the young bunches, they deform the immature banana fingers thereby
inhibiting further development. In severe cases, entire bunches may be
destroyed.

Sigatoka is a major debilitating leaf-spotting disease. Necrotic and dead leaf tissue contain reproductive structures that continue to produce ascospores even when the leaves are removed from the plant. Sanitation consisting of regular removal and destruction of leaf litter has been recommended (O'Connor, 1969; Kranz et al., 1977).

Rhino beetles damage banana pseudostems by burrowing into the stems and eating these from inside. Though a relatively minor pest, they cause sporadic damage.

Tolai banana farmers have developed several control procedures to cope with these pests. Their solution for the control of the scab moth may best be considered in the light of the known ecology and life history of the scab moth. Paine (1964) investigated the ecology of N. octesema and came to several conclusions that corroborate traditional Tolai crop protection practices. Thus, it is known that the scab moth female is attracted to nectar and other olfactory stimuli that direct her to lay her eggs on the petioles and flags of the youngest leaves in close proximity to the base of young banana hands. When the larvae hatch, they crawl under the bracts, where they feed on the newly-formed banana fingers, and are relatively protected against predators, sun and drying heat.

Following on the work of Paine (1964), banana farmers were advised to treat their crops with DDT or a similar insecticide (O'Connor, 1969; Feakin, 1971). The treatment with these insecticides was based on knowledge of the ecology of N. octesema and the growth habit of banana bracts. The recommended application procedure was to use a rubber "puffball" attached to a spout which has to be inserted behind the bract as soon as it starts to lift away from the banana hand.

In contrast, the Tolai solution is simple: since the bracts serve no purpose, as can be deduced from observing banana varieties with non-persistent bracts, they can be removed completely rather than "lifted carefully while inserting the insecticide dust applicator". Because bananas are parthenocarpic, which the Tolai farmers deduced from the obvious fact that bananas usually do not produce seed, the flowers, which attract the female moth of N. octesema, can be removed. These practices take care of scab moth control, but what about the other two pests mentioned?

In a well-kept Tolai banana garden, there are other management practices aimed specifically at achieving pest control. Lower banana leaves that are browning off and can no longer contribute to yield are removed. This is done by pulling these older leaves off the main pseudostem together with the leaf sheath which is attached to the pseudostem as a frayed and fibrous covering. The exposed pseudostem emerges with a smooth surface that is almost as slippery as glass.

The removed leaves contain large numbers of Sigatoka lesions and fruiting bodies that would otherwise have contributed to ensuring survival of Mycosphaerella as an endemic pathogen in banana gardens. Furthermore,

these old leaves and other debris camouflage the scab moth female which
is exactly the same colour as dead leaf tissue. This provides
protection for the scab moth from her natural enemies. The Tolai burn
these old leaves in the evening because they are of the opinion that the
scab moth flies at this time. They contend that the smoke from the
burning leaves will cause olfactory confusion in the moth as she
searches for ovipositing sites. Indeed, they are so aware of the
olfactory stimulus that some go further to collect green vegetation of
specific plants in the forest to burn with the old banana leaves to make
the smoke even more acrid and effective as a means of leading the scab
moth astray.

Leaves are burned in the banana garden among the plants. Burning sites
are consciously rotated so that the fire can control weeds. The ash is
subsequently worked into the soil as fertiliser. At the same time,
other surface debris that contain Sigatoka inoculum and provide shelter
for the scab moth female are raked into the fire.

Rhino beetles, according to at least one Tolai farmer, are "... good at
flying but not so good at landing." This refers to the fact that the
adult rhino beetle in flight is rather cumbersome and that it is unable
to alight on a smooth banana stem. On the other hand, when the stem is
covered in the frayed sheaths of old banana leaves the "hooks" on the
front legs of the beetle enable it to grab on to the stem upon
collision. Pseudostems covered with frayed leaf debris also enable the
beetle to climb up the stem and to burrow into the upper, softer
portions of the pseudostem. In contrast, clean, smooth pseudostems have
two pest-reducing properties. Firstly, the stems at ground level "...
are tougher to eat". Secondly, the available surface area, or the
possible number of feeding sites where entry can be successful, is
reduced drastically.

The relative merits of these practices and significance of these pests
are not important to the context of this paper. Rhino beetle damage may
be unimportant and the leaf-stripping may not be "cost-effective".
Similarly, the Sigatoka disease may be an unimportant problem or the
contribution of sanitation to reducing disease severity may be insignificant. Even the obvious good sense displayed by Tolai farmers when
they control scab moth and the potential value of their control methods
in other areas where scab moth is a pest, are incidental to the theme of
this paper. These anecdotes are important in two other ways: they
present proof of an ecologically-based body of crop protection knowledge
among certain peasant farmers, and they shown how easily such knowledge
can be overlooked and how easy it is to remain locked into a dependence
on pesticides when it is assumed that their application has merely to be
rationalised or optimised.

One may question whether it is a concidence that Sigatoka negra, caused
by M. fijiensis var. difformis, considered to be indigenous to East New
Britain, does not cause the catastrophic crop losses there that it has
done in Central America (Stover, 1980). Of course, if bananas are indigenous to East New Britain, or at least if it is a significant centre

of diversity, then natural selection could have contributed to the apparent absence of Sigatoka negra-induced catastrophies in PNG. However, if Tolai banana farmers display such an explicit understanding of cause and effect, would it be far-fetching also to give them credit for a more deliberate role in the process of selecting for resistance to Sigatoka disease?

We may speculate that a banana industry in East New Britain could have evolved readily on the basis of indigenous understanding of this crop. Instead, a cacao industry was established and it has made an important contribution to the economy of the Gazelle Peninsula and PNG in general. I do not aim to detract from this. Rather it is mentioned as an example where rural planners had an excellent opportunity to bring about a transition from a subsistence farming pattern to a cash farming system without imposing foreign agricultural knowledge. The proximity of PNG to Japan, its traditional economic ties with Australia, established shipping facilities as well as the ready demand for, and cash value of, bananas, would have made the crop an ideal candidate for an agricultural development project.

SUBSISTENCE FARMING SCIENCE
A central thesis of this presentation is that the kind of knowledge displayed by Tolai banana farmers does not reflect a passive, traditional dogma unrelated to cause-and-effect understanding. They do not control scab moth without understanding the underlying ecological reasons why their actions are effective. To be sure, many farmers follow the methods slavishly, because "...their fathers did it that way." However, a small proportion of farmers know not only what they should do to avoid pest damage but also why it works. Thus, I contend that this example is merely a particular instance of subsistence farming science, and that knowledge of this kind is ubiquitous in societies where food production is based on a traditional crop with a long history. If this is the case, then it represents a valuable but usually ignored opportunity, not only as a basis for rational pest control, but also as a means for deciding which kinds of rural development programmes should be chosen in the first place. This view is sufficiently different from accepted opinions to warrant a digression.

It has been suggested that subsistence farmers are submerged in their ecosystems; that they are unable to "wonder at" the system and therefore unable to conceive of the possibility of "manipulating it" deliberately. This passivity and the stoicism that it implies, coupled to the apparently slavish adherence of peasant farmers to inherited practices, may be some of the considerations that have given rise to the widespread and popular belief that subsistence farming methods are governed by traditional dogma rather than cause-and-effect understanding.

Although this may be true of the majority of peasant farmers in any one community, the author's own experience in PNG is that approximately 5% of these farmers have a fundamental understanding of the underlying processes involved. These farmers are the custodians of subsistence

farming science and their farming practices reflect conscious and
deliberate experimentation conducted in their daily lives.

The minority status of farmers with such understanding may be one reason
why they and the knowledge they have, escape attention. Furthermore,
attempts to assess whether there is a rational, systematic and formulated knowledge, i.e. by dictionary definition a "science", underlying
subsistence farming procedures, are further hampered in peasant
societies because crop and cropping patterns are often intertwined with
mysticism and religion.

Panoff (1969) gave an example of how subsistence crops are conceived of
as having a soul and where the society's calendar of religious festivals
is dictated by cropping cycles and environmental influences. Where such
a high degree of animism prevails, it is not difficult to see how outsiders could conclude that farmers in such societies would be reluctant
to tamper with a system so closely associated with gods. However, even
in the example of Maenge horticulture discussed by Panoff (loc. cit.),
the religious and scientific views are not mutually exclusive. Instead
they have been amalgamated so that there is a religiously acceptable
procedure for experimenting with new varieties.

Similarly, Putter (1978) gave an example of religious beliefs and
cultural values intertwined with the perception of a plant disease
problem and specifically with an explanation for blight epidemics,
caused by Phytophthora colocasiae, on the subsistence root crop taro
(Colocasia esculenta). Subsequently, Putter (unpublished) found that
taro farmers on Kar-Kar island in PNG, associate fluctuations in blight
severity with the activities of the god who rules the local volcano.
Thus, whenever the volcano on the island is active, taro blight is more
severe and this is something they consider to be an inevitable consequence of the vagaries of the gods. That part of the deduction that
concerns the fluctuation in blight severity as volcanic activity varies,
was based on experience and observation. The process that inhibits the
farmers from using this knowledge in a rational disease control programme lies in the fact that this information about the disease is con-
founded with superstition.

An analysis of the weather patterns on the island when the volcano is
active revealed that the average daily temperature and relative humidity
are, respectively, lower and higher than when the volcano is dormant.
Precipitation also increases during periods of volcanic activity and
these local weather changes are favourable for P. colocasiae (Putter,
1976). Armed with a rational explanation for the pattern of association
between religious beliefs and disease development patterns, the local
farmers were "guided" to an understanding of how the evil gods achieve
their negative effects. This knowledge enabled the farmers to take
counter measures without generating any anxiety about going against an
inimical god. The grafting of new ideas on to existing ones may demand
an unorthodox compromise between the western scientist's technical views
and religious, superstitious and cultural values of the community.

At least one message should be clear from the Tolai account: listen to farmers and involve them as collaborators in every phase of rural development projects. Concomitantly, it suggests that we should also alter what we are prepared to hear. Few extension workers will dispute the value of listening to farmers to ensure that their "real" and "felt" needs are identified correctly. However, this kind of listening accentuates the negative by emphasising how "miserable" the peasant farmer is. On the other hand, if the aim of listening is to learn from a position of open-mindedness and to give credit to the potential contribution of subsistence farming science, then the rural improvement process stands a chance of developing into a dynamic, two-way partnership in which knowledge exchange can become an important driving ethic.

FARMER PARTICIPATION IN THE EVOLUTION OF RATIONAL PEST MANAGEMENT STRATEGIES

The value of farmer participation is not limited to developing countries. Farmers in developed agriculture will usually respond when their participation is solicited. The concept of farming as a business and as a process of manipulating complex processes via scientifically understood ideas, has given some farmers in developed countries the notion that many decisions and observations are better left to the "experts". For example, the decision whether or not to spray is usually based on pest-predator ratios that were not monitored by the farmers themselves. Agricultural research and extension programmes attempt to attend to every demand and enquiry of the farmer. An expert is always at the other end of the telephone and when a farmer does "wonder" about something he will ask an expert rather than experiment to find out for himself. Of course, this is a vast over-simplification, but coupled to other factors such as the economic imperative to make a profit, farmers in developing countries are increasingly dependent on the advice they receive from their expert services.

In order to test how individuals in a farming community may be brought together to develop a rational basis for pest management and to evaluate the role of a potato late blight forecasting system, Putter (1982) developed in Natal, South Africa, a disease monitoring system that relied totally on farmer participation in both the design and implementation of the project.

Although an effective potato late blight forecasting system had been available for many years and was economically successful in reducing fungicide applications (Putter, 1968), farmers appeared unenthusiastic about it. Indeed, after 20 years, the service had not been evaluated to determine how many farmers used it (Putter, 1982). As part of the programme to investigate the role of this forecasting system, several meetings were held with farmers. At one of these, the idea arose that an unsprayed farm next to a sprayed farm might influence the disease development pattern in that area. This was raised by a farmer who pointed out that the efficiency of the warning system can only be

maximised if all farmers used it diligently. Further discussions
convinced the farmer group that the ratio of the areas of treated to
untreated fields as well as the spatial arrangement of the fields in
this ratio will determine disease activity in the community as a whole.
This led to the idea of a pest management situation analogous to
situations where life-boat theory applies, i.e. farmers who are collectively at risk to one another's pests and pest management routines, find
themselves in a 'life-boat' situation with regard to pest management.
Thus, the group intuitively arrived at the concept that a pest problem
at the community level is a bounded event within which they are mutually
interdependent.

Putter (1982, 1983) subsequently developed the idea of a "pathotope"
from the Greek pathos for suffering and topos for place. Thus, pathotope is to pathosystem (sensu: Robinson, 1976), what ecotope (sensu:
Whittaker et al., 1973) is to ecosystem. A pathotope defines a pest
situation at the community level and is an area within which farmers are
collectively at risk to each other's disease to a given level of
probability. Only the programme of involving the farmers in pathotope
management as a basis for rational pest control and not the quantitative
details of the pathotope concept, will be discussed here.

Pathotopes were tentatively delimited in terms of planting dates,
irrigation practices, varieties cultivated and source of seed. Next, a
map of each pathotope was printed on continuous stationery in four
colours showing roads, rivers, farm boundaries, towns and railway lines.
Every map was bordered by a grid reference to enable each farmer to fix
his position on the map. Once a week each farmer completed and dis-
patched a card that contained information about his crop protection
practices. These data were received and processed by an Apple microcomputer which printed the information at the relative positions on the
maps. The completed maps were mailed to each participating farmer
thereby providing a regular, visible account of what his neighbours were
experiencing in their war against pests. Among other things, this enabled him to draw lines connecting reports of pest incidence on his map
to see if there was evidence of an advancing "front" heading in his
direction.

There are two major advantages of this system. Firstly, the farmers
soon realised the benefits of synchronising their control efforts. In
this situation, potato blight epidemics apparently occurred along a well
defined pathway similar to the Puccinia pathway in North America. Thus,
potato late blight migrates from one pathotope to the next. In turn,
and assuming that every farm in a particular pathotope has an equal
chance of receiving the intial inoculum, it makes obvious good sense for
all the farms to be covered with a protectant spray when invasion is
imminent. The definition of "imminent invasion" was worked out with
farmers in terms of the degree of risk they were prepared to take.

The whole issue of risk aversion associated with disease forecasting,
and especially with potato late blight forecasting (McKenzie, 1984), was
therefore approached differently. Putter (1982) decided that with a

high value crop such as potatoes, farmers would probably opt for a very conservative, preventative programme thereby foregoing the saving of a few early fungicide applications. Therefore, a procedure was adopted which would enable farmers to decide for themselves to what extent they were prepared to place their crops at risk to late blight attack for the sake of saving a few initial applications of fungicide. They coped with the problem by deciding to spray a few days earlier than the previous, earliest recorded date of first blight incidence in their pathotope and to do so all on the same day. In this way, the decision was made by farmers and every pathotope community could decide how it was going to approach the problem.

Secondly, the data about pests were collected and stored by grid reference. This facilitated risk analysis on a geographical basis. It may be possible after several years of data collection, to determine that late blight frequently starts in a particular area.

Had the farmers been approached for their co-operation to contribute data to a data base to enable a risk calculation to be made at some time in the future, they would have been reluctant to take the trouble. A data base aimed at calculating probabilities of blight attack based on records of disease incidence in previous years, will require a considerable amount of historic data. To ask farmers to contribute to such a data base without their being able to perceive any immediate, practical benefit to their current, day-to-day decisions would be an extension officer's nightmare. With the self-administered pathotope approach the farmers were collecting data that had immediate significance to them. Effortlessly and without knowing it, they were contributing to a data base that could help them in important ways in the future.

Other advantages of the approach include the mapping of data in a spatial arrangement that is far more meaningful than the same data in tabular form. Because farmers could see the advantage of complete participation by all potato growers in a pathotope, they took it on themselves to solicit participation from those farmers who initially did not choose to join in. In effect, a self-sustaining, unofficial pest management authority run by farmers for farmers was created. They were placed in charge of their own pest management decisions with a degree of catalysis provided by an outside "expert".

When a pathotope is managed on this basis, forecasting assumes a subservient role to the group-dictated decision of the degree of risk to be tolerated. Seen from the pathotope point of view, the gain in synchronised application that could reduce blight severity in other pathotopes further along the late blight pathway and the value of a community united into a concrete pest management group on rational grounds, far outweighs the benefits of saving a few sprays from following a forecasting system. The pathotope as a basis for managing pests in terms of their activity at the community level, with the word community used both in its ecological and social sense, can reasonably be expected to lead to other, more rational pest management decisions.

CONCLUSIONS
        Both examples of pest management discussed in this paper,
the one from a peasant society practising subsistence farming, the other
from a group of farmers in a developed country producing a high value
cash crop in a sophisticated production system, have convinced the
author that farmers can be used much more effectively in the search for
rational pest control strategies. The key to success is to involve them
from the outset: in the definition of the problem and the objectives to
be reached, and when potential ways of finding solutions are explored.
A farmer who is treated as a person with a potential, rational
contribution to make to the problem that affects his livelihood becomes
a friend and an ally and ceases to be a subject who has to be motivated
and manipulated by experts operating in authoritarian, "top-down"
advisory hierarchies.

REFERENCES
Allen, W. (1976). The African Husbandman. London: Oliver and Boyd.
Anderson, E. (1952). Plants, Man and Life, pp. 245. Noston: Little
        Brown and Co.
Argent, G.C.G. (1976). Wild bananas of Papua New Guinea. Notes of the
        Royal Botanic Garden, Edinburgh 35, 77-114.
Barrau, J. (1955). Native Subsistence Agriculture in New Guinea, pp.
        92. South Pacific Commission, Noumea, New Caledonia.
Barrau, J. (1958). Subsistence Agriculture in Melanesia. Bernice P.
        Bishop Museum, Honolulu.
Barrau, J. (1961). Subsistence Agriculture in Polynesia and Micronesia.
        Bernice P. Bishop Museum, Honolulu.
Conklin, H.C. (1957). Hanunoo Agriculture. A Report on an Integral
        System of Shifting Cultivation in the Philippines. Rome:
        FAO.
de Janvry, A. (1981). The Agrarian Question and Reformism in Latin
        America, pp. 311. Baltimore: Johns Hopkins University
        Press.
Enke, S. (1964). Economics for Development. London: Dennis Dobson.
Feakin, Susan, D. (1971). Pest Control in Bananas, PANS Manual No. 1,
        pp. 128. London: PANS.
Kranz, J., Schmutterer, H. and Koch, W. (1977). Diseases, Pests and
        Weeds in Tropical Crops, pp. 666. Berlin and Hamburg:
        Verlag Paul Parey.
Kimber, A.J. (1972). The sweet potato in subsistence agriculture.
        Papua New Guinea Journal of Agriculture 3-4, 80-97.
McKenzie, D.R. (1984). Blitecast in retrospect – a look at what we
        learned. FAO Plant Protection Bulletin 31(2), 45-49.
Matteson, P.C., Altieri, M. A., Cagne, W. C. (1984). Modification of
        small farmer practices for better pest management. Annual
        Review of Entomology 29. 383-402.
O'Connor, B.A. (1969). Exotic Plant Pests and Diseases. Noumea New
        Caledonia: South Pacific Commission.
Paine, R.W. (1964). The banana scab moth Nacoleia octesema (Meyrick):
        its distribution, ecology and control, Technical Paper 145,
        pp. 70. New Caledonia: South Pacific Commission.

Panoft, F. (1969). Some facets of Maenge horticulture. Oceania 40(1),
      20–31.
Putter, C.A.J. (1968). The economic importance of late blight and
      potato leaf roll virus on potatoes in the Natal Midlands.
      Research Report of the Department of Agriculture and
      Fisheries, South Africa.
Putter, C.A.J. (1976). The Phenology and Epidemiology of Phytophthora
      colcasiae Racib., on Taro in East New Britain Province,
      Papua New Guinea. M.Sc. Thesis, University of Papua New
      Guinea.
Putter, C.A.J. (1978). Pathosystems management under subsistence farming
      conditions. 3rd International Congress of Plant Pathology,
      Munich, 16–23.
Putter, C.A.J. (1980a). The management of epidemic levels of endemic
      disease under tropical subsistence farming conditions. In
      Comparative Epidemiology: A Tool for Better Disease
      Management, eds J. Palti and J. Kranz, 72–93. Wageningen:
      Pudoc, Centre for Agricultural Publishing and Documentation.
Putter, C.A.J. (1980b). An Epidemiological Analysis of the Phytophthora
      and Alternaria Blight Pathosystem in the Natal Midlands.
      Ph.D. Thesis, University of Natal, South Africa.
Putter, C.A.J. (1983). The "Pathotope": a proposed new pathosystems
      management concept. 4th International Congress of Plant
      Pathology, Melbourne, Australia.
Robinson, R.A. (1976). Plant Pathosystems, pp. 184. Heidelberg, Berlin:
      Springer-Verlag.
Spencer, J.E. (1966). Shifting Cultivation in Southeastern Asia.
      University of California Press.
Stover, R.H. (1980). Sigatoka leaf spots of banana and plantain. Plant
      Disease 64, 750–756.
Walters, C.L. (1963). Survey of Indigenous Agriculture and Ancillary
      surveys, pp. 22. Bureau of Statistics, Papua New Guinea.
Wittaker, R.M., Levin, S.A. and Root, T.B. (1973). Niche, habitat and
      ecotope. American Naturalist 107, 321–338.

20. FORECASTING ANNUAL CROP PESTS: ADVANTAGES AND LIMITATIONS

Y. Robert
INRA Laboratoire de Zoologie,
BP 29 F-35650 Le Rheu, France

INTRODUCTION
Although the time between sowing and harvest is relatively short, annual crops are particularly prone to attack by an array of pests at any stage of their phenology. This results in crop losses both in quality and in quantity, and these are not always accurately assessed or rightly attributed. It has been accepted widely that pesticides could solve every problem, especially since the 1950s when new chemicals became available. The trend towards routine insurance treatments, however, has had some well-known environmental consequences; for example, aldrin could be detected in the soil 15 years after its application. Changes in agricultural practices and husbandry, in cultivars and in the technical abilities of farmers, and the increasing problem of pesticide resistance, have prompted the production and use of many novel agrochemicals. Who would have imagined 15 years ago that so many thousands of hectares of cereal crops would be sprayed! To avoid any irretrievable ecological damage in the future, and to satisfy increasing economic constraints on productivity and profitability, it has become obvious that there should be more rational use of pesticides. This implies that the target pests should be identified, their bioecology and actual contribution to yield losses precisely assessed, and the economics of control measures studied. Practically, this also means that an economic injury level should be correctly calculated for a range of situations and that the related critical pest population density should be anticipated and predicted early enough to allow the grower to take action only if necessary. To this end, existing warning schemes have been improved and new ones established. They are based upon three components: monitoring, forecasting and communication (Carter et al., 1982). Forecasting has been given much publicity as it is often seen as the means of achieving a more rational use of pesticides (Norton and Way, 1983). This has stimulated basic and applied research and has improved pest control strategies at all levels from conceiving to applying them. This paper describes some of the main advantages and limitations of forecasting annual crop pests and is illustrated by case studies.

FORECASTING AS A COMPONENT OF WARNING SCHEMES
Any decision to develop an annual crop pest forecasting system is usually taken, (1) because of problems arising from chemical pest control, (2) to try and overcome these problems in the future, and (3) to improve chemical control and to optimise it at lower costs.

However, effective biological control, whatever procedures are used to achieve it, also necessitates prediction of likely and significant trends in target pest populations. Thus, forecasting must be considered as a contribution to the overall process of achieving an optimum timing of control by appropriate measures. This is important because the need for forecasting, and the level of forecast information, depends largely on the objectives of the many decision makers involved (Norton and Way, 1983). Generally, the forecasting of annual crop pests is concerned with anticipating whether and when the pest is likely to invade the crop, what will be its destiny and when its density will exceed the economic threshold. As a result, the use of forecasting techniques varies widely both in space and time as do the control methods themselves.

Spatial scale
Nation or region wide. In some cases it is necessary to forecast likely introductions of pests into a country, for example, the passive introduction of quarantine pests by international trade or tourism. The European and Mediterranean Plant Protection Organisation (EPPO) publishes regularly data sheets on quarantine organisms. However, it seems unrealistic to think that introductions (always referred to as "accidental") can be completely prevented even if quarantine regulations are strictly applied. In some cases, the likelihood of introduction can be forecast when the origin of the plant carrying the pest is known (for example, potato cyst nematodes) but more often, the objective will be to forecast the ability of the pest to survive in a defined area in which it does not now occur (Baker, 1972). The advantage of such a forecast is obvious as it will allow early restrictive action to be taken with some opportunity to eradicate the pest quickly. In fact, relevant data from research done in the country of origin are not always available or they are not relevant to the climatic and environmental conditions that prevail in the importing country; besides, genetical considerations governing the adaptive ability of animals to survive in differing situations are rarely fully assessed and understood. International co-operation is required but is sometimes hampered by commercial interests.

Such co-operation is also required to forecast invasion by active, usually air-borne, pests, for example, the Colorado beetle (Leptinotarsa decemlineata). This species which is harmful to potato and tomato plants is spread over Continental Europe, but has never been allowed to establish in the Channel Islands or in the British Isles as a result of decisions to implement protection taken in 1947 (Portier, 1983). An INRA Field Station was operating for many years in the Cotentin peninsula, and biological and ecological characteristics of the insect were assessed as well as the conditions required for adults to fly and disperse and their ability to survive in sea water. Insect populations and climatic conditions have been continuously monitored since 1954 and forecasts are released by the Plant Protection Service (Advisory Service) in France: if 30% of caged adults have taken flight in one of the 5 Survey Stations, and at the same time the direction of the wind is SE-NW, daily sunshine is more than 6 hours, and maximum temperatures are between 20°C and 30°C, the situation is considered critical and a warning telegram is sent to the Channel Islands. This is repeated each time suitable conditions for invasion occur. Since 1965, no such invasion has occurred and preventative insecticide treatments have been given up in the Channel Islands since 1966. The advantages of this long-term warning system are clear. Apart from direct profit to farmers in the Islands because of the absence of this noxious insect, a considerable amount of scientific and economic data have been accumulated which are of general significance for the chemical and biological control of this pest.

Aphid-borne viruses may also be carried between countries, and southerly winds have been claimed to carry viruliferous alate aphids from the Continent into Swedish sugar beet crops where they disseminate BMYV (Wiktelius, 1977). This possible role of wind is well known and according to Broadbent (1969) "there is circumstantial evidence that persistent viruses such as potato leaf roll are sometimes introduced into Britain from the Continent of Europe". This hypothesis was supported by analysis of back tracks which suggested that in 1947, enormous numbers of the peach potato aphid, Myzus persicae, caught on 15,29 and 30 July might has been carried WNW over the Straits of Dover (Johnson, 1967); however they might also have been blown... from early potatoes in Eastern England (Broadbent and Heathcote, 1961). Similar evidence was given by Cochrane (1980) and Dewar et al. (1980) to explain a mass flight of the rose grain aphid (Metopolophium dirhodum) in 1979 which others suggested was blown from France (Schaefers et al., 1979). Such "circumstantial" evidence raises questions about the soundness and reliability of using some of the climatic factors even if they seem to account for observed events. The European network of 12.2 m suction traps - EURAPHID (Taylor, 1981) - which is intended to assess aphid migration and dispersal and to help produce sound forecasts, is expected to give an insight into this crucial problem.

At a regional or national scale, forecasting may be able to predict which species may become damaging as cropping practices change. For example, the release of new maize cultivars adapted to northern areas has allowed this crop to spread considerably in France; grain maize area has doubled since 1960 and that of fodder maize more than trebled. Besides the European Corn Borer (ECB), Ostrinia nubilalis, which has become very common and damaging in Alsace and in the Paris basin, the bird cherry aphid, Rhopalosiphum padi L., is now important in Brittany where it can spread Barley Yellow Dwarf Virus and Potato Virus Y (Robert, 1978). It appears that maize has favoured the development of this aphid by acting as a "bridge" between spring and, autumn whilst mild winters ensure a good survival. As has been stressed by Norton and Way (1983), one of the main difficulties in forecasting when new problems are likely to arise is that pest control considerations are ignored when changes in agricultural practices are planned, and these changes are made for political and economic reasons. It was not anticipated that increasing the maize area would have implications for seed potato production through aphids and viruses!

Field or region scale.   Forecasting at field level is most
important for the farmer, who needs to know if and when his crop is at
risk; for the adviser, who may feel he is responsible for any success or
failure of control; for the research scientist, who is involved in
determining the best means of forecasting; and for the agrochemical
suppliers who may then plan their pesticide recommendations more
appropriately. A precise understanding of population dynamics is
required to anticipate when an economic threshold is to be exceeded but
this goal is rarely in sight within reasonable time. Even if they claim
to encompass pest population dynamics, many current studies contain no
more than a description of population fluctuations of pests already
present in the crop and whether they are endogenous or exogenous.
However, such an approach was tried in 1970 under the auspices of the
International Biological Programme (IBP) in a collaborative investigation carried out on M. persicae (Barbagallo et al., 1972) and based on
previous work on Brevicoryne brassicae in Australia (Hughes, 1963).
Although it has provided interesting insights into aphid population
study, the method used was shown to have some serious limitations
(Carter et al., 1978). However, the assumption of the model, referring
to a physiological scale for aphid development in relation to temperatures, was attractive for forecasting.

Since then other attempts have been made; they rely on computers able to
solve quickly large numbers of equations (Rabbinge et al., 1979; Whalon
and Smilowitz, 1979a,b; and Carter et al., 1982) and they refer mainly
to potato aphids as virus vectors and to cereal aphids which have become
of great concern in the past ten years. In all cases they require that
pest monitoring is done under field conditions and in real time, in
order to update estimates of populations used in models (Whalon and
Smilowitz, 1979a) and also to validate the models. The practical
success of the predictions is not always up to expectations mainly
because there are still many gaps in our knowledge of the relations
between for example, pest and environment, and plant and predators. So
far, these models are still experimental, they apply to limited
individual fields and are rather cumbersome although some are already
compatible with microcomputer delivery systems (Whalon and Smilowitz,
1979a). Simplified summary models have been derived as exemplified by
EPIPRE (Rabbinge and Rijsdijk, 1983; see also Wilson, Chapter 24) leading to the development of decision rules on whether and when to apply
pesticides. EPIPRE is designed to seek information in order to give
back specific recommendations on a field to field basis and it relies on
participating farmers. Information is circulated between different
levels of decision-making, i.e. between farmers and the Central Computer
System, and involves many pests and diseases at the same time. One
resulting benefit has been to farmers more aware of their problems.
Rabbinge and Rijsdijk (1983) reported that 400 fields participated in
EPIPRE in 1978 and 1100 in 1981 although recommendations were not free;
38% of them were treated in accordance with recommendations in 1980 and
54% in 1981, the ones with the lower yields being more inclined to
follow advice.

An increasing willingness of farmers to co-operate actively
with the research and advisory services has also been reported from
France (De la Rocque and Lechapt, 1980; De la Rocque and Piquemal,
1980). Following the 1975 outbreak of the grain aphid (Sitobion avenae
F.) in the French mainland, a network was established in 1976 to monitor
and forecast pests and diseases on a field to field basis over a large
area; farmers, extension officers, teachers from different institutions,
schools and firms, advisers (Plant Protection Service), and scientists
(INRA) all contributed to this Survey for many years. The proportion of
farmers increased from small numbers in 1976 to 40% in 1979 when 1838
fields were checked weekly for pests and diseases in 82 out of the 95
French Departments (Table 1). At the same time, the participants were
provided with training which has greatly improved their general knowledge of identification, biology and damage caused.

Table 1. Proportion of different participants in the
French cereal pest and disease monitoring network from
1977 to 1979.

	1977	1978	1979
Farmers (%)	20	28.2	39.7
Extension officers (%)	60	50.2	42.8
Advisers (%)	7	10.2	8.8
Teachers and scientists (%)	13	11.3	8.7
Total participants	500	609	877

This willing participation is no doubt a pre-requisite to establishing a
sounder perception of the hazards from pests and to being more receptive
to warnings released by scientists and advisers, which seems far from
being the rule (Mumford, 1981). It also increased awareness that
different pesticides, even if they belong to the same group (e.g.
organophosphates, carbamates, pyrethroids) or are contact or systemic in
action, do not result in a similar efficiency with respect to the target
pest and possible side effects. Moreover, researchers are less isolated
as they have direct access to the field situation and can be updated
accordingly. The EPIPRE recommendations, when followed, have resulted
in fewer treatments but in yields similar to those where farmers played
safe and used propylactic treatments (Rabbinge and Rijsdijk, 1983).
Nevertheless, there is a underlying danger that certain growers who

become more acquainted with pests and ways to control them, may over-
estimate the hazards from pests that have not yet required treatment,
just because they visit their fields more often! A last limitation to
such short-term field-to-field forecasting system, which takes into
account changes in pest populations over a few days or weeks, arises
from the delay between the release of the advice and the decision being
taken. In the case of EPIPRE, it takes an average of 4 days for the
farmer to receive a reply to his field observations; this can be too
short to prepare for action. This defect deserves more attention.

Temporal scale

An annual cropping season does not usually spread over the
entire year but lasts from 3 to 9 months. The physio-phenological
changes of the plants must be very rapid and pest population build-up
can reach outbreak proportions within a short period; for example, in
Britain it has been found that a minimum count of 5 aphids ($S.avenae$)
per ear of wheat at flowering, followed by an increase in numbers a few
days later, would justify spraying (George and Gair, 1979). In France,
this critical number has been tentatively determined as being 10. For
practical reasons, such short-term, tactical forecasting as intended for
aphids or other pests can be difficult to operate and the time between
forecast, decision and actual intervention depends largely on the farmer
being organised to have his crop treated; he may not find the pesticide
required available at the time he tries to purchase it, he or his con-
tractor may be unable to treat the whole of the crops within reasonable
time, or he simply did not receive advice in good time or was not able
to check crop and pest changes. Conversely, if earlier predictions can
be made about whether a pest is likely to infest or colonise a crop and
when, and is liable to build-up to damaging populations, and when, the
above limitations may be partly overcome. This strategic view applies
both to endogenous and exogenous pests. In the former case, since it is
possible to know before planting if pests are already present and at
what density, the likelihood of their populations building-up can be
predicted provided a good knowledge of their population dynamics is
available; thus, decisions can be taken to introduce preventive control,
or not to grow the crop in the particular field. In the latter case, a
comprehensive knowledge of the bioecology and behaviour of the
particular pest may lead to a system of successive forecasts starting
many months before and becoming more and more accurate according to key
events taking place during the pest's life cycle.

A well known example of collaborative work was started in 1968 in
England on forecasting the need for chemical control of the bean aphid,
Aphis fabae Scop., on spring-sown field beans, Vicia faba L. (Way and
Cammell, 1973; and Way et al., 1971, 1981). When more than 5% of plants
are colonised as a result of spring migration, damage to field beans is
"possible" or "probable"; when less than 5% are infested, damage is
unlikely to occur. There are highly significant regressions of this
percentage on the sizes of: (1) the previous autumn remigration; (2)
overwintering egg populations on the spindle tree; (3) fundatrigeniae
spindle populations; and (4) spring aerial migration. They account for
28, 54, 54, and 64% of the variance respectively (Way et al., 1981).

This allows growers to assess very early the potential size of populations liable to invade their crops. The recommended spraying date is determined 1-2 weeks in advance and made available to 16 different forecasting areas. As stressed by Way and Cammell (1980), it is unlikely that this kind of forecasting would attain success with other species.

Other approaches to strategic forecasting have been sought; a number of systems rely on weather variations and are mainly relevant to aphids as pests in their own right or as virus vectors. Since overwintering is important for aphid survival, especially that of anholocyclic species, relationships were sought between accumulated temperature over critical periods during winter and early spring, number of days with frost, rain or other meteorological parameters, and either the time of first emigration flight, the size of spring and autumn flights expressed by number of alatae caught in traps or present in crops (Sparrow, 1974; Rautapää, 1976, Robert and Rouzé-Jouan, 1976, 1978; and Turl, 1980, 1982; and A'Brook, 1981, 1983) or the size of apterous populations at a certain growth stage of the crop or at their maximum (Vickerman, 1977; Watson and Carter, 1983; Pierre and Dedryver, 1984). All critical periods and significant weather factors differ from one species to the other and for one single species according to geographical co-ordinates. This clearly stresses the need for a better knowledge of pest bioecology to understand these differences and is well illustrated by the simulation modelling of predictions elaborated by Pierre and Dedryver's model in Brittany; they are suitable for the West of France but show some discrepancy with observed populations in the Paris basin and near Bordeaux. However, it seems reasonable to think that forecasting some aphid species using prior weather will be possible when more data has been accumulated (A'Brook, 1983).

Illustrations of recent trends in tactical and strategic forecasting can now be given with some contrasting case studies.

CASE STUDIES
European corn borer
In France, the grain-maize crop area has increased considerably during the past 25 years reaching about 1.64 million hectares in 1983 with a mean yield of 6.0 t ha^{-1}. The incidence of the European Corn Borer (ECB), O. nubilalis, has also increased. The plants fail because the larvae feed on leaves and tunnel into plant stalks and ear shanks. The weight of individual ear kernels decreases significantly. In Alsace, Stengel (1969) has shown yield losses to be approximately 15% when stalks are broken above the ears, 33% when broken below the ears, and 50% when ear shanks are broken, depending on cultivar. Another component of yield loss is the difficulty of collecting all the ears by mechanical harvesting because of broken stalks lying on the ground with dropped ears. Overall yield losses can reach 3 t ha^{-1}, i.e. 50% of total yield when 3 or 4 larvae per plant are present at harvest (Stengel, 1982). However, losses usually fluctuate between 5 and 30% according to the level of infestation, cultivar tolerance, overall level

of productivity and climatic conditions (Anglade, 1971). With less than
one larvae per plant, no yield decrease is recorded. In the South of
France, this species can complete two generations a year, in the Paris
basin one complete and one partial generation, and in Alsace only one.
Whilst in the South of France, the population levels are generally kept
at low levels (in 8 years out of 10 on average) by a parasitic fly,
Lydella thompsonii Hert. (Diptera, Tachynidae), the same does not occur
elsewhere. Farmers have for long relied on insecticides such as DDT and
organophosphates and more recently pyrethroids – the application timing
of which has often been imprecise in relation to insect phenology,
economic injury level and environmental consequences on beneficial
insects. The recommended time for treatment of maize, namely at 50%
panicle showing in the sheath, proves adequate in the South but not in
Alsace (Stengel, 1970). Therefore, a more precise forecasting scheme
has been put forward; it is suitable for this area and is a two-step
system adapted to evolving means of control (Stengel, 1982).

The Alsatian monovoltine strain develops slowly and enters diapause at
long light phases (15–16 hours) with about 24% of the population showing
an obligatory, not photoperiod dependent, diapause (Stengel and
Schubert, 1982). Parallel to that, its flight extends over 7–10 weeks
and has been shown to begin after 350-day degrees above $10^{\circ}C$ have been
accumulated; 50% of emergence occurs at 520 day-degrees and 75% at 600
day-degrees. In other words, 3 weeks after the beginning of emergence
(as shown in outdoor emergence cages where 2000 maize stalks with 2–3
larvae in each were gathered in the previous autumn), 50–70% of adults
have emerged, 50% of the total egg masses have been laid, and the most
popular cultivar, INRA 258, has reached 50% male flowering (outset of
pollen shedding) – this proved to be the most susceptible stage to
larvae attack. Thus, it is advisable to apply insecticides 21 days
after first adult emergence and granules are recommended as they
accumulate in the leaf axils through which the larvae bore. The farmer
himself can monitor adult emergence from caged stalks and decide whether
to apply chemicals by a scouting technique; this has proved to be
profitable when an average of 10–12% of plants show an egg mass (from 6%
of early cultivars FAO 100–200 to 15% of late cultivars FAO 300–400) as
observed on 5 x 100 plants per field. This short-term procedure has
many advantages. The farmer himself participates in the decision;
accumulated day-degrees can be substituted for phenological observations, and it is easier for him to do. Many emergence cages have been
set up by farmers, the advisory services and the Maize Technical
Institute in various Alsatian situations; information is centralised to
an automatic answering telephone service. But its main limitation is
that the "treatment window" is rather narrow, and effective application
according to injury level has to be decided within 2 to 3 days. Therefore, medium-term forecasting (Norton and Way, 1983) was sought and this
has led to the defining of risk areas (Stengel, 1982). In effect, the
potential fecundity of females can be forecast as early as the previous
autumn as it apparently depends upon certain climatic conditions occurring before larvae enter diapause according to:

$$y = 342 - 2.51 \, x \quad ; \quad x = \frac{I \, P1}{1000}$$

where y = potential fecundity of females the following year, I = sunshine and P1 = rainfall during the previous months of August and
September. Also, the more fecund are more prone to lay eggs. As a
consequence, less than 0.5 larvae per plant will result in no risk. A
larval population of 0.5 to 1.2 individuals per plant may be damaging if
the potential fecundity of the females is high. Above 1.2 larvae per
plant, damage will occur in the following year. This medium-term forecasting is especially relevant to biological control of the European
Corn Borer with Trichogramma maidis, which has proved efficient in
Alsace where it was introduced in 1974 (Voegele et al., 1975; Stengel,
1982). Such an early assessment of the areas where damage is probable
or certain allows mass rearing to be modulated accordingly and in
regions where Trichogramma cannot be used, the farmer is made aware very
soon of the need to buy insecticides, or better Bacillus thuringiensis
insecticidal granules, which are as efficient and not harmful to the
pest's enemies (Schubert and Stengel, 1974; Stengel, 1982). Then good
knowledge of the pest's biology allow entomogenous insects to be released by the time egg-laying begins, i.e. flight start and then twice
at a 10 day-interval.

Such forecasting is not yet available in the Paris basin. Insecticides
are usually applied early (when the panicle first shows in the sheath)
at a period when plants have not completed their growth. Following
borer and rose grain aphid (M. dirhodum) outbreaks in 1979, broad-
spectrum insecticides were used extensively against the borer in 1980.
This resulted in only a limited decrease in borer larval populations and
in an unexpected bird-cherry aphid (R. padi) outbreak for which treatments with liquid pyrethroids were shown to be mainly responsible as
compared to microgranules. Failure in borer control was attributed to
poor timing because treatment was too early in relation to egg laying,
and R. padi was not affected since these aphids were hidden in leaf
sheaths whilst their enemies were killed. This shows clearly that forecasting one crop pest should be included into a comprehensive integrated
control scheme which takes into account not only the target pest but
also possible effects on its natural enemies and on alternative pests
and their enemies.

Potato virus vector aphids
In France, seed potato crops are grown on about 13,500 ha of
which 7,400 ha are located in Brittany. Virus diseases are a limiting
factor affecting seed potato entry for certification. A careful roguing
of infected plants as soon as they emerge, and burning of the haulms,
are basic means for preventing viruses from being spread, and from in-
fecting tubers. Since the two main viruses, potato virus Y (PVY) and
potato leafroll virus (PLRV), are both transmitted by aphids, chemical
control may be a third possibility. A rational use of insecticides (and

of mineral oils to lessen non-persistent virus transmission) is
especially required since M. persicae can readily become resistant. In
Brittany, it is necessary to have: (1) an early forecast of timing and
size of the spring flight of the main entirely anholocyclic aphid
vectors, Aulacorthum solani, Macrosiphum euphorbiae, M. persicae, R.
padi and some others to decide if granules would be worthwhile or if
delayed spraying would be adequate (Robert, 1976); and (2) an estimate
of the size of the summer dissemination flight as early as possible in
time for decisions about haulm burning.

Figure 1. Alate Aulacorthum solani (o——o) and Myzus persicae (●——●)
caught in yellow trays (60 x 60 cm) in Moustoir-Remungol (M) in 1968 and
Rosporden (R) in 1967 and 1968 in relation to potato plant emergence
(↥), flowering (⊢----⊣) and haulm killing (↧).

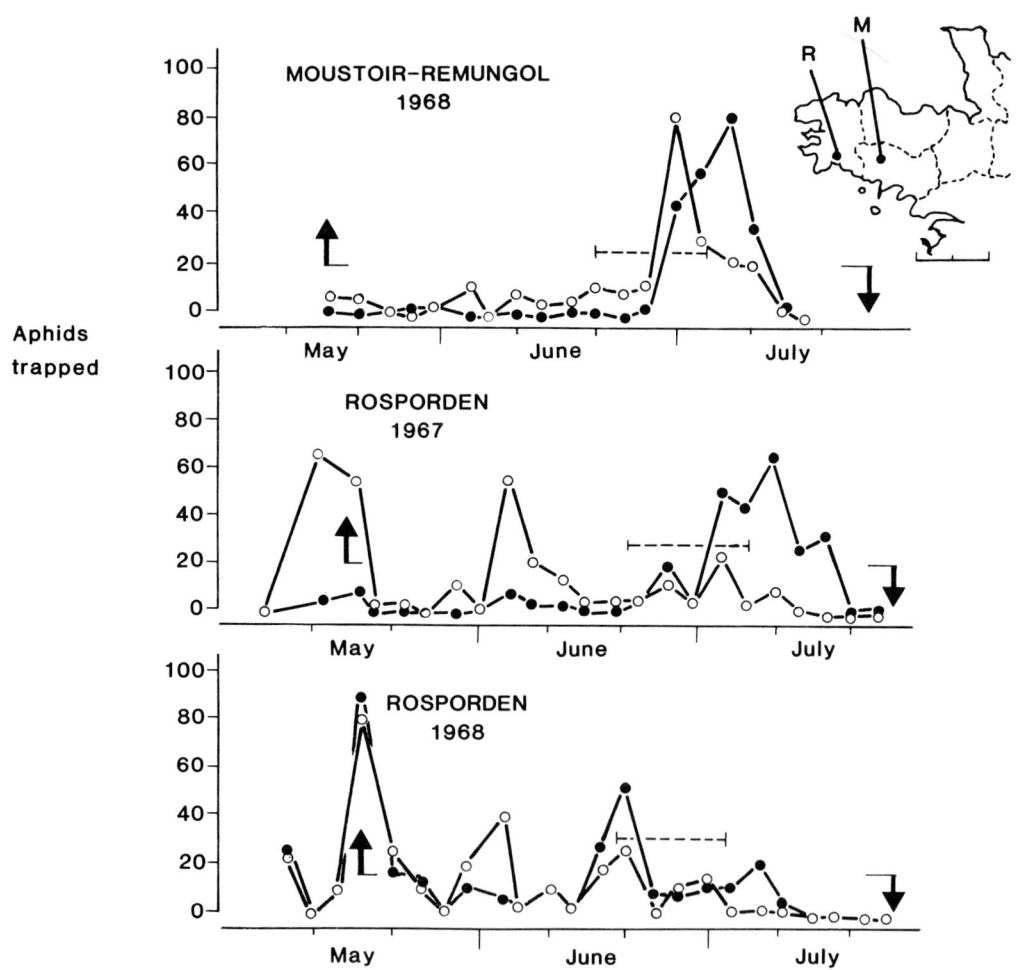

Aphid trapping in large yellow trays (60 x 60 cm) over 17 years (Robert and Rouzé-Jouan, 1978) has shown that in contrast to what can be observed in more northern latitudes, initial potato aphid colonisation by A. solani and M. persicae, taking place in early May, may be on occasions fairly important and poses a real threat since the younger the plants, the more susceptible they are to virus infection. The size of this flight has been shown to be correlated with accumulated temperature over the latter half of April but occurs only for maximum temperatures above $15^{\circ}C$ (Robert et al., 1974; Robert and Rouzé-Jouan, 1978). For the summer dissemination flight, when it exists, at the end of June and the beginning of July, it has been shown similarly that the warmer the latter half of June, the earlier the dissemination flight. An outstanding and probably unique characteristic of the situation is that the non-occurrence of this flight can be easily forecast: the more numerous the spring contamination flight, the less the dissemination flight and vice versa (Figure 1). This is mainly due to the effect of parasitoid Aphidiids, and on some occasions of fungal parasites which results in a very efficient natural limitation of aphid populations as soon as they build up (Robert and Rabasse, 1977; Robert, 1979). Forecasting the size of dissemination flight as decided from the size of contamination flight can be confirmed early in the cropping season by a simple demographic analysis of the incidence of parasitism by dissecting aphids at the beginning of population build-up (Robert, 1981).

With regard to PYV dissemination by R. padi, as experienced from 1974 to 1976 in Europe (Robert, 1978), the number of alate aphids trapped over the cropping period is significantly correlated to mean temperatures accumulated in January, February and March, i.e. the "milder" the winter, the more alate R. padi fly in the spring. All these findings have allowed comprehensive forecasts of the risks arising from aphids to be made when epidemiological data are introduced. A close collaboration between research and potato-related organisations ensures that they are sent to growers. The obvious advantage so far has been to produce high quality seed potatoes with the least use of inecticides and disturbance to the environment.

CONCLUSIONS
Much has been said about Integrated Pest Control which can still be considered as an ideal way of controlling pests using all kinds of compatible methods with the least cost and harm to the environment. Forecasting annual crop pests is one of the basic needs required for decision-making, but so far its use has been mainly to help promote supervised control based on pesticide use. However, it has contributed to some decrease in routine treatments, since the number and timing of treatments can be scheduled to give better cash returns. It has also considerably increased the interest and concern of farmers and practitioners since in most cases, forecasting requires previous local monitoring.

Computerisation of models will attract more attention in the future in relation to implementing and improving forecasting schemes since they contribute to a better study of system analysis, but complex models will probably remain more as research techniques leading to the development of simpler summary models (Carter et al., 1982). It will require long-term research investment to check that simplification does not result in loss of reliability, because this is probably one of the main limitations to the adoption of forecasting by growers.

Another problem is the difficulty in forecasting the weather sufficiently in advance and up to now, all the models must take this important limitation into account. To try and overcome it, predictions based on regression analysis are released very early to prepare growers for action to be taken later, but as reliable forecast information cannot be channelled towards growers much in advance unless improvements in communications are made, present investigations are directed towards the use of media and modern computer technology.

Finally, the crucial question is to decide whether we are right in releasing on a large scale, forecasts which are not thoroughly reliable before waiting for improvements which come when more biological data and checks are available. The reply to that question may not be the same if it is put to growers, advisers, chemical suppliers, or applied research scientists.

ACKNOWLEDGEMENTS
I wish to thank Dr B.D. Smith, Long Ashton Research Station, University of Bristol, for his kind help in making my English more readable.

REFERENCES
A'Brook, J. (1981). Some observations in west Wales on the relationships between number of alate aphids and weather. Annals of Applied Biology 97, 11-15.
A'Brook, J. (1983). Forecasting the incidence of aphids using weather data. EPPO Bulletin 13, 229-233.
Anglade, P. (1971). Bulletin of Technical Information, Ministry of Agriculture 264-265, 951-958.
Baker, C.R.B. (1972). An approach to determining potential pest distribution. OEPP/EPPO Bulletin 3, 5-22.
Barbagallo, S., Inserra, R. and Foster, G.N. (1972). Entomologica 8, 21-33.
Broadbent, L. (1969). Disease control through vector control. In Viruses, Vectors and Vegetation, ed. K. Maramorosch, 593-630. New York: Wiley and Sons.
Broadbent, L. and Heathcote, G.D. (1961). Winged aphids trapped in potato fields 1942-1959. Entomologia Experimentalis et Applicata 4, 226-237.
Carter, N. Aikman, D.P. and Dixon, A.F.G. (1978). Journal of Animal Ecology 47, 667-687.
Carter, N., Dixon, A.F.G. and Rabbinge, R. (1982). Cereal Aphid Populations: Biology, Simulation and Prediction, pp. 99.

The Netherlands. Wageningen: Pudoc, Centre for Agricultural Publishing and Documentation.

Cochrane, J. (1980). Meteorological aspects of the numbers and dist-
ribution of the rose-grain aphid, Metopolophium dirhodum
(Wlk.), over south east England in July, 1979. Plant
Pathology 29, 1-8.

De la Rocque, B. and Lechapt, G. (1980). Phytoma-Defense des Cultures
319, 20-22.

De la Rocque, B. and Piquemal, J.P. (1980). Perspectives agricoles 35,
42-52.

Dewar, A.M., Woiwod, I. and Choppin de Janvry, E. (1980). Aerial
migrations of the rose-grain aphid, Metopolophium dirhodum
(Wlk.) over Europe in 1979. Plant Pathology 29, 101-109.

George, K.S. and Gair, R. (1979). Crop loss assessment on winter wheat
attacked by the grain aphid, Sitobion avenae (F.) 1974-77.
Plant Pathology 28, 143-149.

Hughes, R.D. (1963). Journal of Animal Ecology 32, 393-424.

Johnson, C.G. (1967). Netherlands Journal of Plant Pathology 73 (suppl.
1), 21-43.

Mumford, J.D. (1981). A study of sugar beet growers' pest control
decisions. Annals of Applied Biology 97, 243-252.

Norton, G.A. and Way, M.J. (1983). Forecasting and crop protection
decision making - realities and future needs. Proceedings
of the 10th International Congress Plant Protection,
Brighton 1, 131-138. British Crop Protection Council
Publications.

Pierre, J.S. and Dedryver, C.A. (1984). Acta Oecologica Oecologia
Applicata 5, 153-172.

Portier, G. (1983). Phytoma 353, 22-23.

Rabbinge, R. and Rijsdijk, F.H. (1983). EPIPRE: a disease and pest
management system for winter wheat, taking account of micrometeorological factors. EPPO Bulletin 13, 297-305.

Rabbinge, R., Ankersmit, G.W. and Pak, G.A. (1979). Epidemiology and
simulation of population development of Sitobion avenae in
winter wheat. Netherlands Journal of Plant Pathology 85,
197-220.

Rautapää, J. (1976). Population dynamics of cereal aphids and method of
predicting population trends. Annales Agriculturae Fenniae
15, 272-293.

Robert, Y. (1976). Phytiatrie-Phytopharmacie 25, 187-200.

Robert, Y. (1978). Comptes rendus du 7ème Conference European
Association for Potato Research, Varsovie, 242-243.

Robert, Y. (1979). Annales de Zoologie Ecologie Animale 11, 371-388.

Robert, Y. (1981). In Colloque Franco-soviétique: Lutte Biologique et
Intégrée contre les Purcerons, Rennes 1979, 7-16. Paris:
INRA.

Robert, Y. and Rabasse, J.M. (1977). Role écologique de Digitalis
purpurea dans la limitation naturelle des populations du
puceron strié de la pomme de terre Aulacorthum solani per
Aphidus urticae dans l'ouest de la France. Entomophagia 22,
373-382.

Robert, Y. and Rouzé-Jouan, J. (1976). Neuf ans de piege de pucerons
    des cereals: Acyrthosiphon (Metopolophium) dirhodum Wlk., A.
    (M.) festucae Wlk., Macrosiphum (Sitobion) avenae F., M.
    (S.) fragariae Wlk. et Rhopalosiphum padi L. en Bretagne.
    Revue de Zoologie Agricole et de Pathologie Végétale 75,
    67-80.
Robert, Y. and Rouzé-Jouan, J.M. (1978). Annales de Zoologie Ecologie
    Animale 10, 171-185.
Robert, Y., Bonnemaison, L. and Quemener, J. (1974). In Maladies et
    Parasites Animaux de la Pomme de Terre Brochure 32, 16-25.
    Paris: Institut Technique de la Pomme de Terre.
Schaefers, G., Bent, G. and Cannon, R. (1979). The Green Invasion. New
    Scientist 83, 440-441.
Schubert, G. and Stengel, M. (1974). Comparaison de l'activité
    insecticide de quatre formulations à base de Bacillus
    thuringiensis et d'un ester phosphorique sur la pyrale du
    mais (Ostrinia nubilalis Hübn.). Revue de Zoologie Agricole
    Pathologie Végétale 73, 47-52.
Sparrow, L.A.D. (1974). Observations on aphid populations on
    spring-sown cereals and their epidemiology. Annals Applied
    Biology 77, 79-84.
Stengel, M. (1969). Influence de l'attaque de la pyrale (Ostrinia
    nubilalis Hübn., Lepidoptera, Pyralidae) sur le poids et la
    qualité des grainds de mais en fonction du type de dégâts
    dans les cultures en Alsace (France). Revue de Zoologie
    Agricole Pathologie Végétale 10-12, 101-112.
Stengel, M. (1970). Une méthod de prévision des dégâts de la pyrale du
    mais (Ostrinia nubilalis Hübner, Lepidoptera, Pyralidae).
    Mise au point de la lutte dans les cultures de la Plaine
    d'Alsace. Annales de Zoologie Ecologie Animale 2, 309-325.
Stengel, M. (1982). Essai de mise au point de la prévision des dégâts
    pour la lutte contre la pyrale du mais (Ostrinia nubilalis)
    en Alsace (Est de la France). Entomophaga 27, 105-114.
Stengel, M. and Schubert, G. (1982). Comparative study of the growth
    rate and the photoperiod sensitivity of two strains of the
    European corn borer (Ostrinia nubilalis Hubn., Lepiodoptera,
    Pyralidae) and their hybrids. Agronomie 2, 989-994.
Taylor, L.R. (1981). In Euraphid Rothamsted 1980, ed. L.R. Taylor, 3.
    Harpenden: Rothamsted Experimental Station.
Turl, L.A.D. (1980). An approach to forecasting the incidence of potato
    and cereal aphids in Scotland. EPPO Bulletin 10, 135-141.
Turl, L.A.D. (1982). In Euraphid Gembloux 1982, ed. J. Bernard, 79-82.
    Gembloux: Centre Recherche Agronomique.
Vickerman, G.P. (1977). Monitoring and forecasting insect pests in
    cereals. Proceedings of the 1977 British Crop Protection
    Council Conference - Pests and Diseases 1, 227-234. British
    Crop Protection Council Publications.
Voegele, J., Stengel, M., Schubert G., Daumal, J. and Pizzol, J.
    (1975). Annales de Zoologie Ecologie Animale 7, 535-551.
Watson, S.J. and Carter, N. (1983). Weather and modelling cereal aphid
    population in Norfolk (UK). EPPO Bulletin 13, 223-227.

Way, M.J. and Cammell, M.E. (1973). The problem of pest and disease
        forecasting - possibilities and limitations as exemplified
        by work on the bean aphid, Aphis fabae. Proceedings of the
        7th British Insecticide and Fungicide Conference 3, 933-954.
        British Crop Protection Council Publications.
Way, M.J. and Cammell, M.E. (1980). Constraints in establishing
        meaningful economic thresholds in relation to decision-
        making on pesticide application. EPPO Bulletin 10, 201-206.
Way, M.J., Cammell, M.E., Alford, D.V., Gould, H.J., Graham. C.W., Lane,
        A., Light, W.I., St. G. Rayner, J.M., Heathcote, G.D.,
        Fletcher, K.E. and Seal, K. (1977). Use of forecasting in
        chemical control of black bean aphid, Aphis fabae Scop. on
        spring-sown field beans, Vicia faba. Plant Pathology 26,
        1-7.
Way, M.J., Cammell, M.E., Taylor, L.R. and Woiwod, I.P. (1981). The use
        of egg counts and suction trap samples to forecast the
        infestations of spring-sown field beans, Vicia faba, by the
        black bean aphid, Aphis fabae. Annals of Applied Biology
        98, 21-34.
Whalon, M.E. and Smilowitz, Z. (1979a). GPA-CAST, a computer fore-
        casting system for predicting populations and implementing
        control of the green peach aphid on potatoes. Environmental
        Entomology 8, 908-913.
Whalon, M.E. and Smilowitz, Z. (1979b). Temperature-dependent model for
        predicting field populations of green peach aphid, Myzus
        persicae (Homoptera: Aphididae). Canadian Entomologist 111,
        1025-1032.
Wiktelius, S. (1977). The importance of southerly winds and other
        weather data on the incidence of sugar beet yellowing
        viruses in Southern Sweden. Swedish Journal of Agricultural
        Research 7, 89-95.

21. PRACTICE AND PROGRESS IN PEST FORECASTING

T. Lewis
Rothamsted Experimental Station,
Harpenden, Hertfordshire, AL5 2JQ, England

INTRODUCTION - HISTORICAL PERSPECTIVE
"Insect devastation costs millions of dollars annually .....
wise and timely application of well known remedies would
save much of this constant loss" (Essig, 1894).

Similar statements to this, written nearly a century ago,
have been repeated in many texts on applied entomology, pest control and
pesticide use. Yet the apparently clear objective of "wise and timely"
use of pest control measures in crop protection is often exceedingly
difficult to achieve. Since the beginning of recorded history, unexpected outbreaks of insect pests have devastated crops and caused
untold human misery. Locusts have plagued agriculture in Africa, Asia
and China where nearly one half of the world's population dwells. Leaf-
cutting ants sporadically ruin crops and grassland over vast areas of
Central and South America. In temperate regions, species with short
life-cycles, especially rapidly-multiplying aphids, often appear unexpectedly to damage arable and fruit crops directly or spread epidemics
of virus diseases.

These few examples establish the universality of sporadic pest occurrence. The causes of the outbreaks vary: most arise from complex
responses of pests to weather, others from changing cropping patterns
and husbandry, and a few from what appears to be merely capricious
behaviour. Nevertheless, they all share a common factor, namely, an
unexpected or unnoticed population build-up over weeks, months or years,
followed by relatively sudden infestation of a single or variety of
crops, often over vast areas, for which farmers have been unprepared.

Faced by such unpredictable constraints on crop production, farmers have
traditionally preferred to ignore pests until they reached damaging
levels, partly in the hope that no action would be necessary, and partly
because until 50 years ago there was little that could usefully be done
even with foreknowledge. This situation has now changed. Pressures for
more locally-produced food in developing countries, and for higher
yields, better quality and environmental protection in developed
countries, plus the availability of highly effective control agents,
have given farmers the incentive and means to control most pests,
providing they have enough warning of their impending occurrence.

It is salutory to be reminded that the estimated average world-wide losses due to insects as a percentage of the potential yield amount to 14% for cereals (mainly wheat, barley, rice and maize), 17% for root/carbohydrate crops (potatoes, sugar beet, sugar cane and vegetables) and 11% for oil seed and fibre crops (mainly soybeans, groundnuts, cotton, rape and sunflower seed) (Anon., 1975; Walker, 1975). All these crops, and many others, are attacked by some pests for which monitoring and forecasting methods are now available. For example, aphids and armyworms on wheat, plant-hoppers and leaf-rollers on rice, locusts and armyworm on maize, and aphids on potatoes, beans, brassicas and vegetables can all be detected before damaging populations occur, though this information is not always used to advantage.

Against this brief historical background, this Chapter attempts to categorise forecasts to show their relevance to different planning situations in pest control. Some practices and significant developments in pest detection, data collection and information dissemination are then examined as a guide to improving current forecasting particularly of pests of temperate arable crops.

CATEGORIES OF FORECAST

A "forecast" is perhaps more familiar in the context of weather prediction, when it is usually qualified by a time interval such as monthly (long-term), 24-hourly, or for the ensuing day or night. It generally covers three broad aspects: the type of weather, the geographical region to which it applies, and when it will occur at a specified location within that region. Pest forecasting is similarly concerned with three aspects of insect occurrence, namely the size of future populations, where they are likely to develop or appear, and when numbers above specified limits (ranging from mere presence to populations sufficient to cause economic damage) are likely to occur. Prediction of insect-borne virus disease of plants is likewise concerned with the amount, extent and timing of infection.

However, to be useful, prediction of pest occurrence needs to extend over longer periods than are usually available for weather forecasts, although the unreliability of the latter will also have a major effect on the accuracy of pest predictions. Four broad categories of "forecasts" in relation to pest occurrence are identifiable, depending on (1) the period of warning given; (2) the type of agricultural response required; (3) the speed with which information needs to be disseminated to enable it to be used effectively; and (4) the geographical scale of coverage (Table 1).

Warnings are based on current awareness and estimates of the imminent pest situation, and are given in the expectation that control measures are, or will soon be, necessary. At the very least, they should alert farmers to make daily checks on the pest situation in specified crops. They are particularly relevant to rapidly-spreading air-borne migratory pests such as aphids and plant-hoppers, but can only be useful where there is a system for collecting and disseminating information quickly.

Their quality will depend greatly on the degree of detail in the data available to the forecaster and on his experience. They may be relevant to localities or in more favourable circumstances to large farms.

Table 1. Categories of pest forecasts.

Type of forecast	Period of warning	Agricultural response	Speed of dissemination needed	Geographical scale
Current awareness/ warnings	1-4 days	Operational	1 day	Farms/local
Medium-term	5-21 days	Operational	1-3 days	Local/regional
Long-term	Months or seasons	Tactical	1-2 weeks	Regional/national
Annual probability	Periods of years	Strategic	Weeks-months	Regional/national

Medium-term forecasts of 5-21 days may not be as geographically or temporally precise as warnings, but they are nevertheless of great value because they allow growers to make a considered assessment of pest or virus-vector build up, permitting more time to plan spray schedules in relation to available machinery and suitable weather, and perhaps to restrict or even withhold treatments if careful examination shows crops not to be at a vulnerable stage or only patchily infested. To be fully effective the forecaster needs to be able to communicate quickly with farmers.

Long-term forecasts of a month or season ahead, or spanning seasons (autumn-spring) are useful to allow tactical planning, for example to negotiate sales contracts, buy seed, have it dressed with insecticide, or make bulk purchases of pesticides. At present they are most likely to be useful for pests with long developmental periods whose progress towards the damaging stage can be monitored in relation to weather, and to migrant pests in large regions where consistent seasonal weather patterns occur from year to year.

Annual probability forecasts, by which the expected frequency of a pest outbreak in a given area can be estimated, are of no immediate value to farmers. This approach has received relatively little attention from entomologists, but it is to some extent used intuitively, and could be developed to assess the long-term risk to particular crops from

specified pests in a region. It could be used to indicate the scale of
investment and likely returns to agricultural merchants and pesticide
producers of providing pesticides, marketing facilities and appropriate
control equipment in particular areas.

Generally, the shorter term the forecast, the more restricted the area
to which it is applicable, but the definitions of 'local' and 'regional'
forecasts must be flexible because of variations in local geography and
topography. A regional forecast is likely to cover a far larger area in
a flat plain than a synoptic forecast in a mountainous area, and local
forecasts can sometimes differ from field to field depending on aspect,
shelter from prevailing winds, and soil type.

The distinctive features and requirements of each category of forecast
are best illustrated through examples of systems in use; opportunities
for improvement will then become apparent.

CURRENT AWARENESS AND WARNINGS
Pest event scheduling
Several schemes in the United States have been devised to
predict pest phenology by linking research information, field observations on pest development and weather data in as up-to-date a manner as
possible. One such "biological scheduling" system run by Michigan State
University is based on field observations for about 150 pests, and on
more detailed data on pest development, obtained by both research and
field studies, for a further 12 pests. These data are linked by
computer to current weather data to provide useful predictions of when
field observers should search for, and expect to find, a particular
stage of a pest attacking a given crop anywhere in the State (Gage et
al., 1982).

Briefly, observations made by "pest management field assistants",
ranging from professional county agents to farmers and students, form a
data file containing details on pest, life stage, day-degree base, day-
degree occurrence range, and crop or host. These data, when combined
with current weather data, are used to estimate when a pest life stage
is, or will be, present in a particular place. For a few carefully
studied pests, developmental parameters provide the length of each life
stage in day-degrees between appropriate lower and upper thresholds.
The users can request via computer more detailed information such as the
distribution of stages and events for a single pest species throughout
the State, or for all pests and stages currently present at one site.
The information is either tabulated or presented graphically and is
widely available through the computer, fulfilling the requirement for
rapid information dissemination essential in current awareness systems.

The system is designed to provide early warnings once a week to enable
field personnel to examine crops and estimate pest density and damage in
order to facilitate management decisions, though, of course, once alert-
ed they still need to identify the pest, assess its density and damage
potential, and interpret the observations to make decisions on control.

By covering many pests and crops the scheme has stimulated an awareness of the importance of timing in pest monitoring to improve the accuracy of forecasts; and by encouraging regular feedback on events to expand its historical data base, it is self-improving. A disadvantage is the requirement for many (about 250) trained field personnel, which is beyond the staffing capabilities of many other organisations.

Aphid forecasting - Rothamsted Insect Survey

One of the aims of the aphid monitoring system established by the Rothamsted Insect Survey throughout Great Britain is to provide early warning of crop infestation by migrant aphids with a minimal requirement for field inspection because there are too few staff in the agricultural extension services to be able to examine even a small proportion of the major arable crops grown. As with the Michigan system, the objective of this aspect of the Survey's programme is to alert advisers and farmers to inspect crops in regions identified at risk. However, the input of data is not obtained from field inspectors but from 23 suction traps distributed throughout the UK (Taylor, 1977a, 1979).

Details of the trapping, analysis and information dissemination system in use are given by Woiwod et al. (1984). Briefly, the trap samples are collected daily during the main aphid flight period (early April to mid-November) and at weekly intervals for the rest of the year. Catches are sorted and identified to species or species groups at two sites, by specialists who receive samples twice weekly by mail and Datapost. Thereafter the urgent early warning advisory requirements are met by rapid entry of the data into a computer from which summaries of many types are extracted. The principal immediate output is a weekly Aphid Bulletin listing the totals of 33 pest species or species groups record-ed at each trapping station, plus other summaries comparing the current data for each of 8 pest species in ten regions with data for the same time during the previous year and with the long-term average (Tatchell, 1982). This information in the form of an interpretative commentary accompanying the Bulletin is circulated by post, telex and telephone weekly to the agricultural industry and advisers, who can then inspect crops as and where necessary. Recent transfer of the database to a computer with a virtual memory system will allow the rapid production of current aphid density maps (Woiwod and Tatchell, 1984). One important feature of the system is that continuous information on aphids numbers has now been collected for 20 years, providing a range of experiences with which to compare current data, and incidentally a historical data base which, when linked to suitable economic thresholds established by complementary research, allows medium- and even long-term forecasts for some species.

The desirable 4-day period of warning for such current awareness systems requiring rapid dissemination of up-to-date information is not totally achieved by either the Rothamsted or Michigan systems in their present form. The reasons for this shortcoming in the Rothamsted system has been the absence of widespread farmer ownership of computer and viewdata systems to receive the information, and the time taken to transport,

sort and identify the trap catches. The former is now being resolved as
farmers are becoming more aware of the advantages of modern communication systems, and an attempt is being made to shorten the collecting
and processing period by using upwardly looking radar to provide information on gross air-borne movement to anticipate the need for urgent
examination of trap catches.

A prospective radar system

With the availability of modern high-speed electronics and
micro-computers, it is now possible to sample the radar signals from
insects as small as aphids and to process them immediately. This
ability can both improve the radar performance, and provide an opportunity to design a fully automatic system for pest monitoring capable of
measuring the numbers of insects flying at different heights, their
speed and direction of flight and also capable of recognising and categ-
orising various insect taxa.

The radar system currently being developed at Rothamsted should provide
such immediate information, particularly about aphids flying at heights
between 12 - 300 m. Eventually a number of such fully automatic radars
will be sited in the UK and linked directly to Rothamsted by telephone
(Bent, 1984).

The prototype radar is a vertically pointing dual antenna system, using
a high power pulsed transmitter operating at a wavelength of 3.2 cm.
The transmitter is connected directly to one of the parabolic antennae
via a rotatable centre feed dipole aerial to produce a 3.6^{o} pencil beam.
A second receiver parabolic antenna is mounted close to the transmitter
antenna and offset at a small angle to the horizontal. A motor rotates
this antenna about its vertical axis, thereby conically scanning the
receiver radiation pattern.

As an insect flies over the radar site and traverses the region in which
the transmitter and receiver radiation patterns overlap, the intensity
of the radar echo signal produced depends upon the target's range, its
position within the beams and the radiation scattering properties of the
target, which are defined in terms of the radar cross-section. The
target position can be determined by measuring the amplitude and phasing
of the radar echo at this frequency, and the target's radar cross-
section provides information on the weight and body shape of the insect.
Returning radar echos from some insects also contain additional information associated with the beating wings and this can be used to identify
the insect targets. In the prototype system, the radar signals are
electronically sampled at eight height ranges between 12 and 300 m and
fed directly into a computer for immediate analysis and categorisation.

The sampled radar signals are processed by a dedicated multi-processor
computer that controls and monitors the operation of the radar, acquires
the data from the radar's electronic sampling circuits and performs the
rapid calculations ($100,000$ s^{-1}) required to categorise the targets flying over the radar site. As the radar signals are acquired the targets
are classified and their height, speed and direction are stored in the

computer memory. Over a period of 15 to 30 minutes this produces an
average insect density-height profile for the different categories of
insect, together with heading and speed distribution. After a specified
time the dedicated radar computer will be contacted via a telephone-
modem and the data in the memory bank transferred to a central machine
for further collation and analysis.

The radars together with their electronic sampling circuits and dedicated computers are housed in mobile trailers, with the intention of siting
them at selected locations in the UK. Each radar will be contacted by
the central computer and the data transferred to Rothamsted. This
information will be used together with information from the suction trap
network, field counts etc. and further processed to produce up-to-the
minute maps, trends and forecasts, and should be a significant step towards a novel early warning system requiring minimal staffing.

MEDIUM-TERM FORECASTS
Many of the features of this category of forecast are
illustrated by systems operating for moth pests in China and the UK.
They provide a complete contrast in scale, and staffing. Neither uses
the sophisticated computing necessary for extensive current awareness
systems, but, at different levels, the input of biological and weather
data is considerable.

Moth forecasting in China
The Chinese systems for forecasting armyworm (Mythimna
separata) infestation on cereals, and rice leaf roller (Cnaphalocrocis
medinalis) on rice, provide medium-term forecasts over an enormous geographical range. Current practices are based on uniquely comprehensive
systems for observing the arrival of adults of these migratory pests,
and their oviposition, by a range of simple but effective traps at many
thousands of recording points (Li and Chen, pers. comm.). The systems
are backed by very detailed knowledge of life stage development and by
an excellent synoptic meteorological network (Reed, 1980). The success
of the systems also arises from the ability of many field workers to
recognise the pests, a knowledgeable decision-making hierachy and good
communication between observers, weather and pest forecasters, and
growers.

The detailed coverage achieved is illustrated by the monitoring system
operated for armyworm in Jilin province, where, to detect adult
immigrants, shown by marking and recapture experiments to be capable of
travelling up to 1200 km, there are 100 light traps, 100 attractant
traps baited with either molasses or wine depending on the time of the
year, and 1000 straw bundles on which the adults alight; in addition
1000 bundles, each of three millet stalks, are used to detect oviposition. The traps are examined daily and data on the number, sex, ovarian
development and estimated age of individuals are sent to the Jilin
Academy of Agriculture, Gongzhuling by post every 3 days. If extraordinarily large catches of more than 1000 individuals occur, the
information is sent by telegram. Using the data and weather information

from the synoptic meteorological centre in Changchun, six forecasts of moth abundance are issued each season and distributed to 500 centres for local action. Control is recommended when larval populations on wheat reach 1 per 100 tillers. For different provinces, the number of egg clusters per 100 bundles is used to predict larval numbers up to two weeks ahead. The present system has evolved over 18 years.

A similarly intensive system is operated in southern provinces to detect rice leaf roller. Detailed counts of all instars, pre-pupae, pupae and adults are made in April and, on the basis of these, the date of appearance, the size of the damaging second generation, oviposition period and optimum spraying date are forecast up to 20 days ahead of the event. Sudden outbreaks of this pest appear typically over wide areas covering 6-12° of latitude. Marking and recapture of moths on an enormous scale (5 million moths were marked and released 1977 to 1980 at 39 stations in 8 provinces) have revealed the origins and routes of migrants. Ovarian dissections have shown that each year young adults migrate soon after emergence in five main surges northwards corresponding to successive generations during March to August, followed by three migrations southwards from August to October. These movements are associated with three particular synoptic patterns: frontal systems in which moths often appear in areas before the trough and behind the front, usually accompanied by rain and downward currents; anticyclones, when moths migrate up the eastern plain in summer as areas south of the Yangtse river are influenced by sub-tropical high pressure; and typhoons, the western fringes of which generally disperse insects southwards along the east coast (Chang et al., 1980). Table 2 shows the basis of a forecast in 1981 for the appearance of the second generation in Guandong province.

Table 2. Rice leaf roller, Guandong Province, China: basis of 1981 forecast.

1980:	1st generation 300 - 950 adults ha^{-1}
1981:	1st generation 3800 - 5800 adults ha^{-1}
	(This large increase noted)
20 April 1981:	Survey and counted all larval instars, pupae and adults
25 April 1981:	Precise report available
Forecast:	30 April - 3 May : beginning of adult peak
	6 - 9 May : peak adults
	6 - 9 May : beginning of egg hatch
	12 - 15 May : peak egg hatch
	20 - 22 May : end egg hatch
Recommend control:	12 - 15 May

With such a wealth of basic developmental, distribution and meteorological data, it is only the lack of computerised collation and mapping techniques that prevents Chinese forecasting systems reaching an even higher standard, though the heavy staffing requirement probably precludes adoption of similar systems in more developed countries.

Pheromonal monitoring for forecasts

No survey of current forecasting practices would be complete without mention of the role of pheromones in pest detection, though to convert knowledge of the presence of a pest detected by pheromone traps into a useful forecast for control, requires much additional developmental information. Such chemical monitoring, usually by attractants dispensed from simple and inexpensive traps, is cheap and can be used easily by farmers to produce a detailed pattern of pest presence on an individual field. This is useful when accurately-timed spraying is essential, especially when the maturation of crops in an area is not synchronised. It has the great advantage of producing largely "self-sorted" catches because many of the lures, particularly pheromones, attract a limited taxonomic group, usually a single species. Because chemical lures are usually so intensely attractive, very sparse pest populations, which would probably be overlooked by other methods, can be detected, allowing medium-term forecasts of impending infestations and time to plan control. The main disadvantages of chemical monitoring are that it usually measures an activity, such as feeding or sexual urge, rather than direct abundance, and many of the materials so far available do not attract the larval stage that actually damages the crop or the females which lay the eggs, making the interpretation of catches in terms of crop infestation difficult.

The monitoring system for pea moth operated in the UK illustrates a well-researched approach to a pheromonal-based scheme that alerts the grower to the presence of the pest in a crop 10-13 days before control may be necessary, and later provides accurate assessment of the optimal first spraying date. To be effective against this pest, insecticidal sprays must be timed to kill newly-hatched larvae before they enter pods, but the critical period varies between years and even fields. Two sticky traps, each containing an analogue of the moth's natural pheromone (Wall and Greenway, 1981) are placed in each pea field on adjacent headlands on the side of the prevailing wind. Traps are examined every 2 days and when 10 or more moths are caught in either trap on two consecutive sampling occasions, calculations for rate of egg development are started to predict first larval hatch (Lewis and Sturgeon, 1978). After detecting a threshold catch, a grower knows that control measures are likely to be required, and can then calculate the best spray date himself from max./min. temperature data recorded in each field or from a nearby meteorological station, or, alternatively, contact a 24-hour phone-in service run by local advisory services which will tell him when to spray. From 1980-83, growers who have operated this system and have detected a threshold catch and sprayed accordingly have decreased mean damage levels 4-fold, whereas those who sprayed without achieving a threshold have reduced damage only sporadically; the sensitivity and usefulness of the system are thus confirmed (Wall, Garthwaite and Blood-Smythe pers. comm.).

LONG-TERM FORECASTS
The more distant in time the event to be forecast, the more
uncertain the accuracy, especially when the interval spans one or more
seasons. Nevertheless, for many crop protection decisions long-term
forecasts can often be helpful.

Black bean aphid forecasting in UK
An excellent illustration of the way in which forecast
accuracy decreases as the forecasting period increases is provided by
the very well studied forecasting system for Aphis fabae on spring-sown
field beans in the UK (Way et al., 1981). In southern England, the
species migrates to spindle trees (Euonymus europaeus) in autumn, where
overwintering eggs are laid. From these, fundatrigeniae hatch in the
following spring to produce migrants that fly to infest spring-sown
beans. Estimates of the sizes of populations involved in these four
"stages" in the annual cycle, made either by egg counts or suction trap
catches, have been used to forecast infestation levels in field bean
crops from 2 weeks to 8 months ahead. The forecasts of crop infestation
become progressively more accurate from the autumn to the following
spring migration, accounting for 28%, 54%, 54% and 64% of the variance
respectively. Autumn trap catches provide forecasts 8 months ahead of
notably large or small populations on field beans, but otherwise lack
precision. They can be used to anticipate years in which the spring
migration will be large and treatment probably essential. Egg sampling
in winter provides a more accurate forecast approximately 5 months
before crop infestation, and confirms the areas at particular risk,
alerting pesticide merchants and growers to the likely need for widespread insecticidal applications. Forecasts made from peak spring
population counts on spindle in May and from the spring migration in May
to mid-June may be seen in the context of forecasting categories listed
above (Table 1) as reasonably accurate medium-term forecasts. Table 3
shows the reliability of area forecasts based on autumn suction trap
catches and overwintering eggs over 8 years, confirming the promise of
the system.

It is important to stress the considerable research effort expended by
about 20 entomologists on the basic biology of this pest, trap sampling,
methodology, widespread and intensive inspection of host plants, and
data analysis and interpretation, to produce this limited but useful
degree of reliability in a long-term forecast for a single species
attacking a minor crop. This underlines some of the reasons why similar
forecasts are unlikely to become commonplace in the near future.

Wheat bulb fly forecasting in UK
An interesting and unusual approach to long-term forecasting
based on land use in relation to the annual cycle of a pest was attempted for wheat bulb fly in eastern England by Kempton et al. (1974)
following a suggestion by Raw (1967). Much of the difficulty in controlling the pest is due to its peculiar life cycle. There is only one
generation a year; eggs are laid in July and August on bare soil or
fallow land or on exposed soil beneath root crops such as potatoes, but
they do not hatch until 6-7 months later in January or February. Only

larvae hatching in fields already sown to wheat, barley or rye survive, boring into the base of tillers where they feed until pupation in May. Depending somewhat on the soil type, crops sown where egg populations exceed 1.25 M ha^{-1} are likely to require insecticidal protection. Improved methods of forecasting egg population before seed orders are placed would help merchants to prepare enough treated seed in years when the pest was scarce.

Table 3. Accuracy of area forecasts of black bean aphid (Aphis fabae) based on suction trap catches in autumn and eggs on spindle (Euonymus europaeus) in winter, 1969-75 and 1976-77 (after Way et al., 1981)

Forecasting procedure	Damage	No. of areas	No. correct	No. incorrect
Autumn suction trap catches	Unlikely (0-15 alatae/trap mid-Sept. - early Nov.)	26	22	4
	Possible and probable (>15 alatae/trap mid-Sept. - early Nov.)	21	19	2
Overwintering eggs on E. europaeus	Unlikely (0-1 eggs/ 100 buds)	26	23	3
	Possible and probable (>1 egg/100 buds)	26	24	2

Analysis of 20 years' records (1952-71) showed that mean egg populations over large areas vary up to 5-fold from year to year, and that much of the variation could be explained by variation in land use, which controls the area of oviposition sites, and in weather, which affects plant growth and the survival of the immature stages of the egg-laying flies. For example, a multiple regression with coefficients estimated from data over this 20-year period accounted for 71% variation in mean egg density between years in the extensive wheat-growing area of the Isle of Ely, providing a potential method for forecasting several months ahead the likely need to control the pest.

The considerable effort required to produce the biological and statistical background for this type of forecast is again worth emphasising; the samples for this analysis were taken from nearly 2500 fields over 20 years. Unfortunately, a weakness of this approach, as with all that depend for success on consistent land use, is that agricultural practice has now changed, and areas of fallow for oviposition are rare. Most bare land available for oviposition is exposed after harvest of vining pea crops, adding a new variable to the situation, with the consequent need for a further period of long-term recording.

ANNUAL PROBABILITY FORECASTS

The abundance of some species, particularly those living near the edge of their climatic range, is likely to fluctuate more erratically from year to year than that of species well adapted to the climate. Two approaches have been used widely to assess the probability with which outbreaks of a pest are likely to occur in a particular region. The experimental approach is to measure in laboratory or field the range of physical factors a species can tolerate, and then to determine whether it could survive in the chosen region by comparing specific tolerances with the range of climatic factors experienced there. In places where these tolerances are exceeded, outbreaks are most unlikely. Unfortunately, the biological data for most potential pest species are incomplete and laborious to collect; furthermore, this approach presupposes that all the constraints on a pest's development are known and can be equated with corresponding climatic variables, which is rarely so.

An empirical but more reliable method is to measure the frequency with which infestations have established over a period of years to provide for a given pest a forecast of the probability of its occurrence. This approach has the advantage of being based on direct evidence of actual occurrences or observed damage, so it incorporates the elements of climate critical to a species without having to identify each individually. Its drawback is that it is possible only when long-term records have been kept, and, as mentioned, such instances are regrettably few.

From the farmer's viewpoint, annual probability forecasts of pest occurrence, however obtained, are unlikely to be an overriding factor determining whether a particular crop is grown, except where there is long experience of its failure due to frequent direct or indirect pest attack. (In the UK, the avoidance of seed potato crops in lowland areas suitable for the rapid growth and spread of the virus vector, Myzus persicae, is an example). A farmer is more likely to decide to grow a crop for agronomic and economic reasons, and then to use warnings and medium-term forecasts as an aid to protecting it from damaging infestations.

There are no standard methods of producing annual probability forecasts. Usually they are derived from long-term data collected by diverse methods and often for other purposes. Examples relating to one pest and one aphid-borne disease liable to occur on crops in the UK illustrate how differing sets of data, each collected over at least 30 years, might be used to estimate the risk of outbreaks in a single year.

Colorado beetle in England

Leptinotarsa decemlineata is a non-endemic species for which the climate in the UK is barely suitable. Since it became firmly established in Europe in the 1920s, single beetles have been found in most years in Britain, mostly associated with imported vegetable produce, with perhaps a few flying directly from Continental Europe. Some winters are mild enough for the diapausing beetles to survive in soil and

warm spring days are suitable for dispersal. Oviposition and egg hatch
may be depressed in cool periods and heavy rains may kill young larvae,
but in hot, dry years the beetle could complete one generation and perhaps a partial second. Since measures to prevent the pest's establishment were tightened in 1945, a detailed record of the number of dead and
living beetles recorded in the UK, and more importantly the number of
breeding colonies recorded annually, has been kept. Over a 33-year
period from 1946-1978, the beetle was found to breed in only eight
years, and in three of these only in one or two places (Bartlett, 1979).
This indicates that there is only a 0.15 to 0.24 probability of the
insect establishing on this northern edge of its range given the geographical barrier of the English Channel, the climate and the stringent
eradication policy pursued. It would seem unnecessary even to prepare
to take precautions in the UK in most years.

'Virus Yellows' in sugar beet

A second example is provided by the long-term data collected
on the incidence of 'virus yellows', caused by beet yellows virus (BYV)
and beet mild yellowing virus (BMYV) transmitted to sugar beet crops by
aphids. The successful survival of Myzus persicae, the main vector, and
its early availability to infect crops in spring is dependent on the
severity of frosts and of surface moisture during the previous winter
(Watson et al., 1975; Taylor, 1977b; and Harrington, pers. comm.).

Figure 1. The incidence of sugar beet yellows
virus at the end of August in about 1000 English
beet fields chosen at random (after Watson et
al., 1975; and Heathcote, pers. comm.).

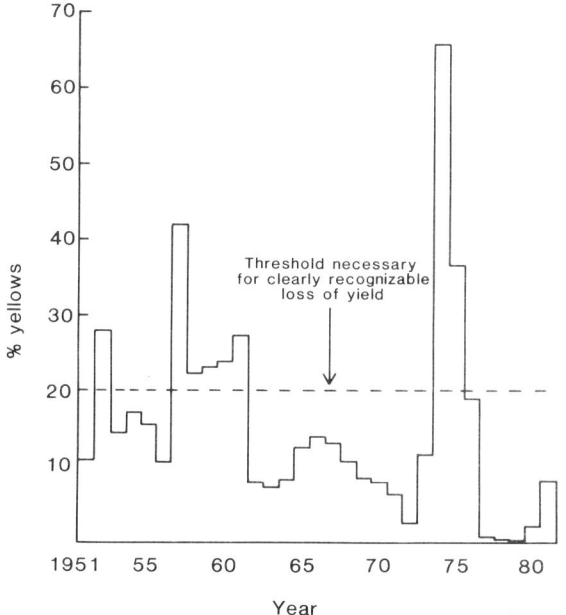

In England, all sugar beet is grown under contract and records of disease incidence are obtained from annual surveys made by field staff of the British Sugar Corporation. Figure 1 shows the annual percentage yellows infection at the end of the August from a sample of about 1000 fields from 1951-1981.

The amount of infection apparent at the end of August gives a reasonable indication of the likely loss of yield. Below 20% infection, no substantial losses occur; above that figure, each 5% increase in infection equates very approximately to a 1.3% loss of sugar. It thus appears that on a national basis, yellows would only have depressed yields in 8 out of the 30 years covered. Sugar losses of more than 10% would only have occurred in three years and of more than 5% in only six years (Heathcote, 1978 and pers. comm.). However, this interpretation underlines one of the main disadvantages of this type of probability forecast, namely that they are usually compiled from such general information that important local differences are swamped. In this instance, incidence of yellows varied greatly depending on locality, often being, for example, four times more prevalent in southern East Anglia than in Yorkshire; in 1980, the 2% national average covered a range from 0.1 to 19%. Thus, in an area or on a farm where climate was conducive to spread, and vectors and local sources were abundant, the risk of infection in a given year could well be much greater than that indicated by the overall forecast.

ACKNOWLEDGEMENTS
I am pleased to acknowledge the help of many Rothamsted staff in preparing this paper, especially R. Bardner, G. Bent, G. Heathcote, G.M.. Tatchell, C. Wall and I.P. Woiwod.

REFERENCES
Anon. (1975). Estimated losses and production costs attributed to insects and related arthropods, 1973. Co-operative Plant Pest Report, US Department of Agriculture 1, 22-36.

Bartlett, P.W. (1979). Preventing the establishment of Colorado beetle in England and Wales. In Plant Health. The Scientific Basis for Administrative Control of Plant Diseases and Pests, eds D.L. Ebbels and J.E. King, 247-257.

Bent, G. (1984). Developments in detection of airborne aphids with radar. Proceedings British Crop Protection Conference, Pests and Diseases, Brighton 1984, 665-674. British Crop Protection Council Publications.

Chang, S.S., Lo, Z.C., Keng, C.G., Li, G.Z., Chen, X.L. and Wu, X.W. (1980). Studies on the migration of rice leaf roller Cnaphalocrocis medinalis Guenee. Acta Entomoligica Sinica 23, 130-140.

Essig, E.O. (1911). Injurious and Beneficial Insects of California. Supplement, The Monthly Bulletin, California State Commission of Horticulture, Sacramento, pp. 541.

Gage, S.H., Whalon, M.E. and Miller, D.J. (1982). Pest event scheduling system for biological monitoring and pest management. Environmental Entomology 11, 1127-1133.

Heathcote, G.D. (1978). Review of losses caused by virus yellows in English sugar beet crops and the cost of partial control by insecticides. Plant Pathology 27, 12-17.

Kempton, R.A., Bardner, R., Fletcher, K.E., Jones, Margaret G. and Maskell, F.E. (1974). Fluctuations in wheat bulb fly egg populations in Eastern England. Annals of Applied Biology 77, 102-107.

Lewis, T. and Sturgeon, D.M. (1978). Early warning of egg hatch in pea moth (Cydia nigricana). Annals of Applied Biology 88, 199-210.

Raw, F. (1967). Some aspects of the wheat bulb fly problem. Annals of Applied Biology 59, 155-173.

Reed, R.J. (1980). Meteorology in China. In Science in Contemporary China, ed. L.A. Orleons, 213-235. Stanford, California, USA: Stanford University Press.

Taylor, L.R. (1977a). Aphid forecasting and the Rothamsted Insect Survey. Journal of the Royal Agricultural Society of England 138, 75-79.

Taylor, L.R. (1977b). Migration and the spatial dynamics of an aphid, Myzus persicae. Journal of Animal Ecology 46, 411-423.

Taylor, L.R. (1979). The Rothamsted Insect Survey - an approach to the theory and practice of synoptic pest forecasting in agriculture. In Movement of Highly Mobile Insects: Concepts and Methodology of Research, eds R.L. Rabb and G.G. Kennedy, 148-185. Raleigh, North Carolina, USA: North Carolina State University Press.

Walker, P.T. (1975). Pest control problems (pre-harvest) causing major losses in world food supplies. FAO Plant Protection Bulletin 23, 70-77.

Wall, C. and Greenway, A.R. (1981). An effective lure for use in pheromone traps for monitoring pea moth, Cydia nigricana (F.). Plant Pathology 30, 75-76.

Watson, M.A., Heathcote, G.D., Lauckner, F.B. and Sowray, P.A. (1975). The use of weather data and counts of aphids in the field to predict the incidence of yellowing viruses of sugar beet crops in England in relation to the use of pesticides. Annals of Applied Biology 81, 181-198.

Way, M.J., Cammell, M.E., Taylor, L.R. and Woiwod, I.P. (1981). The use of egg counts and suction trap samples to forecast the infestation of spring-sown field beans, Vicia faba, by the black bean aphid, Aphis fabae. Annals of Applied Biology 98, 21-34.

Woiwod, I.P. and Tatchell, G.M. (1984). Computer mapping of aphid abundance. Proceedings British Crop Protection Conference: Pests and Diseases, Brighton 1984, 675-683. British Crop Protection Council Publications.

Woiwod, I.P., Tatchell, G.M. and Barret, Angela M. (1984). A system for the rapid collection, analysis and dissemination of aphid-monitoring data from suction traps. Crop Protection 3, 273-288.

22. PROGRESS TOWARDS RATIONAL WEED CONTROL STRATEGIES

G.W. Cussans, R.D. Cousens and B.J. Wilson
Long Ashton Research Station, University of Bristol,
Bristol, BS18 9AF, England

INTRODUCTION
This paper concentrates on defining the needs for weed
control, and does not consider herbicide behaviour. In practice,
herbicide performance is variable and any system of rational weed
control should include prediction and optimisation of that performance.

One simple, practical observation may be worthwhile. Herbicides may be
applied pre-emergence or up to quite advanced post-emergence stages to
control, either single species or a range of species. Concepts of
planned, rational weed control would be infinitely easier to apply to
the use of late-applied chemicals against a single species, rather than
where a pre-emergence application is used to kill a wide range of
species.

THE OBJECTIVES OF WEED CONTROL
These may be to eradicate the weed completely or to contain
it at a frequency or a degree of vigour which will not interfere with
crops to an unacceptable degree. It is possible to consider eradication
because most weeds are comparatively spot-bound organisms with limited
mechanisms of dispersal (except by human agency). Seeds of some weeds
are dispersed by the wind and although such plants are relatively uncommon, they are potentially able to fill niches vacated by less mobile
species. Epilobium spp. have become a problem in this way in some
fruit-growing areas. Nonetheless, some weed species can be eradicated.
Corncockle (Agrostemma githago) and darnel (Lolium temulentum) have been
eliminated from UK agriculture and others, notably the common wild-oat
(Avena fatua) may have been eliminated from individual farms. However,
we consider that these are rather special cases; that eradication is a
word which may often be used with insufficient care (see Cussans, 1980);
and that the main practical aim of weed control is containment.

SHORT-TERM CONTAINMENT OF WEEDS
In order to achieve this, we must begin with an understanding of how much damage is caused by weeds. Theoretically the
simplest rationalisation would be to prescribe simple decision rules for
whether or not to spray, based on an economic threshold level. This is
a weed density at which the cost of damage caused by the weed population
equals the cost of control. Many people have considered this approach,

but others have postulated an <u>absolute threshold</u>, a weed density below
which no competition occurs.

This latter concept takes the widespread assumption that the crop yield
response to increasing weed density is of sigmoid form (Fig. 1). If
this were so it might be possible to suggest an absolute threshold as
shown in Figure 1. Even with the curve shown, there would not be a true
absolute threshold; the line for crop loss passes through the origin.
However, such a concept might be acceptable for practical purposes if
the responses were sigmoidal. In fact, there is no evidence whatever to
support the existence of this form of response. When 195 data sets were
examined (Cousens <u>et al</u>., 1984), the general form of the competition
response curve was best described by a rectangular hyperbola (Fig. 2).
This indicates that, far from having no effect, individual weed plants
have most effect at low densities.

A reasonable simplification, at low densities only, is to assume a
linear response and indeed much experimental data can be described in
this way.

However, these are retrospective analyses rather than predictions, for
we still have to contend with the plasticity of weed growth. Individual
plants may attain very different sizes at maturity. It is likely,
therefore, that yield loss will be better related to weed biomass than
to plant numbers. Where this has been studied there has indeed been a
good relationship. In a few cases, a non-linear hyperbolic relationship
has been found but in most cases there has been a linear relationship
between weed biomass and crop yield loss. An example is given in
Figure 3.

In many cases (see Wilson and Peters, 1982), the relationship between
weed biomass and loss of crop biomass is 1:1. This is not unexpected
where the species concerned is one such as wild-oat (<u>A. fatua</u>) which is
similar to the crop in its pattern of growth. In such cases the weed
replaces crop. However, the slope of this curve may not be constant
from year to year and is certainly not the same for every weed species.

Summarising what is known and what remains to be done: for single
species populations, we know the underlying shape of the response curve
to seedling population and we know that the crop yield loss is directly
and simply related to weed biomass. We also know in qualitative terms
many of the factors controlling the development of weed biomass. This
final outcome is influenced by crop and weed density, by the characteristic pattern of growth for the individual species, the time of weed
seedling emergence relative to that of the crop and by other factors,
notably soil moisture. More work is still necessary to allow prediction
of weed biomass development and hence competition based on early counts
of seedlings, which is all that is possible before spray decisions must
be made.

Figure 1. A commonly accepted model of the response of crop yield to increasing weed density which is sometimes used erroneously to adduce an absolute threshold below which no competition occurs.

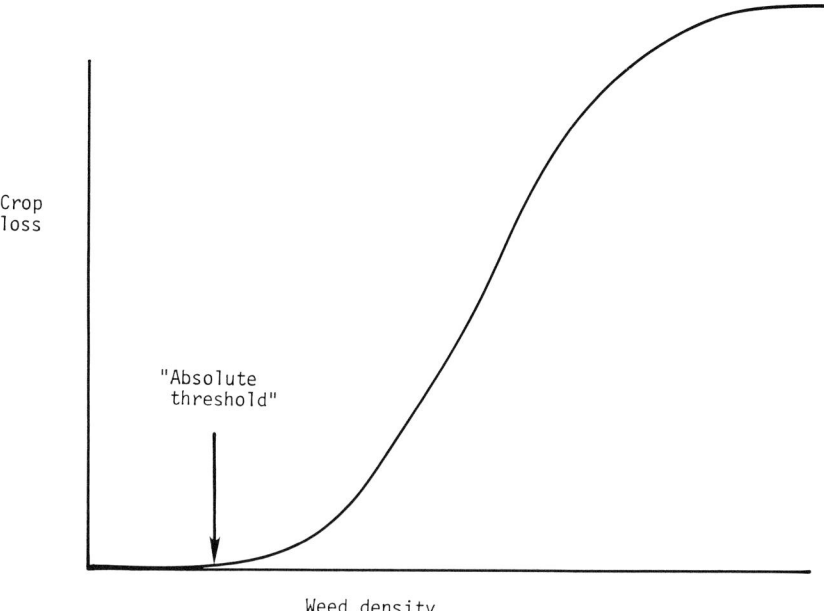

Figure 2. The model of crop yield response which can most often be fitted to data.

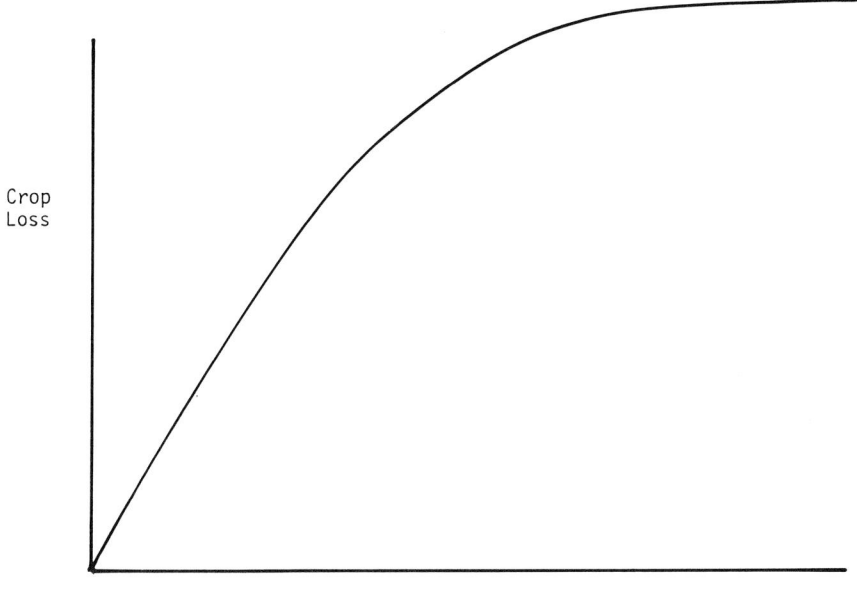

Figure 3.   An example of the relationships between weed
biomass and total crop weight or grain weight in <u>Vicia</u> <u>faba</u>
(from Glasgow 1976).

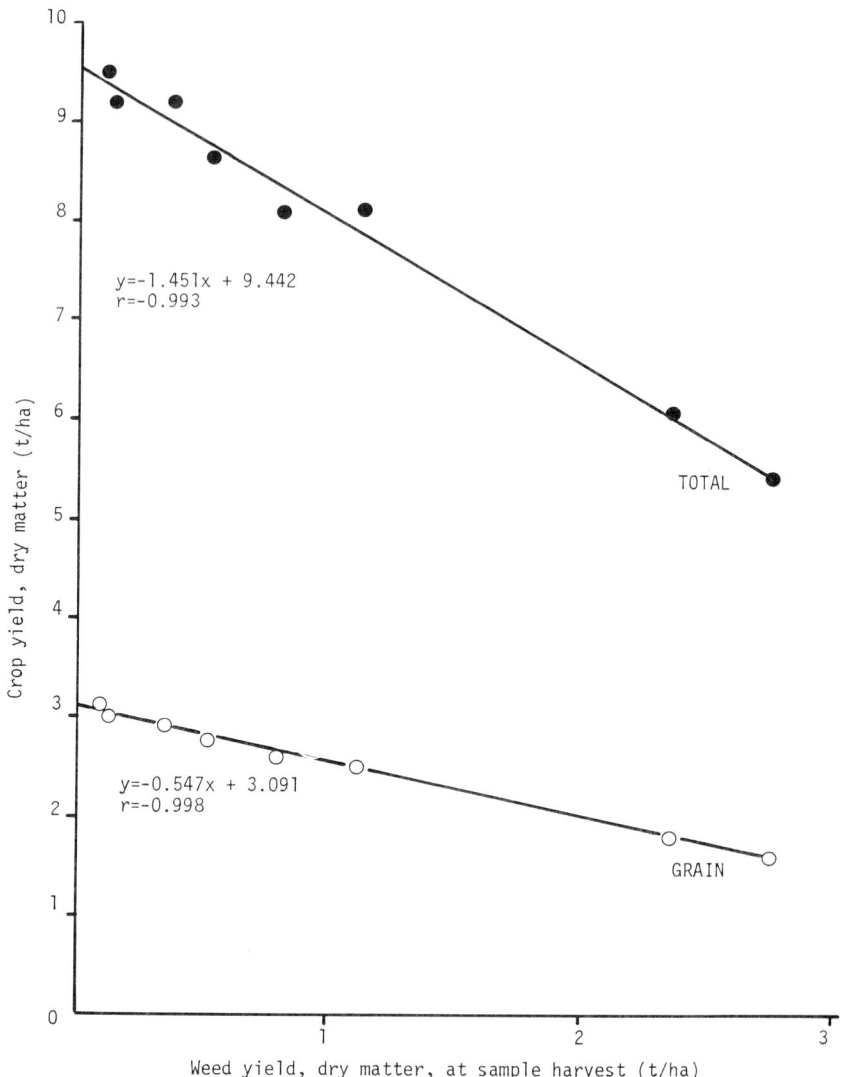

Most of the work done by our own organisation has been with a single
major weed species, such as wild-oat (<u>A. fatua</u>) or black-grass
(<u>Alopecurus</u> <u>myosuroides</u>).   The objective has been to achieve some understanding of the principles involved before moving on to the more
difficult problem of determining the effects of mixtures of weed
species.

However, this situation of mixed weed flora is also being approached in
a simpler, pragmatic way.   If it is assumed that the yield decrease due
to weeds is related linearly to weed dry weight, then the relative

competition efficiency of different weed species should be related to
the dry weight produced by each plant. We are conducting simple growth
analyses of a range of weed species in wheat crops on a number of
private farms. The dry weight of individual weed plants is expressed
relative to that of individual crop plants. This is done for two
reasons: the crop is the one plant species present at all sites; if the
competition is of a replacement type then this "crop equivalent" ratio
will be a direct predictor of yield effects.

Some preliminary results are given in Table 1. It must be emphasised
that these are average results from one season only. We have already
identified some site-to-site variation and must expect year-to-year
variation.

Table 1. Some crop equivalent values for different
weed species.

Galium aparine	1.7	Lamium purpureum	0.2
Papaver rhoeas	0.9	Myosotis arvensis	0.2
Matricaria spp.	0.6	Veronica hederifolia	0.1
Stellaria media	0.5	Viola arvensis	0.1
Veronica persica	0.2	Legousia hybride	0.1
		Aphanes arvensis	0.1

Table 2 shows how all the weeds present were combined into one total
"crop equivalent", which could then be used to predict the proportional
yield loss, if a direct replacement of crop by weed occurred.

This direct pragmatic approach has achieved some success and we plan to
ally it with more fundamental work to improve the quality of our predictions. It has to be remembered that predicting the response to weeds
is only one element of a larger and more complex problem. To predict
the economic outcome of weed control one needs to predict also the crop
yield, the price likely to be achieved and the efficiency of weed
control.

An example of prediction based on some of our data for difenzoquat, a
wild-oat herbicide, is shown in Figure 4. The curves are based on a
herbicide kill of 91% for full dose, and 78% for half dose. A grain
price of £120 t^{-1}, weed-free yield of 6.5 t ha^{-1}, spray application cost
of £4.50 ha^{-1} and herbicide cost (full dose) of £40 ha^{-1} were assumed.
Yield loss was assumed to be related to weed density according to the
relationship:

$$Y_L = \frac{0.75d}{1 + 0.0078d}$$

where, Y_L is percentage yield loss and d is weed density. The cost index is the ratio of the cost of treatment divided by the potential gross return on a clean crop expressed as a percentage. From Figure 4, it can be seen that approximate economic threshold weed densities would be 9 plants m^{-2} for full dose, and 5.5 plants m^{-2} for half-dose. At these densities, yield losses of 6.3% and 4.0% respectively would be incurred if spraying did not take place.

Table 2. "Crop equivalent" (CE) values for weeds in a wheat crop and prediction of proportional yield loss - based on one field experiment.

	Plants m^{-2}	Crop equivalents m^{-2}	
Wheat	275	275	
Galium aparine	12	20.4	
Stellaria media	16	8	Total
Veronica persica	1	0.2	weeds
Myosotis arvensis	34	6.8	= 35.6
Viola arvensis	2	0.2	
Total		310.6	

Prediced proportional yield loss $= \dfrac{\text{Weeds as CE (35.6)} \times 100}{\text{Total CE (310.6)}} = 11.5\%$

Recorded grain yield loss $= 15\%$

In our view this is a valuable research exercise but is probably too complex to be appropriate for practical decision-making. The following factors mitigate against the precise use of spray thresholds:

> 1. In practice, weed populations are immensely variable within fields so a high degree of precision would be unobtainable;
>
> 2. The prediction of crop yield and price is likely, itself, to remain approximate;
>
> 3. Longer-term considerations may also be of great importance.

Figure 4. The relationship between economic threshold weed density and the cost of control, relative to gross crop value, at two levels of control efficacy — for _Avena fatua_.

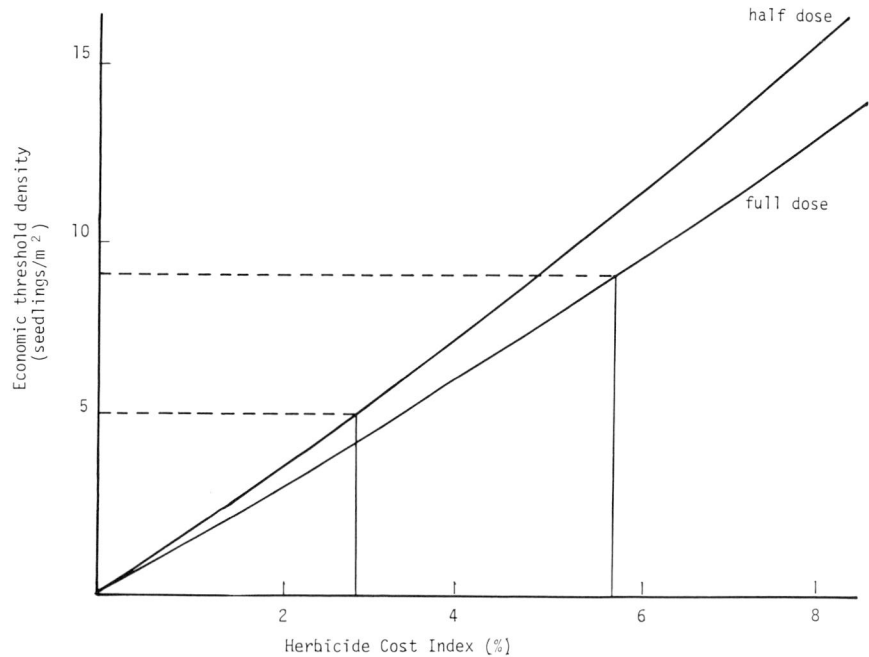

Before the longer-term aspects are discussed, let us introduce another concept, that of a safe decision rule or action threshold. In practice, many fields are treated at weed levels far lower than the economic threshold. The reasons for such caution should be analysed before dismissing it as irrational.

A substantial safety margin may be required:

> 1. To allow for the possibility of poor herbicide performance;
>
> 2. To allow for poor assessment, e.g. a low density in some parts of a field may be a sign of worse infestations in other (unseen) areas;
>
> 3. To avoid population increase of species which are especially difficult to kill.

All of these criteria could be loosely grouped under the heading of risk. Despite the difficulties, quantification of these risk elements will be a vital practical step in rationalising herbicide use.

We have done very little work on suggesting numerical ratios
between the economic threshold and a safe decision rule. We have, however, suggested a ratio of 10:1 for A. myosuroides to allow for the
great proliferacy of this weed and the difficulty and expense of controlling it. With easy-to-control species, the ratio could be as little
as 2:1. Weeds such as Papaver rhoeas, even when they are causing very
little yield loss, may be an affront to many farmers' pride. Such
aesthetic considerations are not irrational but they cannot be quantified and have therefore been ignored in our approach.

LONG-TERM CONTAINMENT OF WEEDS
          It may be understandable, if not accurate, to visualise the
soil as providing an infinite reservoir of the seeds of some broad-
leaved weeds. Indeed, the seeds of many weeds are so long-lived and the
seed reserve sufficiently large that the consequences of control (or
lack of it) in one season are not apparent in the next. Seedling emergence may be more influenced by the timing and nature of seedbed preparation and by the climate than by recent seeding, although clearly this
must influence the population in time.

Such weeds may best be managed by a threshold approach, as described
earlier. Seedling emergence would be treated as almost a random event
and treatment imposed when the level exceeded some decision rule.

In marked contrast to this, all of the grass weeds and some broad-
leaved species have seeds with a much shorter life span, typically
5-9 years. The potential for increase or decline of populations over a
short span of years is much greater and thus the consequences of weed
control in one season may be discerned more readily in the next year or
so. In the case of the grass weeds, it is also common for decisions to
be made relating to one species, and not to an association of different
species all being influenced to some degree by the same herbicide, as
occurs with broad-leaved weeds. Finally, the grass weeds are more
important economically; they compete strongly with the crop and appropriate herbicides tend to be expensive.

For all these reasons, we have spent some time developing a series of
simple population models for some of the major grass weeds in the UK: A.
fatua, A. myosuroides and Bromus sterilis. Initially, these were exponential models, taking no account of density-dependence. This was
appropriate if the models were only used for considerations of low
populations, as would be of practical interest. However, later
variations (Doyle et al., 1985) have introduced density-dependent
fecundity into the basic models. In each case we started with a simple
description of the life cycle, as illustrated for A. myosuroides in
Figure 5.

Values are assigned to each of the transition pathways. These may be
observed from a single experiment but, more often, they are an average
of a number of observations. Herein lies the weakness of such models
for frequently we know that the average value comes from a wide range of

observations. Many of the component elements vary with climate or some other variable but in most cases we do not have an adequate base of data or working hypothesis to introduce these variable factors into the models. We can, however, introduce some variables of great practical importance, notably cultivation method. It is known that the degree of burial of seeds by cultivation has a great deal of influence on grass weed populations (Cussans et al., 1979). Tables 3 and 4 show one way in which these models help us to quantify the effects of tillage. We have used the models to calculate the percentage kill which would be required to maintain static populations of A. myosuroides (Table 3) and A. fatua (Table 4) over a number of years of continuous winter cereal cropping.

Figure 5. A simple population cycle for Alopecurus myosuroides (from Cussans and Moss, 1982).

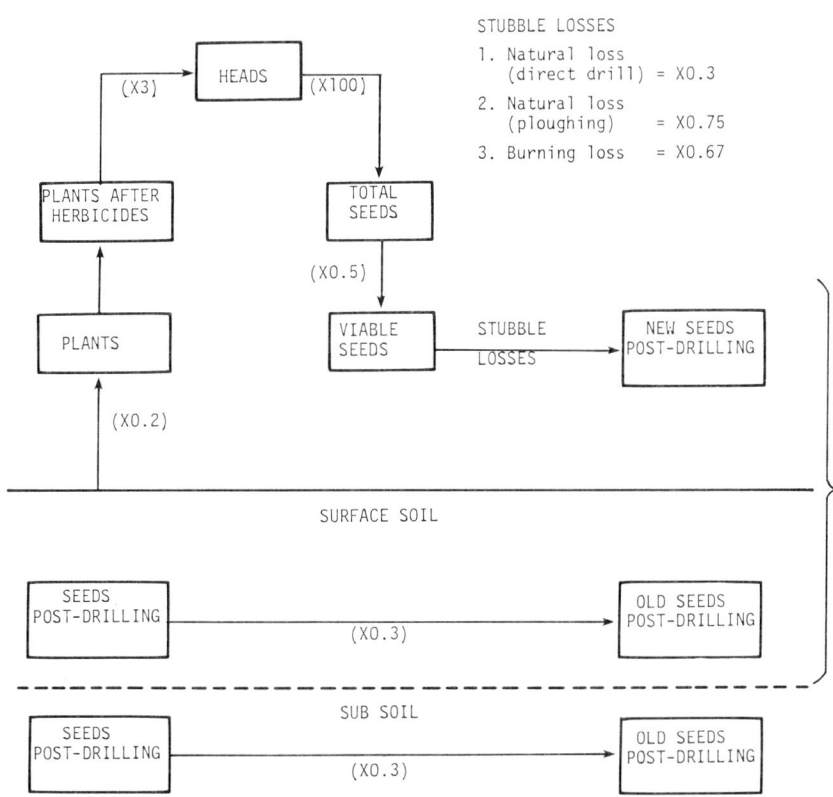

Table 3. The annual percentage kill by herbicides needed to maintain a static population of A. myosuroides in continuous winter cereal cropping (from Cussans and Moss, 1982).

	Straw burnt	Not burnt
Ploughed	50	65
Direct drilled	88	92

Table 4. The annual percentage kill by herbicides needed to maintain a static population of A. fatua in continuous winter cereal cropping (after Wilson et al., 1984.)

	Straw burnt	Not burnt
Ploughed	71	81
Tine cultivated	73	86

We can, of course, use these models in a number of ways, such as simulating the effects of changes in rotation. A further example is shown in Figure 6 where the effects of tillage and straw burning are expressed in terms of the number of years in which spraying would be necessary.

Such models have proved useful as teaching aids and have put some rationale into what we can observe. They "feel" right but must clearly be limited by the reliance on average performance data. However, they have allowed some simple predictions already and must be susceptible to improvement.

FUTURE PROSPECTS
Predicting competition
Competition is a vital element in attempts to forecast weed populations as well of obvious importance in economic analyses. This is because the outcome of competition determines both crop yield and the production of weed seeds or vegetative propagules. This outcome is not easy to predict, but now we have a basic model which is fundamentally

sound and it should be possible to add to it consideration of some of
the major climatic and edaphic variables. This must be seen as a priority area for on any objective analysis it seems absurd that we know so
much about killing weeds and comparatively little about the damage they
do. However, if prediction of competition were to be recognised as a
priority target, it should be possible to make substantial advances
within a relatively short span of time, perhaps five years.

Figure 6. A model for <u>Avena</u> <u>fatua</u> showing the frequency
with which a spray threshold would be exceeded in continous
cereal cropping.

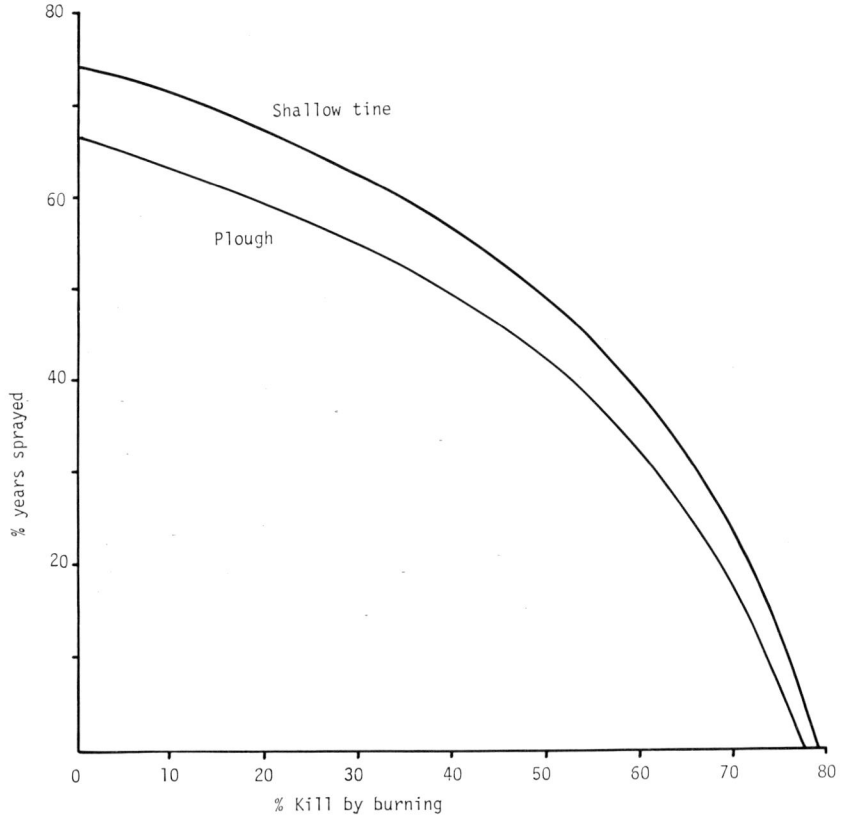

Seed analysis
Because weeds are basically sedentary, it should also be
possible to forecast populations by determing the seed content of the
soil, and some workers have indeed tried this approach. It suffers from
two problems. The first is the simple arithmetic of seed numbers
relative to bulk of soil. It is common to base advisory treatment
thresholds or decision rules for insect pests on numbers of eggs per
gram of soil. In contrast to this, the decision rule for wild-oat, <u>A.</u>
<u>fatua</u>, has been set by various authorities between 0.5 and 10 plants
m^{-2}.

Figure 7. A diagram of the germination periods of some
common weed species (from Roberts, 1982).

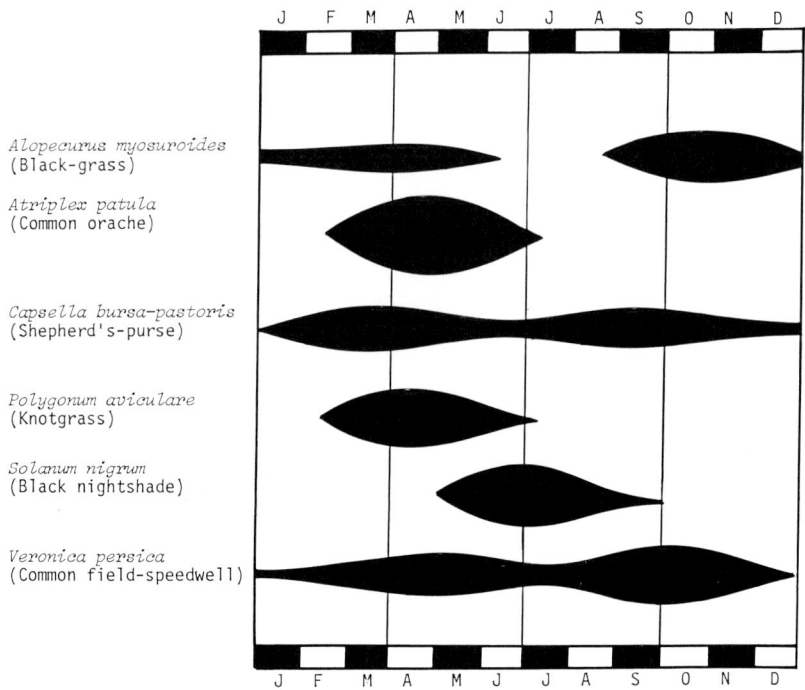

Figure 8. The germination of seeds
of Rumex crispus in response to
light, nitrate supply, fluctuating
temperature and to combinations of
these factors (after Roberts, 1973).

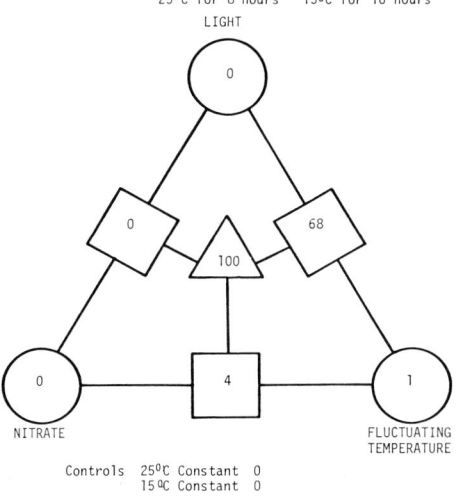

Assuming 10% successful seedling emergence, this corresponds to a range of 5 to 100 seeds m^{-2} of soil to a depth of 12 to 15 cm. This represents a density of around 30 seeds per tonne of soil at the safer of the decision rules. The relative lack of enthusiasm for soil testing among weed specialists may be readily understood, and the few who have ventured into this field of work have handled many tonnes of soil in the furtherance of their art! Having taken the soil samples, seed extraction is by no means straightforward and many workers have rejected extraction and direct identification in favour of germination techniques. These may involve retaining the soil samples for years and identifying seedlings as they emerge. This represents a challenge to modern technology. It may never be possible but surely we should attempt to automate this laborious process. If it were possible, the prospects for weed forecasting would be enormously increased.

However, even if we did know the seed content of the soil, we would not have achieved the goal of forecasting because germination of weeds is notoriously unpredictable. We do have a "coarse grained" ability to predict germination in that seasonal patterns of seedling emergence have been studied over many years. Figure 7 gives a set of examples from which it can be seen that Atriplex patula germinates only in spring, and is therefore only likely to infest spring-sown crops, whereas V. persica may germinate in spring or autumn and is therefore not so restricted. This coarse prediction is very useful and more could be made of it, but it really points the way to the third major need for the future.

Prediction of germination
Germination of weed seeds is governed by a range of factors. Dormancy may be profound when the seeds are first shed, with little germination for the first year or so. This dormancy may itself be inconsistent, Peters (1982a,b) has shown that dormancy of A. fatua can be influenced by temperature and water supply during maturation. Roberts (1973) has shown that single stimuli rarely have as much influence as the interaction of two or three stimuli. Figure 8 shows a diagrammatic representation which Roberts has used to demonstrate one of his experiments. Because field emergence of weeds is so complex, very few workers have attempted seriously to predict it. This, above all, is the challenge for the future. So much laboratory work has been done on some major type species that it seems essential to at least try to marshal the formidable array of isolated facts into a predictive model of field germination. If this could be done, say for A. fatua, then the way really would have been opened for forecasting and therefore for rationalisation of weed control.

REFERENCES
Cousens, R.D, Peters, N.C.B. and Marshall, C.J. (1984). Models of yield loss - weed density relationships. Proceedings 7th COLUMA - EWRS International Symposium on Weed Biology, Ecology and Systematics, 367-374.

Cussans, G.W., Moss, S.R., Pollard, F. and Wilson, B.J. (1979).   Studies
          on the effects of tillage on annual weed populations.
          Proceedings EWRS Symposium on the Influence of Different
          Factors on the Development and Control of Weeds, 115-122.
Cussans, G.W. (1980).   Strategic planning for weed control - A
          researcher's view.   Proceedings 1980 British Crop Protection
          Conference - Weeds, 823-831.   British Crop Protection
          Council Publications.
Cussans, G.W. and Moss, S.R. (1982).   The population dynamics of annual
          grass weeds.   In Decision Making in the Practice of Crop
          Protection, Monograph No. 25, ed. R.B. Austin, 91-98.
          British Crop Protection Council Publications.
Doyle, C.J., Cousens, R.D. and Moss, S.R. (1985).   A model of the
          economics of controlling Alopecurus myosuroides Huds. in
          winter wheat.   Crop Protection 5, 143-150.
Glasgow, J.L., Dicks, J.W. and Hodgson, D.R. (1976).   Competition by,
          and chemical control of, natural weed populations in
          spring-sown field beans (Vicia faba).   Annals of Applied
          Biology 84, 259-269.
Peters, N.C.B. (1982a).   Production and dormancy of wild-oat (Avena
          fatua L.) seed from plants grown under soil water stress.
          Annals of Applied Biology 100, 189-196.
Peters, N.C.B. (1982b).   The dormancy of wild-oat seed (Avena fatua L.)
          from plants grown under various temperature and soil
          moisture conditions.   Weed Research 22, 205-212.
Roberts, E.H. (1973).   Oxidative processes and the control of seed
          germination.   In Seed Ecology.   Proceedings of the 19th
          Easter School in Agricultural Science, University of
          Nottingham, 1972, ed. W. Heydecker, 189-218.   London:
          Butterworths.
Roberts, H.A. (ed.) (1982).   The Weed Control Handbook: Principles. 7th
          Edition, pp. 535.   Oxford: Blackwell Scientific
          Publications.
Wilson, B.J. and Peters, N.C.B. (1982).   Some studies of competition
          between Avena fatua L. and spring barley.   I.   The influence
          of Avena fatua L. on yield of barley.   Weed Research 22,
          143-148.
Wilson, B.J., Cousens, R. and Cussans, G.W. (1984).   Exercises in
          modelling populations of Avena fatua L. to aid strategic
          planning for the long-term control of this weed in cereals.
          Proceedings 7th COLUMA-EWRS International Symposium on Weed
          Biology, Ecology and Systematics, 287-294.

23. ADVISING THE FARMER ON PESTICIDE USE: A VIEWPOINT FROM THE AGROCHEMICAL INDUSTRY

J.O. Walker
BASF United Kingdom Ltd, Hadleigh, Ipswich, IP7 6BQ, England

FACTORS AFFECTING THE DEMAND FOR ADVICE

Advice is required by farmers to improve the efficiency and profitability of their farming business. In the United Kingdom, the need for advice relating to pesticide use has increased because of the accelerating complexity and cost of pest control. Table 1 shows that in wheat production, pesticides increased from 16% of variable input costs in 1970/71 to 40% in 1981/82. The number of pesticides from which the farmer must choose has also increased (Table 2), as have the alternative techniques and equipment for their application. At the same time, private and State-funded research provides an ever-rising flow of data concerning optimum timing, economic thresholds, treatment interactions and other factors relating to pesticide use. Many of these data are inconclusive or even conflicting, which is not surprising in view of the complex biological systems in which treatments have to be tested and compared.

Concurrent with these technical changes relating to pesticides, farming generally is also passing through a phase of exceptional change. The yield of crops is rising (Fig. 1) due to the greater sophistication of husbandry, mechanisation and management, coupled with the introduction of varieties with high yield potential. Farming is also operating in a more complex business environment in terms of financial and marketing regulation, and increasing controls are applied by Statute or voluntary agreement to protect the health and safety of those involved in the industry and to protect consumers and the natural environment. Simultaneously the number of workers employed in agriculture continues to decline and is now very low compared with most countries.

The net result of all these changes is that fewer farmers and employees are producing higher yields by the application of more complex technology. They are doing this in an increasingly complex legal and economic environment, and the latter is becoming less favourable. During the last decade, farm incomes measured in real terms have been generally declining (Fig.2). Now we face a situation where the major products of EEC farming are in surplus and it must be assumed that it will be politically unacceptable to continue producing excess food at prices artificially supported above world market values at a high cost to tax-payers. Reductions in the Common Agricultural Policy support levels for cereals will necessitate a review of all aspects of current systems of crop production.

Table 1. Winter wheat - output, variable costs and gross margins in the UK: 1970/71-1981/82 figures adjusted to 1981/82 values.

| Year | Yield (t ha^{-1}) | Price (£ tonne^{-1}) | Gross output (£) | Costs (£ ha^{-1}) | | | | |
				Seed	Fertiliser	Spray	Total variable	Gross margin
70/71	4.11	136.3	560.2	36.2	45.9	15.4	98.4	461.8
71/72	4.66	124.9	582.5	34.4	45.6	17.6	99.7	482.8
72/73	4.50	134.9	607.2	33.9	52.6	23.6	112.8	494.5
73/74	4.31	174.1	749.8	29.1	47.3	26.8	105.3	644.5
74/75	5.26	134.0	704.9	33.7	41.7	29.8	107.5	597.4
75/76	4.46	134.9	601.6	36.8	54.0	37.4	130.8	470.8
76/77	3.92	140.6	550.9	32.5	46.0	40.3	122.6	428.3
77/78	5.22	114.1	593.5	37.1	47.8	45.2	133.5	460.0
78/79	5.56	129.3	731.0	36.4	56.3	61.6	160.8	570.2
79/80	5.46	122.5	669.1	37.1	61.6	69.1	176.0	493.1
80/81	6.42	117.4	753.2	33.0	64.1	71.9	175.8	577.4
81/82	6.29	112.4	706.7	33.5	69.9	73.7	183.5	523.2

(Source: University of Cambridge, Agricultural Economics Unit, 1981/82.)

Table 2. The number of officially approved agrochemicals (active ingredients) recommended for wheat in the UK.

Year	Herbicides/ Plant Growth Regulators	Insecticides/ Fungicides	Total
1970	78	19	97
1973	100	33	133
1976	138	44	182
1979	172	68	240
1981	174	127	301

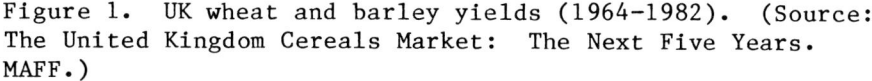

Figure 1.   UK wheat and barley yields (1964-1982).   (Source: The United Kingdom Cereals Market: The Next Five Years. MAFF.)

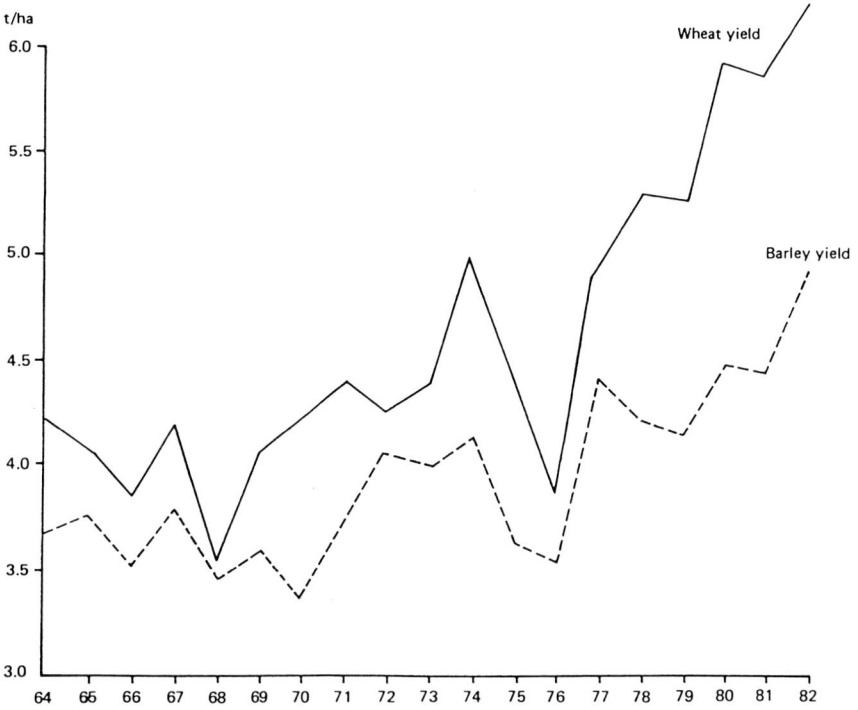

There is yet another dimension to the complex of factors bearing on crop production.   That is the rapidly rising awareness and hostility of the public concerning the effect of modern arable farming systems on the environment.   The trends towards large-scale cereal monoculture have been made possible by the development and use of pesticides to overcome problems such as weed and disease control, which were formerly solved by crop rotation.   Monoculture has been motivated by artifical market price structures and by the economies of scale which justify more sophisticated management and marketing of the crop, and enable high capital investment in specialised equipment. Such systems of production satisfy the profit motives of farming but they do not satisfy the public which views land as a national inheritance and recreational facility as well as a crop factory.

In the technical, social and economic environment outlined I believe it is inevitable that the demand by the farmer for advice will increase. When demand increases in a free market it is natural that a supply develops to satisfy the demand.   This Chapter examines the potential sources of advice, the type and quality of advice required, and the techniques of communication which can be used.   The Chapter is restricted to arable agricultural cropping and concentrates on the cereal crops which represent nearly 70% of the total output of farm crops in the United Kingdom (Anon. 1984).

Figure 2.   UK farm incomes, measured in real terms, 1970-83.

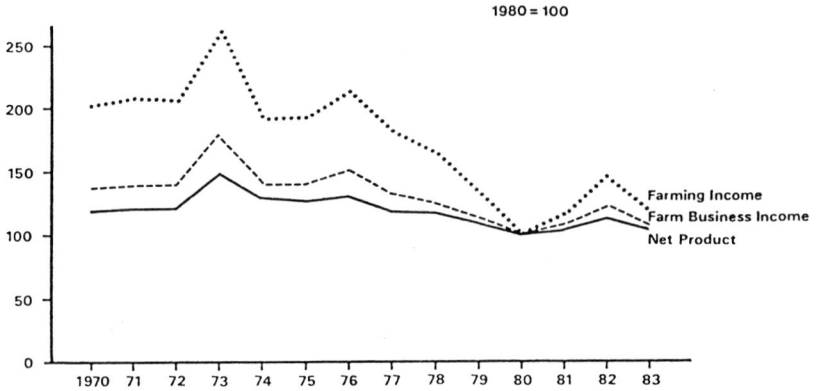

Net Product is a measure of the value added by the
agricultural industry to all the goods and services purchas-
ed from outside agriculture after provision has been made
for depreciation.

Farming Income is the return to farmers and spouses for
their labour, management skills and own capital invested
after providing for depreciation.

Farm Business Income is the return to farmers, spouses,
non-principal partners and directors for their labour and
management skills and on all capital (own or borrowed)
invested in the industry, after providing for depreciation.

(Source: Annual Review of Agriculture, HMSO. (1984).)

THE SCOPE OF ADVICE REQUIRED
In 1977 at a symposium on cereal yields (Anon., 1977), the
speakers were asked to comment on the apparent lack of progress in consistently raising UK cereal yields during the preceding decade;
examination of the relevant data on cereal yields (Fig.1) will explain
the reason for the question at this time. It was also questioned why,
in large numbers of field trials, responses to nitrogen in the UK ceased
at a lower level of input than that reported to be used with advantage
by leading farmers in northern Germany. Speaking from the floor, W.
Fiddian from the National Institute of Agricultural Botany (NIAB)
postulated that the reasons for the lack of response to high nitrogen
could be because the experiments did not take into account the needed
levels of other inputs including nutrients, fungicides and growth
regulators and also possibly husbandry factors.

It is now difficult to believe that the need for experiments to be
conducted on integrated husbandry was put forward as a theoretical
proposal only seven years ago. It is a salutary lesson on the dangers
of thinking too narrowly about future requirements.

My own view is that advice on crop production needed in future must be extremely broadly-based and must embrace the following factors:

Marketing
Crop selection and rotation
Soil management
Crop protection
Machinery
Recording
Finance
Legislation and control

We must not fall into the trap again of researching and advising on one aspect of crop management and ignoring all other factors.

The objective of all advice is succinctly described by Forrest (1983), who stated that the surviving farmers will be those that achieve the highest output of the quality required by the market at the least cost. He also noted that most important decisions are often made before the seed goes into the ground. Such decisions include where the harvested crop is going, what quality is needed and when.

It is not possible to discuss under separate headings the factors required for crop production any more than it is sensible to advise on them in isolation. The demand of the market must be considered against the cost of meeting that demand with the soil and environment available on a farm. The effect on cash flow of producing the required crop must also be considered, together with the needs for machinery, labour and management expertise and experience.

In considering the more specific question of advice on decision-making in the use of pesticides, the following factors should be taken into account:

Crop market requirements and financial objectives
Crop cultivar characteristics
Climatic environment
Crop rotation
Field history
Soil characteristics and nutrient status
Cultivation systems and calendar of operations
Application equipment type and capacity
Technical expertise available

Again I would emphasise the need for a multidisciplinary approach to advising farmers on pesticide use so that crop protection is considered in the context of crop production and business and marketing objectives.

Fortunately this need is now recognised, and Lester and Prew (1983) and Prew et al. (1985) have reported on extensive projects at Rothamsted Experimental Station involving a multidisciplinary approach to studying

factors affecting the yield of cereals and other crops. Factors examined include drilling equipment, sowing date, nitrogen rates and timing,
irrigation, pest and disease control and soil type.

In addition, our cereal cultivars are now tested by NIAB (1983) with and
without standard fungicide programmes. Also, the Agricultural Development and Advisory Service (ADAS) has extended cereal cultivar testing by
examining interactions on a range of soil types at a large number of
sites. Cereal breeders, agrochemical companies, farmer groups and
others are all moving in the same direction, but the scope of multidisciplinary research which can be undertaken by any one organisation is
limited by the huge resources required to deal with the complex
variables. Consequently a great number of experiments and a large proportion of advice are still based on comparisons of single variables,
often made under conditions which do not relate to modern crop
production practices. Also much current experimental work and advice is
based on comparisons where all the variables are changed together so
that the significance of single factors cannot be judged. However, this
approach has up to now provided valuable results because of its relative
simplicity and low cost. In practice it probably forms the basis of the
greater part of the recent advances in cereal yield increases which are
attributable to husbandry factors.

In future, I believe that both multifactorial experiments and empirical
testing of packages of variables selected on the basis of intelligent
judgement will have a role to play in providing a basis for advising
farmers. The multidisciplinary approach to advice means that the time
and cost needed to develop the basic information for a new cereal
cultivar will increase considerably. It will also increase the learning
burden both on advisers and farmers. Associated with such development,
detailed records will be needed on the characteristics of individual
fields. These records should include a history of cropping and yields,
pest incidence, and nutrient levels backed by soil analysis and information on climatic factors and soil characteristics.

The learning burden suggests that there are good arguments for changing
cultivars less frequently than currently. More value should be placed
on the information banks and practical experience acquired with a
cultivar on a specific farm location when decisions are made on possible
substitution by an alternative but unknown cultivar because of a small
advantage indicated on the NIAB comparative yield scales.

SOURCES OF ADVICE FOR THE FARMER
The traditional sources of advice to the farmer on pesticides are shown in Table 3, which is based on data collected in ADAS
surveys. The results of the surveys are confirmed in independent
private market studies undertaken on behalf of the agrochemical
industry. It is concluded that in crops where the supply of pesticide
products (and other inputs) is on a scale to ensure profitability,
commerce is the dominant supplier of advice and particularly the
specialist agrochemical distributor. In contrast, in small-scale

high-value horticultural crops where the market for advice is much
smaller, but where the technical level of advice needed is complex,
non-commercial advisers play a major role.

Table 3. Traditional sources of advice on pesticides for
the farmer. (Source: MAFF Surveys.)

	Arable Crops	Vegetables	Orchard	Hardy Nursery Stock
	1977	1972	1973	1976
Commerce (Distributors)	97	78	70	29
ADAS	19	30	20	42
Other official bodies	1	1		2
Farming literature, conferences	9	3	20	1
Other farmers/ growers	2	8		6
Growers' groups	2	6		4

Since these surveys were done in the 1970s there have been considerable
changes in the market affecting sources of advice. The most notable of
these are discussed below.

1. The agrochemical manufacturer
    The manufacturing industry has enjoyed several decades of
rapid growth in a young market where there were many opportunities to
supply new patented products. Now the majority of the obvious niches in
the market are filled by products which are often difficult to better
technically by a sufficient degree to justify the cost of introducing a
new compound.

The established technically viable products are coming off patent, and
manufacturers without the burden of a large research base are producing
off-patent compounds and competing in the market place, with the result

that prices and profits are severely reduced. At the same time, fewer
marketed compounds are carrying the fixed costs of agrochemical research
and developments, which are increasing due to the need to synthesise and
screen more complex molecules and to satisfy more stringent toxicological and environmental requirements. Pressures on profit are resulting in the need to cut costs, which will mean it will be more difficult
to justify and maintain advisory services to the farmer.

Nevertheless some manufacturers, including my own company (BASF), are
pioneering the development of field trials to evaluate integrated crop
husbandry programmes examining agrochemicals, nutrients and plant growth
regulators in association with other aspects of cultivar management.
Manufacturers are also examining and testing new systems of application
in relation to their products.

2. The agrochemical distributor

The agrochemical distributor was traditionally the corn
merchant who supplied chemicals as they became available. Over the last
two decades, agrochemical distribution has become recognised as a
specialist business requiring a high degree of expertise as a basis from
which to advise the farmer. Consequently, agrochemical distribution is
now carried out as a separate business or as a specialist division of a
merchant business. The more progressive companies carry out extensive
replicated field trials to compare products and to develop integrated
crop husbandry recommendations with agrochemicals and fertilisers as the
main factors.

Progressive distributors are also establishing their own laboratory
facilities for soil and tissue analysis and for pest diagnosis. Some
companies are developing computer services to assist their customers in
their decision-making (see Wilson, Chapter 24). However, the profitability of distributor businesses is also declining, partly as a result
of increased market competition and partly as a result of the increasing
scale of farming which moves more bargaining power into the hands of the
customer. The raising of standards by the adoption of the voluntary
British Agrochemical Supply Industry Scheme (BASIS) and legislation on
health and safety has also raised costs. Thus, the progressive
distributors have made a big commitment to maintain their position as
leading suppliers of advice to farmers, but the cost of sustaining this
could become a problem since traditionally the advice has been supplied
free of charge.

The way ahead appears to require a separate charge for advice, the
actual charge depending on the amount and depth of advice required by
the farmer.

3. Agricultural Development and Advisory Service (ADAS) of the Ministry of Agriculture, Fisheries and Food

ADAS has the responsibility of providing the main official
advisory service to farmers in England and Wales; the Agricultural
Colleges have fulfilled a similar function in Scotland. ADAS has an
apparently excellent structure for the provision of advice, having over

20 experimental farms at which the products of research can be tested and demonstrated in practice before authoritative advice is provided on new treatments to the farmer. The six regional centres of ADAS also provide excellent specialist information on the principal aspects of farm management and crop husbandry, including crop protection; they are supported by scientific and analytical laboratory facilities. At the International Congress of Plant Protection (ICPP) in 1983 it was stated by the Chief Agricultural Officer for ADAS that the number of field advisers conveying ADAS advice direct to the farmer had increased, and stood at 1036 at that time. ADAS services have historically been free to the farmer but charges for analytical and diagnostic services are now made, and will be extended to other services in the near future.

The use of the ADAS services by farmers (Table 3) has been surprisingly low, but it must be noted that the number of 1036 ADAS farm advisers is considerably exceeded by the number of agrochemical distributors' staff providing an on-farm service for arable crops. It is clear also that much of the information provided by the regional centres finds its way to the farmer via other routes, for example, through publications, media, open days and conferences.

4. Consultants

In a study in 1982 by BASF it was concluded that consultants influence about 0.5 million ha of cereal production in the UK and it appears that the number of consultants has continued to grow since that time.

There is however a need to define more clearly what is meant by a consultant, since in the UK the term is generally applied to all those who are members of the 'Association of Independent Crop Consultants'. The word 'independent' really applies to their independence from product sales. In practice, a majority seem to be employed by farming groups which reflect an evolution of the farm business to a point where a sub-division of the farm management is needed to provide specialist technical management. The medium-sized farms cannot afford this level of specialisation and so they form into groups which employ a full-time technical expert or consultant funded from an acreage levy. Generally one consultant services about 4,000 hectares of cereals, and the cost per hectare at about £6 is relatively low. Other consultants are completely independent and self-employed but they may nevertheless work with a relatively fixed group of clients. The advantages of continuity of the consultant - client relationship are obvious if information banks and experience on the characteristics of individual fields are developed by the consultant. Large farming companies with an area of over 4,000 hectares are likely to employ their own technical experts.

The advantage of the consultant to the farmer is that he is independent of the sale of particular products, although some consultants do negotiate for the supply of products on behalf of clients. Also their livelihood is directly dependent on the employer or customer perception of the quality of advice in relation to business profitability.

The main limitation of consultants appears to be that, for economy, they have few or no staff, and they therefore lack research facilities to help keep abreast of new developments. Set against these limitations many of the best consultants appear to have excellent personal relationships with ADAS and Agricultural and Food Research Council (AFRC) research stations which can provide the practical research and literature updating required. Indeed, it appears that at least some of the official bodies see the consultants as a valuable outlet and communication medium for their research.

In an open discussion of advisory services organised at the ICPP in 1983, it was stated that 70-80% of the advice given by leading consultants is now related to crop husbandry problems rather than being confined to the narrower field of pesticide recommendations.

5. Farmer-controlled experimental husbandry centres
This concept in the UK is best exemplified by the Cotswold Cereal Centre (under the title of the Arable Research Centres), near Cirencester, Gloucestershire. This already has sister establishments under the same management board. There are other independent centres which work on a similar basis. The Norfolk Agricultural Station can be considered as the forerunner to this approach. These organisations have the common facet that they carry out extensive experimental programmes, employ permanent scientific staff and are financed mainly by controlling farmers. Other organisations such as commercial companies, and research and educational establishments also contribute to the management and funding. The approach has most of the advantages of a consultant serving a farming group with the added advantage of an experimental base to maintain and extend levels of expertise. The farmer control should ensure that the work programme is related to practical farming needs. Often 'one-off' results are obtained from single experiments; whilst these can be of considerable interest, the need to consider results from a number of experiments done under different conditions over several years must be stressed.

The longer experience of Norfolk Agricultural Station, which is now funded by MAFF as well as by member farmers is, I believe, encouraging, and represents a most valuable way of turning the products of State research into practical advice which is valued by the farmer.

A similar approach is used by German farmers in Schleswig-Holstein, which is well-known for its leading achievements in profitable crop production. Here, the County Agricultural Department (Landwirtschafts-kammer) runs a group of experimental husbandry centres which are jointly financed and managed by a partnership of State and farmers. One of the notable points about the centres which carry out extensive replicated experiments, is the large number serving this relatively small area. The farmer finance is based on an obligatory hectarage levy and in return he has a strong voice in the management of the research pro-grammes. There is also a high level of communication of results to the farmer by way of demonstration days and written reports which include detailed advice on the management of each cultivar grown in the area.

Farmers are also provided with plans which enable them to inspect the sites whenever they wish.

COMMUNICATION MEDIA
I have already emphasised that the farmer has experienced a huge increase in the volume of information he needs to cope with to take advantage of the benefits which are available from research into all aspects of crop husbandry. I have also drawn attention to the fact that the level of labour on UK arable farms is low, typically one person per 70 ha, but in many cases much lower. Since the complexity of farm management is increasing in all areas, including marketing, machinery and legislation, any further intensification of crop husbandry will produce serious management problems. I have already indicated that such intensification will be necessary in order to maximise profitability and competitiveness by integrating decisions on all aspects of crop husbandry, and relating them to the needs of individual crops in individual field environments. Choice of media for communicating advice will consequently become more critical.

There are many well-established routes for communicating advice. The printed word must be the most common form ranging from research reports, through MAFF and commercially-produced literature to the farming journals and newspapers. The spoken word is also very widely used, and again ranges from research symposia to farmer meetings organised by industrial companies and State organisations, to radio and television broadcasts and telephone messages. Demonstrations are also widely used as multi-wide bases for communication.

The development of the specialist farm consultant for which the farmer is willing to pay illustrates the farmer's awareness of the difficulty of handling the rising tide of information needed for crop husbandry decisions. However, to move further towards integrated decision-making it seems to be essential to harness the advantages of computer technology. This technology can speed the transfer of new information and enable the farmer or his adviser to draw on huge banks of data to meet his specific needs. It can also provide a powerful decision-making aid by storing farm, field and crop information and relating this to the farmer's marketing and cropping objectives.

Information has been available to the farmer for some years via viewdata services based on Prestel, and currently there are approximately 5,000 frames of agricultural information available from about 30 organisations which means the data base is becoming more attractive. At present a relatively small but growing number of farmers have the necessary equipment to operate a two-way interactive service which requires a microcomputer terminal on-farm linked to the national data base via telephone lines. It is difficult for the individual subscriber to justify acquiring the necessary terminal equipment until there is a worthwhile data base to receive; conversely it is difficult for data base providers to justify the cost until there are a large number of subscribers. Despite these problems, such interactice services, for example Prestel-Farmlink in the UK and Teletel in France, are now developing. Potential information providers, which should include all those currently providing advice to farmers, will feel obliged to participate in these services, and there are indications that many farmers will install the necessary terminals and use them.

CASP - a Computerised Advisory Service
As an example of the type of advisory service which is beginning to be provided to the farmer via an agricultural viewdata service, I will describe a computerised advisory service known as CASP, developed by BASF and now made available to farmers via specialist agro-chemical distributors.

The service has been developed to meet some of the objectives which I have put forward as requirements to improve the efficiency of arable crop production in the future. Design objectives of the CASP computer program are as follows;

1. Store data on farm characteristics, including identity location and cropping.

2. Store data on individual field characteristics such as identity, size, soil characteristics, nutrient status, cropping history and pest history, including pest range and level.

3. Store data on agrochemical and nutrient treatments, including biological range and levels of effect, timing, rate, environmental interactions, compatibility and sequential treatment characteristics and cost.

4. Store data on crop cultivar charcteristics.

5. Correlate cropping plan with field data and crop cultivar characteristics to provide a target yield for a specific crop in a specific field.

6. Correlate target yield with field characteristics, including climatic, pest and nutrient status to provide a nutrient and pesticide management plan for the whole growing season. The management plan is to take account of gross margin objectives by minimising inputs in relation to the need for achieving target yields. A gross margin analysis is to be provided.

7. Enable the amendment of all data banks and other variables by keyboard entry.

8. Enable the user to amend the management plan at any time by the keyboard input of new or alternative data.

The concepts of providing an integrated decision-making system for pesticide and nutrient recommendations are illustrated in Figure 3.

Figure 3.   CASP - the design basis for computer programs integrating decision-making for crop husbandry inputs.

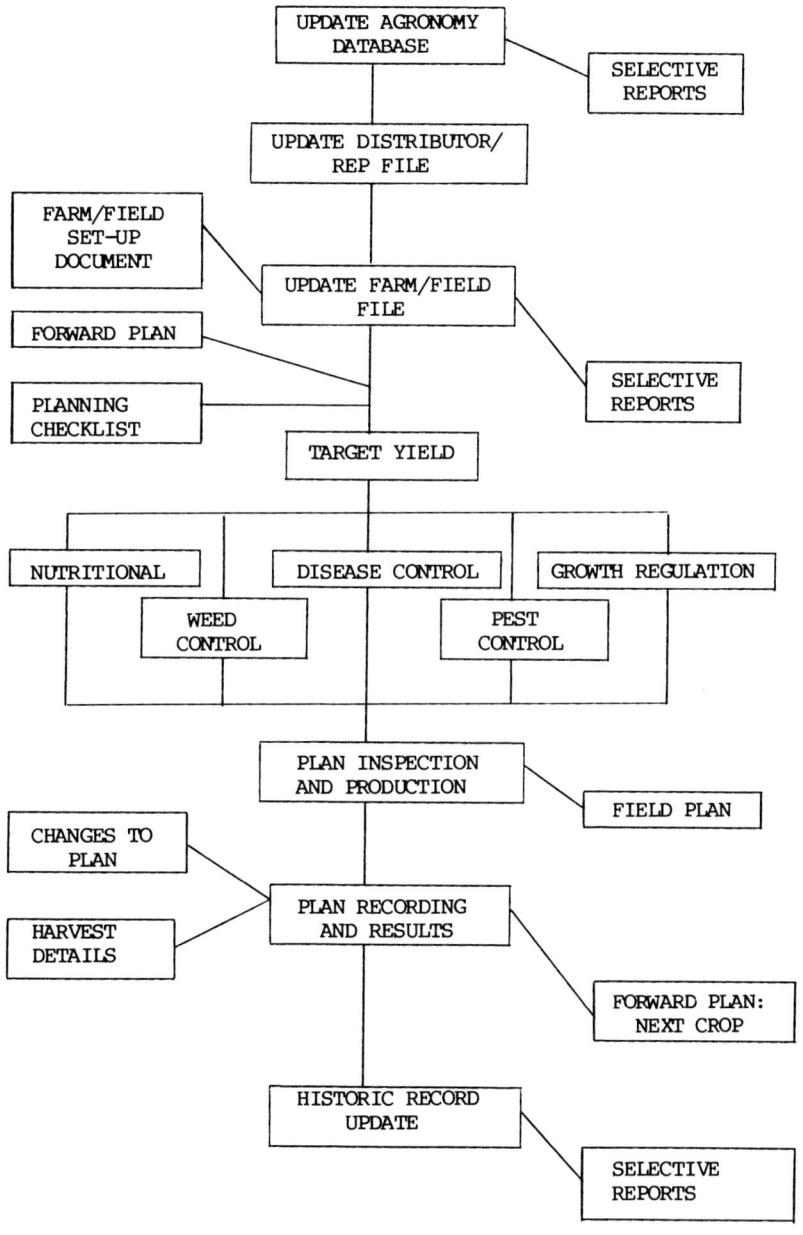

The overall benefit of the system is that it enables the storage of large amounts of data concerning the crop, the specific environment in which it is to be grown, and the characteristics of pesticide and nutrient inputs. These data are then correlated with the objective of optimising gross margins by tailoring the technical attributes and cost of inputs to the yield potential, with marketing objectives and with an analysis of the pest problems and nutrient deficiency risk to which the crop will be exposed.

Once a farm has been set up on the system, the keyboard data inputs needed for a new crop are relatively few. Also the action plan provided for the farmer is relatively simple, as illustrated in Table 4.

At present the CASP service is operated from selected agrochemical distributor offices, but the system has been designed for linking into viewdata services should this be required. When this development takes place, the lessee of the system will still be responsible for updating the data files relating to crops, products and decision-making cor-relations; the user will simply input information at his terminal key-board concerning the crop and cultivar he intends to plant and note any additions to his field data base such as previous year's yield. A printed plan for the new season could then be produced on the farmer's desk.

The approach used in CASP in relation to pesticide application is based on the use of risk analysis to create a plan at the beginning of the growing season, but this is supplemented with in-season decision-making programmes to deal with problems which arise while the crop is growing.

The system has been designed so that the data banks for CASP farms can also be used to provide an in-season warning service to alert growers of particular risks in specific crops and fields. The service could at present be activated by observation of the development of a specific pest problem such as a new strain of disease, and warnings could then be sent to growers advising action on the specific cultivars and fields at risk. When practical forecasting models are available these too can be used to provide warnings, and to avoid the current large degree of reliance on routine or calendar sprays.

DISCUSSION
The view has been put forward that arable farming has enter-ed an era where economic pressures will make decision-making on arable husbandry techniques much more critical due to economic forces reducing farm profitability. Since pesticide inputs now account for approximate-ly 40% of the variable costs of producing our most important arable crop, winter wheat, optimum advice to farmers in relation to pesticide inputs is obviously critically important.

It is concluded that the opportunities for improving advice to meet the farmer's profit objective lie with the development of integrated advice so that decisions on all crop husbandry inputs, including cultivation

Table 4.   A CASP field plan.

Distributor: A Distributor	Farm: Hall Farm	FIELD PLAN 12
Rep/contact: A N Other	Contact: J Smith	1
Tel Nos :	Tel No : 123 1234	Ref 98-02-011-01-833

		Index	mg/lit	
Soil analysis date 26/08/83				Field: No 5
	Nitrogen N			Crop : WINTER WHEAT /3
Soil type SANDY LOAM	Phosphorus P	5	85.00	Variety: RAPIER /0
SEMI ORGANIC	Potassium K	3	300.00	Undersown
	Magnesium Mg	3	145.00	
Soil series	Copper Cu			Target yield 7.9
	Boron B			Potential yield 8.3
pH value 07.7	Manganese Mn			Estimated Selling
STRONGLY ALKALINE	Sulphur S			price 120.0
	Zinc Zn			Drilling date
Date last limed 00/00	Molybdenum Mo			(PROVISIONAL) 23/09/83
	Calcium Ca			
Line requirement	Iron Fe			

Growth Stage	Type	Code/Product Name	Application Rate	Area	Notes	Field Price
00-00	N	N034 POTASH SOLID	65.00K	C		29.25
00-14	N	N030 NITROGEN SOLID	25.00K	C		26.25
12-23	H {	H097 IPU 54	4.50L	C		105.00
	H {	H095 IOX-BROMOX 20/20	0.70L			17.93
12-24	H	H009 DIFENZOQUAT 63	0.74K	C		69.70
21-22	N	N030 NITROGEN SOLID	55.00K	C	Apply N early to mid February	57.75
22-25	N	N030 NITROGEN SOLID	85.00K	C	Apply N early to mid March	89.25
24-31	G	CHLORMEQUAT 64	1.40L	C	If late 31 use only 1.10	14.15
30-30	F {	F001 CARBENDAZIM 50	0.50L	C		13.50
	F {	F040 MANEB 80	1.00K	C		3.60
30-31	N	N030 NITROGEN SOLID	80.00K	C	Apply N early to mid April	84.00
32-37	F {	F006 TRIADIMEFON 25	0.25K	C		18.00
	{	F040 MANEB 80	1.00K			3.60
50-59	F {	F022 FENPROPIMORPH 75	0.75L	C		31.45
	{	F012 CHLOROTHALONIL 50	1.50L			36.45
	{	F005 CARBENDAZIM 50	0.30L			8.10

Estimated sale of crop	2844.00
Total treatment price	607.98
Gross margin	2236.02

techniques and timing, cultivar attributes and the characteristics of
individual fields, are all correlated with decisions on crop management.
The decision must also be correlated with marketing objectives.

The establishment of correlation factors in a system with so many
variables is extremely complex, although it is encouraging to note that
research is now proceeding on this basis. Due to the complexity of the
problem it must however be assumed that our knowledge will remain
imperfect for the foreseeable future and, in practice, 'judgement' will
have to be substituted for proven information in many situations. Field
tests with fixed packages of inputs, although of limited scientific
value, may be worthwhile to provide practical guidance.

It is suggested that a less rapid change of cultivars grown would also
assist in justifying the cost of testing cultivars in integrated programmes. It would also enable greater use to be made of the farmers'
and advisers' cultivar husbandry knowledge and experience.

The application of decision-making which takes account of large numbers
of variables will be complex, and I believe that it is inevitable that
computer technology must be employed widely to cope with the large data
bases and correlations necessary.

In the near future, decision-making computer programs should be regarded
as an aid to management rather than a substitute, and specialist technical advice will remain vitally important. Indeed, I believe the
computer should be viewed as a tool to assist the adviser or technical
specialist to relieve him of burdens, and to enable him to spend more
time looking at the crop and to consider wider aspects of decision-
making such as the choice of cultivation techniques and application
systems.

I conclude therefore that the task of advising will become more complex,
but if this is true we must consider the financial pressures I have
described on those who have traditionally given advice free of charge.
The answer, I believe, is that if advice is worth having, the customer
will pay for it. The development of paid consultants confirms this.

Forecasting who will provide this advice is difficult. I believe the
consultant or specialist adviser will continue to grow in importance as
a natural development of specialisation of management within the farm
business. The creation of consultant groups will also develop, possibly
associated with farmer-funded centres such as the Arable Research
Centres. This will provide a practical experimental base with finances
shared by individual consultants and farmers and possibly by State
services, while the advantage of the self-motivated individual adviser
can be retained. Consultants will however need to be supported by the
costly research base which can only be provided by the State or large-
scale private business. The role of the State depends on politics,
which makes prediction even more difficult, but I do believe the
scientific base provided by UK public-sector research is essential. It
is almost certain that research will become more complex and costly to

achieve results as the law of diminishing returns takes effect as crop yields progress nearer to the potential optimum.

The agrochemical manufacturer will, I believe, draw back from on-farm advice and concentrate resources on channelling information and services through the distributor and others who can provide a broader-based service to the grower. The distributor who at present plays an important role will in turn need to integrate his advice on pesticides with other aspects of crop husbandry. I think this is likely to be achieved via integration of pesticide business with a broader range of farm merchanting services. The ultimate development could well involve partnership or integration with farm businesses.

REFERENCES

Anon. (1977). Proceedings of a Symposium on Cereal Yield. Hadleigh: BASF.

Anon. (1983). National Institute of Agricultural Botany "Annual Report for 1983. Cambridge: NIAB.

Anon. (1984). Annual Review of Agriculture 1984. London: HMSO.

Forrest, J.B. (1983). Towards the '90s: Cereals and Oilseed Rape. Symposium held at The Royal Agricultural College, Cirencester.

Lester, E. and Prew, R.D. (1983). The Yield of Cereals. Monograph Series No.1. National Agricultural Centre, Royal Agricultural Society of England.

Lester, E. and Prew, R.D. (1983). A multidisciplinary approach for evaluating the factors which affect cereal yields. In The Yield of Cereals, ed. D.W. Wright, 79-84. London: Royal Agricultural Society of England.

Prew R.D., Church, B.M., Dewar, A.M., Lacey, J., Magan, N., Penny, A., Plumb, R.T., Thorne, G.N., Todd, A.D. and Williams, T.D. (1985). Some factors limiting the growth and yield of winter wheat and their variation in two seasons. Journal of Agricultural Science, Cambridge 104, 135-162.

24. COMMERCIAL IMPLEMENTATION OF FORECAST METHODS

J. Wilson
Comput-a-Crop, Beeches, Mill Lane, Scamblesby,
Louth, LN11 9XP, England

INTRODUCTION

Better crop protection has been a major factor in the change of agriculture away from mixed farming and rotation towards "arable only" ventures and continuous cereal production. The very name "crop protection" implies that crops can be protected successfully from pest and disease attack, and farmers and advisers alike once hoped for the elimination of pests and pathogens. With the high cost of control measures, it is now acknowledged that complete prevention can give way to suppression of symptoms for many pests and diseases without loss of profitability. The consequent reduction in the release of toxic sprays into the environment is a welcome spin-off from less frequent pesticide use. The economic effect of an outbreak of disease on an individual variety varies from field to field and considerable improvements in profitability may become possible from taking field conditions into account. At recent grain prices, some 30% of prophylactic sprays of cereal crops in the UK are likely to be uneconomic, amounting to half a million hectares (Cook 1983).

The emphasis in spray decision-making is moving from absolute crop protection, at whatever price, towards the management of disease at an economically tolerable level of incidence or severity. The benefits to be derived from basing disease management decisions on field conditions will always be limited by the degree of expertise of the decision maker. There is now a growing market for decision aids for all those who work with field data, from consultants and advisers to individual farmers. These products are founded on computerised data bases and their introduction is treading new ground, by introducing skills, techniques and procedures as radically different from prophylactic treatment as chemical crop protection is from disease avoidance by rotation. The integration of such decision aids will take some time. Their success will depend on how easy they are to use, the quality of field data collected and how much support and training is provided.

THE TRADITIONAL BASIS FOR ADVICE

Crop protection practice has three major components – problem definition, advice and product application. Farmers choose the control method which they are persuaded is appropriate to their needs. Seldom do they choose to leave pests and diseases untreated. They have relied on the expertise of others to a very large extent to identify disease and pest levels and to advise on chemical choice, although ultimately farmers are themselves reponsible for safe, adequate and timely spraying. Arable farmers' traditional skills have encompassed the preparation of land, the choice of the right crop for a given step in the rotation, crop planting, weed control and harvest, but not pest and disease control.

The introduction of synthetic pesticides into agriculture has brought new choices and new constraints. Today's farmers are the first generation to grow up with an agrochemical armoury and have been subject to rapid change in technology which affects every area of traditional husbandry. With their product support teams which are necessary for the commercial success of sophisticated chemicals, and which are financially justified by the high added value of petrochemical by-products, manufacturers have largely controlled crop protection practice by bombarding the grower with advice-based product promotions.

Farmers have been ill-equipped to distinguish between an economically justifiable control measure and those sprays which serve only to enhance the supplier's profit margin. The crop protection industry has developed and controls a body of technical data which has tantalisingly been just out of reach of the farmer. The mystique associated with artificial control measures has produced an elite group of consultants, whilst it is the practicalities of spray application that have been embraced by farmers and adopted by a highly skilled work force as one more step in farm mechanisation.

Consultants, to safeguard their reputations, and government advisers, to safeguard the nation's food production, initially advised control of pests and diseases on the premises that control of insects affecting aerial parts of the plant can be left safely until the pest is visible, and that soil pests and plant diseases should be treated largely prophylactically since the symptoms only become visible after the pathogen has done some damage to plant tissue.

Government and independent scientists and advisers have performed field experiments to establish biological and economic criteria on which to base spray decisions. The definition of components of disease development and crop response by empirical research programmes has led to predictions of loss or damage which justify disease control advice. It is only when a large body of data is available to illustrate disease development in the presence or absence of spray treatment, as well as crop development in the presence or absence of disease, that the economic effect of a pathogen and its control can be assessed. Prediction of levels of damage from disease then becomes feasible.

TRANSITION TO NEW METHODS: THE FARMERS' RESPONSE
Government advisory bodies, consultants and manufacturers have for some time included the probability of damage in their equation when formulating advice to farmers. Decisions on crop protection have been based on regional disease forecasts relating to the known

susceptibility of crop varieties, information which farmers take on
trust. In the light of a high regional probability of disease development, growers have come to expect crop damage and yield loss in the
absence of crop protection measures. They have relied on the expertise
of outside advisers to assess the need to spray and then discussed
choice of product and timing of application and negotiated a price.
With the reduction in cereal prices and the increasing level of farm
borrowing to finance production, farmers are recognising now that all
areas of expenditure, including pest and disease control, must be
examined. Farmers have viewed spraying as an insurance against unforeseen losses, especially in high-risk continuous cereals, but also in
high value, high quality fruit and vegetables. With the high cost and
toxic nature of many chemicals, they are keen to reduce sprays, by
reducing the dose applied, or by spraying less frequently. However,
probabilistic models based on regional data and historical averages will
very rarely advise against spraying.

Disease management can most accurately be based on individual field
data, a task beyond individual advisers and requiring computer processing to integrate the components of risk analysis with variables recorded
about the growing crop.

For on-farm disease forecasting, the information to complement regional
and varietal data is obtained from disease recognition in the field and
an interpretation of incidence and severity. This information must be
based on knowledge. A new expertise and confidence in the practice of
disease recognition is required before farmers will embrace on-farm
disease forecasts as part of their decision-making toolkit. A record of
other relevant factors such as crop growth stage, weather history and
crop and soil nutrient status must be maintained routinely and
consistently.

Due to lack of formal training in the recognition of disease symptoms,
and also the high cost of crop losses, farmers may be nervous of using a
system that to an extent relies on them developing expertise in disease
assessment. They may also keep insufficiently detailed or up to date
field records.

Once a farmer has opted to use forecast techniques for decisions on
disease management, decision-making can extend beyond price-based
product choice to fundamental choices of whether or not to spray when
disease is present. It can seek confirmation from established, trusted
advisers that they are not taking undue risks. Many would prefer to
filter the advice given by a forecasting system through their consultant
or adviser, to share the responsibility of decision-making. These
factors must be taken into account in the design of commercial disease
forecasting systems.

CHARGING FOR FORECAST SYSTEMS
Farmers are used to getting free advice on product choice
and use, but expect to pay consultants who are prepared to advise

against spraying. They expect free training in the use of
manufacturers' products from their merchants but are prepared to pay for
courses in new husbandry techniques and to keep up to date with the
latest research in crop management. Therefore, given the availability
of field data and the opportunity to learn the necessary new procedures,
farmers will pay for computer-based disease forecasting as a management
tool, provided only that the advice generated is independent of the
chemical manufacturers.

There is only one price range at which such novel products can be sold,
and that is one which the market will stand. Market research and test
marketing help to establish a figure. Pricing may be set at amounts per
farm, per field, or per hectare and buyers are encouraged to take into
account any other benefits derived in addition to the advice received.
The greatest benefit to both supplier and farmer is acquisition of data
in the course of day-to-day business, whether for farmer decision-making
or supplier market intelligence and product development. Without a very
rapid growth in the sales of such service-based products, their benefit
to the suppliers must not be isolated from the effect their introduction
has on developing the market for other goods and services. As use of
such systems depends on access to a computerised data base, the information provided by the user may be captured as test data in order to
evaluate and to help develop new and improved products, for the benefit
of both subscriber and supplier. Access to a computer-based product
will encourage farmers to record their data consistently and to use them
for other management decisions made through additional computerised
products.

COMPONENTS OF COMMERCIAL DESIGN AS ILLUSTRATED BY EPIPRE - A DISEASE AND PEST MANAGEMENT SYSTEM

The design constraints of a computer-based forecast system
extend beyond but do include those imposed on the development of single-
purpose commercial computer software. The criteria in common are those
of modular program design for ease of maintenance, clear programmer and
user documentation, good quality test data to aid debugging, user-
friendly input and output formats, and good system audit and validation
procedures for security of data. The four components of the software of
the EPIPRE system are data recording, data analysis and processing, data
archiving and output of advice (see also Tait, Chapter 17). These are
echoed in the other components of the system, namely subscriber training, in-season technical support and field trials, and out-of-season
data analysis, to test system enhancements and to promote the product.

Difficulties which were predicted in implementing the system centred
mainly around user experience. As unforeseen strengths and weaknesses
are discovered, EPIPRE software or procedures may be modified accordingly. The subscriber, whether farmer or consultant, has to supply an
accurate history of each site to be monitored, take plant samples at
intervals defined by the software in response to the data provided,
assess disease incidence in the sample and report any fertiliser or
spray applications and rainfall since the previous sample was recorded.

The value of the advice given is limited by the accuracy of the yield
forecast submitted at the start of each season and the standard of field
sampling and recording of growth stage, disease recognition and reporting. EPIPRE has contingency values for the progression of cereal growth
stages and will report if a user's recorded growth stage is impossible
or unlikely. New reporting criteria were built into the system as a
result of the exceptional yields in 1984. A range of advices is
generated to take account of over- and under-estimation of yield
expected. EPIPRE will indicate the possible occurrence of pesticide
resistance from disease counts, takes into account residual values when
generating advice and will not advise consecutive treatments of similar
materials.

The commercialisation of EPIPRE highlights aspects of design which reach
beyond normal product boundaries. Farmers have a free choice of advice
and will be able to obtain services from a growing number of computer-
based systems. While a field history or an in-season field operation is
being recorded, it should be stored within the computer system in a form
retrievable by other software to avoid duplication of effort. Short-
term information derived from data processing for one purpose may
provide suitable input data for another, perhaps longer-term recording
requirement. For instance, a number of the inputs required by EPIPRE,
including varietal information, field histories, weather and yield data,
are available as outputs from other programs and devices. Yet the output from EPIPRE itself is valuable input data for long-term trend
analysis of the incidence, severity and control of diseases, and
provides current disease intelligence.

PRACTICAL IMPLEMENTATION

The greatest variation in the accuracy of disease prediction
and in the implementation of advice originates from human factors.
There are four distinct areas of opinion and skill which influence
success. First, the initial acceptance of the benefits of forecasting;
second, attitudes to disease management advice in the light of farming
experience; third, the ability to learn growth stage and disease
recognition; and finally, the willingness to accept the discipline of
consistent sampling and record-keeping. The worst difficulties encountered must receive attention when developing a system. However,
many subscribers do not have the same doubts or encounter the same
difficulties in using forecast methods, and some of those who do have
problems also find discipline in other areas of business or crop
husbandry hard to accept. The objective is to provide a system which
caters for the needs of most potential users, and those who are un-
receptive can sometimes contribute comments which make a product more
widely acceptable.

Acceptance of forecasting

In the promotion of EPIPRE, the commonest reasons for not
trying the system are that the farmer or consultant has nothing to learn
from such a system as they are already satisfied with their profitability, or can see no benefit to be gained by increased attention to

detail. Other objections are that their prime consideration in spray
decisions is, and will remain, the weather forecast, or that field walking takes more time than they choose to make or have available, and that
management time is at a greater premium than money to buy sprays for
prophylactic application.

Response to advice
Experience of the EPIPRE Bureau has ranged from farmers who
have paid fees, made no observations and then vigorously recommended the
system to other farmers, to consultants who would not follow the advice
given without a full briefing on program structure, content and elements
of the risk analysis components, which were not available to subscribers. As most farmers expect to spray when disease is present, a
number of subscribers regretted spraying in advance of field walking
which subsequently generated advice from EPIPRE not to spray. Those who
expected the system to save them money by advising fewer sprays were
difficult to convince when a spray was advised.

Crop and disease recognition skills
Most agriculturalists, including farmers, expect disease and
growth stage recognition to be the most difficult part of EPIPRE to
follow efficiently. Growers only monitor each disease during the growth
stage range over which EPIPRE anticipates losses could be caused. This
varies from one disease to another. Over the range of diseases covered,
we have found that farmers have had the hardest job recognising symptoms
of Septoria spp. In common with many commercial agriculturalists and
plant pathologists, farmers confuse these with stress symptoms. The
element of training which tackles this problem incorporates the use of
samples of plants suffering from stress. ADAS Disease Recognition Cards
are supplied for all the relevant diseases, and farmer meetings are
organised during the growing season to enable subscribers to compare
samples in a standing crop. Where there is any doubt about disease
recognition, samples can be returned for identification and encouragement is given to seeking a local second opinion. Commonly, even farmers
who regularly walk their fields are unused to assessing crops for
disease incidence and surprise has been expressed at how easy Erysiphe
graminis is to identify. This sort of comment from experienced farmers
demonstrates that no existing knowledge in this area can be assumed, but
is soon acquired. In second and subsequent years, subscribers rapidly
gain confidence in their disease assessment abilities. Initially, it
has proved more important for subscribers to determine when the sample
is not affected by disease. During both 1982 and 1984, amounts of leaf
disease were low enough in fields monitored by EPIPRE to prompt subscribers to check carefully in other fields not monitored by the system,
exercising their newly-acquired skills, before making a decision on the
need to spray.

Sampling procedures, discipline and consistency
Some farmers who lack confidence in disease recognition have
been the most diligent in following sampling instructions. At every
level of response to EPIPRE, the prevailing impression is that of farmers keen to learn and prepared to seek a greater understanding of their
crops.

The most common source of problems for Bureau users has been their lack of discipline in completing the observation records. It is emphasised at meetings, on visits and in the documentation that, as long as the system is updated with details of completed applications, the advice generated is valid for a current observation. Where no disease has been detected or the disease level has not increased, some subscribers have not recorded their sample counts. In the same circumstances, some have skipped an observation altogether and thus lost the benefit of a prompt of when the crop should next be inspected. Having recorded a sample count, subscribers have been known to file the record card instead of posting it, or delay sending it for several days. EPIPRE has a built-in reminder generator. When a variety of different people have performed observations or consultants try to fit in EPIPRE field monitoring with a regular schedule of visits to other fields, then the consistency of recording has been poor. The sampling procedure itself has been simplified since the system was introduced to the UK and has not been a source of problems.

The routine of field sampling by a given date, of examining plants and recording symptoms found, together with keeping records of spray and fertiliser applications, imposes a discipline on subscribers. On some farms where EPIPRE advice has affected spray decisions, farm staff have watched keenly for effects on yields and have been prepared to record other field details. This has improved the general standard of farm recording. Farmers have contributed ideas and opinions on how the product might be extended. Some have concluded that they could gain greater benefit from the system if the data collected for EPIPRE could form the basis of a field monitoring and recording system; this in turn creates demand for other products.

THE NEED FOR STANDARDS AND PRODUCT MONITORING

Standards of crop protection advice by merchant and manufacturer staff are maintained by product training, and in the UK by the BASIS scheme. Where advice can be derived from an increasing number of computer-based resources, standards are difficult to define. There will always be conflict amongst software producers on data formats and between agricultural scientists on the data which a system requires. Standards would be desirable in system documentation for transfer of data from one system for use in another, in sampling procedures for the maximum benefit to be derived from each sampling event, and in disease severity assessment for each disease, and in data archive format. It is unlikely that any standards will be established, because a major question arises over data ownership and confidentiality. It would probably be in the interest of farmers for government research bodies to have access to site-specific data, and to assess disease incidence and standards of disease control. In the same way as farmers are required to furnish information for annual returns to the Ministry of Agriculture, authority could be given to obtain other farm data. Unless this was available in a consistent format, or information was available for the conversion of data to a single format, then it would be of limited use. In the absence of a standard, the best means of central-

ised monitoring to analyse national and regional trends in disease
incidence and control, would be for system suppliers to divulge results.
This would only be helpful if the design of the system which generated
advice and summarised and archived results was also disclosed. This
places a burden on both system producer and government, without clearly
identified benefits to either.

In product monitoring and assessment, government-funded researchers have
a positive role to play in highlighting areas for development and
improvement, without the distraction of vested interest. The greatest
contribution may be made by the traditional training role of
universities and the extension services, in developing or supplying
teaching aids to farmers, in order to maximise the accuracy and benefits
of forecasting. Although each individual advisory product must be
supported by adequate training in the skills necessary for its use,
there are skills which farmers need to acquire which are common to most
systems and may be more readily learnt independent of system suppliers.
It is most certainly in the national interest for farmers to obtain a
better understanding of their own crops and to become more profit
orientated in their choice of husbandry procedures.

Finally, means of developing new advisory products are limited by the
availability of relevant data. It is noticeable that all advisory
services for disease control currently promoted commercially are confined to winter wheat, not just because it is a major crop but because data
are lacking for the development and validation of techniques and
procedures for other crops.

CONCLUSION
         Recognition by farmers that attention to spray decisions on
an individual field basis can enhance profitability, enables commercial
concerns to sell disease management advice for cereal crops. Spray
decisions will be based increasingly on the economics of maintaining
adequate crop health to safeguard financial returns, instead of on
traditional objectives of total crop protection.

Commercial systems of disease forecasting can help to transfer disease
management expertise from its origins in applied research to individual
farmers. Supplier and customer demands for new products stimulate
development of a range of decision aids. Commercial implementation and
accountability for the quality of advice at field level has highlighted
the weaknesses in farmers' recognition skills and recording procedures.
It stimulates the development of training aids which complement field
assessment techniques.

Practical considerations of finding time for sampling, and of consistent
and regular recording of parameters which farmers can learn to observe,
should be paramount in product design. Advisory systems can only
succeed if they accommodate human variation in behaviour and prompt
users to keep consistent records.

Government-funded research bodies have a key role to play in safeguarding the farmers' and nation's interest against the vested interest of companies who use farm information to exploit the market for their products, and in co-ordinating research into the development of new advisory techniques and procedures. The establishment of sampling, recording and documentation standards would speed-up product development and streamline data logging and analysis to the benefit of the farmer.

REFERENCE
Cook, R. J., (1983). Profit from Cereals Course. Stoneleigh, Warwick: Royal Agricultural Society of England.

SUBJECT INDEX